"An impressive exercise in urban forensics engaging infrastructure as social engineering. Case studies from the United States and Northern Ireland provide comparative analyses of strategies for urban segregation by law and by design. Cunningham provides an invaluable resource for understanding and ameliorating legacies of spatial injustice in our cities."

Richard Plunz, Professor Emeritus, Architecture, Planning and Preservation, Columbia University, New York City

"For all those students and practitioners in urbanism, particularly those addressing the problems of deeply divided cities, where contested issues of identity and inequality overlap, this is a must-read book. It brings novel insight to the perennial problem of how you reconcile community cohesion and social inclusion by remaking urban space for connectedness over segregation. Perceptively, the book acknowledges the historic injustices of race, religion and class that have imprinted themselves in layers of spatial discrimination and containment that continue to obstruct opportunity for the most marginalised. In a very readable style, and with an impressive set of references, the book challenges us to respect this legacy and all its complexity."

Frank Gaffikin, Emeritus Professor, Queen's University Belfast

Human Rights and the Architecture of Conflict

A MASTERPLAN FOR SPATIAL INJUSTICE—FROM BALTIMORE TO BELFAST

Human Rights and the Architecture of Conflict exposes how governments, public officials and private actors on both sides of the Atlantic entrenched racial and ethnic divisions through manipulation of the planning and design of the built environment.

Based on interviews, never-before-seen documents, and field work carried out in Belfast, Chicago, Miami, Washington D.C., nd New York City, this book shows how the planning and design of our built environment impacts the physical, mental, social, economic, political, and environmental well-being of communities. Tim Cunningham, an urban scholar and human rights advocate, reveals how the British Army set about reconfiguring the urban fabric of Belfast as part of a counter-insurgency strategy in the 1970s. His research shows how the techniques used in Northern Ireland during this period mirror earlier processes deployed in U.S. cities under urban renewal and the Interstate Highway Program. A global genealogy of segregation that examines the trajectory of coercive urbanism in the twentieth century, the text highlights the real-life walls and barriers that cleave communities along ethnic and racial lines and the role of architects, planners, developers, and public officials in erecting them. The final chapter considers some contemporary efforts to address the legacy of these practices through restorative architecture and planning initiatives that aim to deliver more cohesive, sustainable, and inclusive urban spaces.

This book is ideal reading for courses in architecture, city planning, community development, geography, human rights, sociology, transitional justice, urban studies, and Irish history.

Human Rights and the Architecture of Conflict

Tim Cunningham

Routledge
Taylor & Francis Group
NEW YORK AND LONDON

Designed cover image: Tim Cunningham

First published 2026
by Routledge
605 Third Avenue, New York, NY 10158

and by Routledge
4 Park Square, Milton Park, Abingdon, Oxon, OX14 4RN

Routledge is an imprint of the Taylor & Francis Group, an informa business

ISBN: 978-1-032-07648-5 (hbk)
ISBN: 978-1-032-07545-7 (pbk)
ISBN: 978-1-003-20812-9 (ebk)

DOI: 10.1201/9781003208129

Typeset in Times
by SPi Technologies India Pvt Ltd (Straive)

Dedication

For Marny, Oisín, and Cara

Contents

About the Author

Tim Cunningham is currently a Visiting Assistant Professor at the Pratt Institute, New York City. Prior to commencing his PhD at the Transitional Justice Institute, Ulster University, he worked for the Committee on the Administration of Justice, an independent human rights organization based in Belfast and was convenor of the Equality Coalition which brought together grassroots organizations representing the most marginalized groups in Northern Ireland. As part of his advocacy work on promoting equality and human rights he met regularly with international experts, including representatives from UN Committees (e.g., CERD, CEDAW, and CESCR) as well as local and international NGOs and officials from the New York City and New York State Comptrollers offices. He has given numerous guest lectures and talks in Ireland, Britain, and the U.S. He was also actively involved in historic preservation as a board member of the Historic Monuments Council for Northern Ireland from 2009-2015. A native of County Armagh, he is currently resident in New Jersey.

Acknowledgments

I would like to thank my PhD supervisors, Professor Rory O'Connell and Dr. Anne Smith, of the Transitional Justice Institute (TJI), Ulster University, for their contribution to this work. Without the PhD, there would have been no book. Their attention to detail was surpassed only by the speed with which they provided feedback on the numerous drafts. Thanks also to Prof. Christine Bell, Dr. Kris Brown and Prof. Fionnuala Ní Aoláin, who supervised earlier stages of my work. Thanks also to Professor Brendan O'Leary at the University of Pennsylvania and Professor Brandon Hamber of Ulster University, who examined my PhD thesis. Their astute observations and critical analysis significantly enhanced my thinking on this issue. I would also like to thank Professor Catherine Turner of Durham Law School and Prof. Kieran McEvoy, School of Law, Queens University Belfast who at various points during the PhD provided invaluable advice. Thanks also to Lisa Thompson, who sent me the email notification about the PhD studentship that started the ball rolling on this project and who kept the TJI ticking over in the early years of my study before she moved on to greater things. I would also like to thank Dr. Jacqueline Reilly, former Head of the Doctoral College, Ulster University, for her assistance along the way as well as my former colleagues at the Transitional Justice Institute, especially Aisling Swaine and Ita Connolly.

Special thanks go out to Professor Frank Gaffikin, Emeritus Professor, School of Natural and Built Environment, Queen's University Belfast, and Ken Sterrett, Senior Lecturer in Planning and Urban Design, School of Natural and Built Environment, Queen's University Belfast, both of whom helped steer the post-graduate interdisciplinary reading group that was a valuable source of information and guidance. The reading group also provided me with the opportunity to publish my first piece of work in this field and provided several opportunities to present and refine the ideas that are contained in this book. Thanks also to the other members of the post-graduate reading group, especially Aisling Shannon Rusk, who provided me with a first-class architectural tour of Belfast, which helped reassure me that I was indeed barking, but at least it was up the right tree.

Thanks also to Professor Richard Plunz, Graduate School of Architecture, Planning and Preservation (GSAPP), Columbia University, who helped clarify my thinking on several issues pertaining to this study and who provided me with the title for my thesis. The work we did together for the GSAPP studio on Belfast came at just the right time. Special thanks also go out to the late Professor Janet Smith of the Nathalie P. Voorhees Center for Neighborhood and Community Improvement, University of Illinois, Chicago, who really did walk the walk and who helped guide my ideas as we meandered around Cabrini Green.

Sarah Schindler at the Sturm College of Law, University of Denver was especially helpful in clarifying my thinking on the transatlantic implications of my original research and was the person who suggested I contact Routledge about writing a book on this. Deborah Archer at NYU also provided valuable insights. I would also like to thank Dr. Brian Feeney of St Mary's University College Belfast, who suggested to me that the Belfast Development Office (BDO) files from the 1980s might be worth looking at. He wasn't wrong. Thanks also to Maureen Reustle at Kean University for her encouragement.

I would like to thank everyone who gave up their valuable time to be interviewed and the many others along the way who provided valuable insight, advice, and assistance, many of whom lived in the communities that I visited and have to deal with the psychic weight and physical isolation of coercive architecture on a daily basis. I would also like to give special thanks to Barbara Bullard with whom I have co-organized a series of walking tours looking at the implications of gentrification in the Bedford-Stuyvesant neighborhood in Brooklyn. Special thanks also go out to my former colleagues at the Committee on the Administration of Justice (CAJ) especially Maggie Beirne, Aideen Gilmore, Liz McAleer, Martin O'Brien, and Paul Mageean, all of whom provided me with an invaluable hands-on education in human rights. Some of the ideas that we grappled with back in the late 1990s, namely, how to understand and help address concentrated patterns of spatial disadvantage in Belfast provide the basis for this book.

Any book-writing project requires the assistance of librarians. I have had the good fortune to be aided by some excellent ones. Joanne Knox at the Ulster University library was especially helpful in the early days, as were the rest of the library staff at Ulster University. Sal Tuszynski at the Pratt Institute proved invaluable, securing endless requests for interlibrary loans with ease. The Center for Teaching and Learning (CTL) at Pratt certainly lives up to its name. Special thanks to Judit, Holly, Maura, and Zach for the wonderful resources that they provide which played an important supporting role for this project over the past few years. Their Virtual Research Caves were especially valuable in offering a safe haven for focused writing. Thanks also to Prof. Sameetah Agha for hiring me at Pratt and offering me the opportunity to put some of my ideas into practice.

Special thanks also to the wonderful staff of the Public Record Office of Northern Ireland (PRONI), who are the custodians of an amazing and rich archive, a tiny fraction of which is shared in this book.

Thanks also to the wonderful staff at Taylor and Francis, without whom there would be no end product. Chloe Layman was the staff member who worked with me on the original proposal, and she has since moved on to greater things. Becky Lane and Emily Irvine provided outstanding editorial assistance.

Most of all, thanks to Marny, Oisin, and Cara for everything.

Acronyms and Abbreviations

Alliance Party Founded in 1970, the Alliance Party sought to present a more moderate and non-sectarian form of unionism that included both Catholic and Protestant representatives. The Alliance Party has grown in recent years becoming the third-largest party in the Northern Ireland Assembly.

BQE Brooklyn-Queens Expressway.

Belfast City Council (BCC) Local government authority for Belfast. During direct rule from Westminster, Belfast City Council, like all local authorities in the region, had limited powers.

British Army Deployed in Northern Ireland as part of Operation Banner from 1969 to 2007.

Department for Social Development (DSD) The DSD was established as a Northern Ireland government department under the power-sharing arrangements that followed the signing of the Agreement in 1998. DSD was responsible for urban regeneration, community development, social legislation, housing, and social security benefit. In May 2016, it was renamed to the Department for Communities as part of a restructuring that reduced the number of Northern Ireland government departments from 12 to 9. The Department for Communities now oversees these functions as one of the nine current Northern Ireland Executive departments.

Democratic Unionist Party (DUP) Founded in 1971 by Ian Paisley, the party emerged as a hardline unionist party in opposition to more conciliatory unionist policies. The DUP became the largest unionist party in the 2000s, overtaking the UUP.

Direct Rule Following the suspension of the Northern Ireland government and parliament in March 1972, the British government in London took direct control of the region's affairs. Under Direct Rule, a British Secretary of State was appointed to oversee Northern Ireland, with U.K. ministers given control of various Northern Ireland departments and the power to legislate for the region through Orders in Council. This system, intended initially as a temporary measure, continued from the early 1970s to the late 1990s due to the ongoing conflict and political deadlock. This system of governance centralized decision-making for Northern Ireland within the British government and resulted in limited power for local political representatives.

FHA Federal Housing Administration

Girdwood Barracks Located in North Belfast, this was a key military installation for British Army operations, which conducted patrols from the base. As part of the peace process and the scaling down of military presence in Northern Ireland, Girdwood Barracks closed in 2005. After its closure, the Girdwood site, comprising 27 acres, became the focus of redevelopment plans. It was seen as a potential symbol of post-conflict transformation and an opportunity to address local needs, particularly the severe housing crisis in North Belfast. However, the redevelopment process was fraught with challenges, which are explored in this book.

Good Friday Agreement (also known as the Belfast Agreement or "The Agreement") Signed on April 10, 1998, the Good Friday/Belfast Agreement is a peace accord that established a framework for power-sharing governance in Northern Ireland, addressing issues of sovereignty, human rights, and demilitarization. It created new political institutions, including the Northern Ireland Assembly and Executive, and set principles for cooperation between Northern Ireland, the Republic of Ireland, and the United Kingdom. Key aspects of the Agreement include power-sharing government representing both unionists and nationalists.

Green (ing) In the Irish context, this term is used to describe the Catholic/Nationalist/Republican community, institutions, beliefs, and territory. For example, land or areas dominated by the Catholic/Nationalist/Republican community might be referred to as green territory. The "greening" of an area in Belfast refers to an increase in the proportion of Catholic residents. This terminology is quite distinct from the use of "green" in most other geographical contexts to refer to environmental practices or concerns.

HOLC Home Owners Loan Corporation. Established under the New Deal by President Roosevelt, the HOLC pioneered a uniform method of property appraisal using color-coded maps. Areas containing African Americans or immigrant communities were colored red and deemed most at risk for mortgage default, giving rise to the term "Redlining."

Irish Republicanism/Republicanism Irish republicanism is a more radical form of nationalism that seeks a united Ireland free from British rule and emphasizes complete Irish independence. Historically, the term has been associated with armed struggle. Republicanism is primarily supported by segments of the Catholic community and is most prominently represented by Sinn Féin, though the party has evolved in the last couple of decades to pursue its goals through peaceful political means.

IRA (Irish Republican Army) The Provisional Irish Republican Army (IRA) (also known as "the Provisionals" or "the Provos") was an Irish republican paramilitary organization that aimed to end British rule in Northern Ireland and create a united Ireland, conducting an armed campaign against British security forces and civilians and bombing economic and military targets. In 2005, the organization announced an end to its armed campaign and instructed its members to pursue their objectives through purely peaceful means.

Loyalism Loyalism is a more militant form of unionism, often associated with working-class Protestant communities. It demonstrates a strong allegiance to the British crown and is sometimes linked to paramilitary groups. Loyalism is represented by smaller parties like the Progressive Unionist Party (PUP) and has historically been connected to paramilitary organizations like the Ulster Volunteer Force.

MHA Miami Housing Authority.

Nationalism/Irish Nationalism Nationalists advocate for the unification of Ireland, seeking to bring together the Republic of Ireland and Northern Ireland into a single state. It emphasizes Irish cultural identity and self-governance. Nationalism is primarily supported by the Catholic community and is represented by parties like Sinn Féin and the Social Democratic and Labour Party (SDLP).

North of Ireland This term is often used by nationalists and republicans as an alternative to "Northern Ireland," reflecting their view that the region is part of a united Ireland rather than a separate entity within the U.K.

Northern Ireland Northern Ireland is a constituent region of the United Kingdom, comprising six counties in the northeast of the island of Ireland. It was created in 1921 when Ireland was partitioned, with the remaining 26 counties of the island becoming the Irish Free State (later the Republic of Ireland).

Northern Ireland Act 1998 The primary legislation that implements the provisions of the Belfast/Good Friday Agreement in U.K. law. The Act establishes the framework for devolved government in Northern Ireland, including the Northern Ireland Assembly and Executive, and incorporates key principles of power-sharing and cross-community consent.

Northern Ireland Housing Executive (NIHE) (Housing Executive) The public housing authority for Northern Ireland, established by the Housing Executive Act (Northern Ireland) 1971. It is Northern Ireland's largest social housing landlord and serves as the enforcing authority for housing regulations. The NIHE's functions include examining housing conditions, developing programs to meet housing needs, managing its own housing stock, and administering the Housing Benefit Scheme to assist low-income residents with rent.

Office of the First Minister and Deputy First Minister (OFMDFM) (now called the Executive Office) Established by the Agreement, this is the central department of the Northern Ireland Executive. It is led jointly by the First Minister and Deputy First Minister, who have equal powers and responsibilities. Their key functions include co-chairing meetings of the Northern Ireland Executive, coordinating the work of the Executive and its response to external relationships, and determining significant or controversial matters for Executive consideration.

The Office is responsible for various policy areas, including economic policy, equality and human rights, European Union issues, and government machinery and public appointments.

Two junior ministers assist the First Minister and Deputy First Minister in their duties.

Orange Colloquial term used in Northern Ireland to describe the Protestant/Unionist/Loyalist community, institutions, beliefs, and territory. For example, land or geographical areas dominated by the Protestant/Unionist/Loyalist community are sometimes referred to as "orange territory."

Orange Order A Protestant fraternal organization known for its annual parades, which have been a source of tension, especially when they pass through Catholic areas.

Power-sharing The system of government established in Northern Ireland after the Good Friday Agreement, designed to ensure representation for both unionist and nationalist communities.

RUC (Royal Ulster Constabulary) The police force of Northern Ireland from 1921 until it was replaced by the Police Service of Northern Ireland (PSNI) in 2001 as part of the reforms that followed the Good Friday/Belfast Peace Agreement. The RUC was overwhelmingly Protestant and members were targeted by the IRA during the conflict.

Sinn Féin Founded in 1905 by Arthur Griffith, the party became prominent after the 1916 Easter Rising but split in 1922 over the Anglo-Irish Treaty. Sinn Féin re-emerged as a significant force in Northern Ireland politics in the 1980s and played a crucial role in the peace process and Good Friday Agreement. The party was closely aligned with the Irish Republican Army (IRA).

Social Democratic and Labour Party (SDLP) Founded in 1970, the SDLP advocated for Irish nationalism through peaceful means. John Hume, who led the party in the 1990s, played a key role in the peace process. The SDLP was the largest nationalist party until overtaken by Sinn Féin in the 2000s.

Troubles Used to describe the period of conflict in Northern Ireland that lasted approximately 30 years, from the late 1960s to 1998.

Ulster Defence Association (UDA) Loyalist paramilitary organization. Operating under the nom de guerre of Ulster Freedom Fighters (UFF), it declared its goal as defending Protestant and loyalist areas and opposing republicanism by engaging in paramilitary activities that often resulted in the deaths of Catholic civilians.

Ulster Unionist Party (UUP) Evolved from the Ulster Unionist Council, founded in 1905 and dominated Northern Ireland politics from 1921 to 1972, forming every government during that period. The UUP played a key role in the Good Friday Agreement negotiations in 1998 but lost its position as the largest unionist party to the DUP in the early 2000s.

Ulster Volunteer Force (UVF) Loyalist paramilitary organization. Sought to maintain Northern Ireland's union with Britain and combat Irish republicanism, primarily targeting Catholic civilians in sectarian attacks.

Unionism Unionism/Ulster Unionism supports Northern Ireland remaining part of the United Kingdom. It seeks to maintain the political, economic, social, and cultural connection with Britain and ensure continued British citizenship of Northern Ireland's residents. Unionism is predominantly supported by the Protestant community and is represented by parties such as the Democratic Unionist Party (DUP) and the Ulster Unionist Party.

Westminster Westminster refers to the U.K. Parliament in London, which has historically held sovereignty over Northern Ireland. The term is often used in discussions about the relationship between Northern Ireland and the rest of the U.K.

1 The Architecture of Inequality

1.1 INTRODUCTION

Faced with explanations about the findings of ground surveys and plot ratios, most people acquiesce and defer to the technical expertise presented to them as a fait accompli. This is exactly what happened to James Roth, who owned a farm on Long Island. Roth was told by planning officials that his land lay in the proposed route for the new Northern State Parkway and that he would have to surrender 14 acres of valuable property needed for the new road. Roth requested that the parkway be rerouted about 400 feet through barren land on his farm, which would have allowed him to continue in business. Officials refused, claiming that the route had been decided based on "engineering considerations" and "could not possibly be changed."[1] Roth agreed to the demands, partly because he had been assured that the engineering study would not allow for any possible alteration of the plan. What Roth did not know was that the Northern State Parkway was originally supposed to have traveled along a route 3 miles to the north of his property. The original route would have taken the Parkway through the 18-hole private golf course belonging to millionaire Otto Kahn. Kahn was as keen as James Roth to protect his land and used his financial largesse to pay a $10,000 bribe to Robert Moses to route the Parkway away from his golf course and the estates of other wealthy landowners. The road ultimately bifurcated and destroyed James Roth's farm. Municipal corruption of this nature is not typically considered to be within the remit of human rights, but the incident is illustrative of a wider principle that is relevant to this book, namely, that the planning and design of the built environment is fundamentally about political power, and as such, it is capable of abuse.

Although never elected to any office, Robert Moses did more to shape New York City than any other individual in history, forcibly relocating approximately half a million people in that city alone to create his own modernist vision for the twentieth-century metropolis. Like many white elites, Moses' vision for the future of his city was a highly racialized one in which affluent commuters would be able to move freely and seamlessly in private vehicles along sleek new parkways that snaked around New York City. Moses' vision did not allow much space, literally or figuratively, for African Americans or poor ethnic communities, and he went to great lengths to render them invisible. In some instances, this involved sending people correspondence

DOI: 10.1201/9781003208129-1

from his own office deliberately designed to mislead the recipients into thinking they were being ordered by the courts to vacate their property. Those who were displaced typically had to seek refuge elsewhere in more spatially isolated districts, with access to fewer facilities and opportunities for work, often paying higher rents. Moses also used his sweeping powers to try and render African Americans invisible from the public spaces within his control. For example, when he opened facilities at Jones Beach, Moses hired a small number of African American lifeguards, deliberately stationing them in a cluster at a swimming pool in the farthest corner of the complex. The aim was to "nudge" African Americans who arrived at the facility into believing that this section was the area they were required to use for swimming.

The magisterial biography "The Power Broker" by Robert Caro, demonstrates the huge social and economic costs of Moses' actions and are discussed further in Chapters 3 and 5 of this book. One unfortunate consequence of Moses' notoriety is that it serves to obscure the actions of similar technocrats, civic officials, and politicians in cities across the U.S. who also worked to deliver a modernist vision predicated upon "whitening" downtown by reshaping urban space to inhibit African American social, economic, physical, and residential mobility. In the decades following World War II, highway routing, public housing site locations, the location and expansion of university campuses, hospitals and medical facilities, sports arenas, retail developments, and civic centers were planned and designed to assist with enshrining segregation in concrete at a time when courts were dismantling the legal architecture of Jim Crow. Unlike Robert Moses, officials in many U.S. cities responded to legal requirements to integrate swimming pools by closing them and filling them with concrete. These efforts would ultimately erupt in flames in a series of freeway revolts and other urban riots across the U.S. as many African American communities sought to resist forced displacement and destruction of their property.

Unfortunately, by the time of these conflagrations, patterns of social exclusion and racial segregation had been established in many of these cities and have largely remained in place to the present day, continuing to shape racial disadvantage in the U.S. Moreover, despite its many failings, the modernist vision espoused by Robert Moses and others in the U.S. was exported across the Atlantic where it took shape under the guidance of Moses' namesake, Sir Robert Matthew in the U.K. Like Moses, Matthew was an unelected planner who had significant influence shaping cities in the U.K. during the 1950s and 1960s period. One city that Matthew was commissioned to redesign was Belfast, where "The Matthew Plan," as outlined in Chapter 3, played no small part in helping to precipitate the violent political conflict that would engulf Northern Ireland in the late 1960s. The experiences of Belfast, as in L.A., Detroit, and many other U.S. urban centers at this time highlight important lessons about the relationship between spatial inequalities, planning, and social conflict that are addressed in this book.

Ignatieff has argued that there is no contradiction between believing that races and religions should live together in a global city while "simultaneously choosing to live apart, side by side, in residential communities self-segregated by race or religion." He suggests it may be that living apart is what makes living together possible but that it is choice in the matter that renders the arrangement "legitimate and possible."[2] This book shows how in many cases individual choices were denied, sometimes by laws that directly prohibited racial or religious groups from living in certain locations but in many other instances by a much more complex set of architectural, planning, and economic factors that conspired to reshape and divide cities on both sides of the Atlantic.

1.2 THE ARCHITECTURE OF SEGREGATION

Notwithstanding the Equal Protection Clauses enshrined in the 14th Amendment of the U.S. Constitution, the country has a long history of using law as an instrument of racial exclusion. Most people are familiar with the Jim Crow laws that spread across the South in the decades following the Civil War restricting African American use of public transportation, schools, lunch counters, water fountains, restrooms, and other public amenities. These laws derived their legitimacy from the Supreme Court ruling in *Plessy v Ferguson* that "separate but equal" facilities were constitutional until a series of legal challenges in the 1950s and 1960s. By this time, rights activists had begun to challenge separate but equal legal doctrine in a number of protests across the South. Rosa Parks famously stood up for her rights by refusing to take a seat at the back of the bus on which she was travelling, while in 1960 six-year-old Ruby Bridges needed federal marshals to escort her to school in the face of white mobs chanting racial slurs. Because northern cities did not use laws to the same extent to racially exclude African Americans from public spaces, racial segregation in the U.S. has come to be viewed, particularly by many outside the country, as a largely Southern phenomenon that ended in the 1960s with the arrival of the Civil Rights Movement under the leadership of Dr. Martin Luther King Jr. and others.

High-profile battles that the American civil rights movement fought in the 1950s and 1960s to desegregate public spaces, while significant, obscured other powerful, but less visible forms of racial exclusion that incorporated architecture and urban planning techniques. After World War II, African Americans in towns and cities across the U.S. were especially impacted by federal laws and practices that prohibited them from renting property, buying homes, or securing mortgages in neighborhoods of their choosing. As the eyes of the world followed the assaults on civil rights protestors, hundreds of thousands of African Americans were forcibly evicted from their homes and businesses under the guise of ostensively progressive federal programs like "urban renewal" and the development of the interstate highway system that eviscerated entire communities and destroyed Black wealth. As Samuel Jordan, co-founder of the Baltimore Transit Equity Coalition, has pointed out, while Rosa Parks and others were refusing to give up their seats, "Baltimore's politicians and business community were planning to demolish Black neighborhoods for an expressway to serve white automobile drivers."[3] The destruction of Baltimore's Black neighborhood was not an unfortunate consequence of secular planning and economic development policies but rather part of an orchestrated attack on African American political, social, economic, and cultural power. Moreover, this pattern was repeated in cities across the country from Miami, Florida, to Chicago, Illinois, and from Pittsburgh, Pennsylvania, to L.A., California.

Ronald Bayor has shown how Atlanta became more residentially segregated at the very time that schools and public spaces were subject to integration efforts. White elites in that city, concerned about their image and mindful of pictures from Little Rock and Birmingham that showed peaceful protesters attacked by billy clubs, dogs, and water cannons, reluctantly met demands to integrate schools and public facilities and were aware that racial residential patterns, the routing of highways, street design, and the siting of public housing developments received far less national attention compared with "Whites Only" notices on lunch counters, restrooms, and water fountains. This allowed for "a quieter resegregation and control of Black spatial mobility," which achieved many of the same objectives as the earlier Jim Crow laws but not the same opprobrium from reformers within the city, and beyond.[4]

This book examines how, during the twentieth century, exclusionary laws and architectural and planning practices in the U.S. were drafted, sponsored, and implemented by federal, state, and local government officials, with the help of private actors, many of whom came from the real estate industry, shaping patterns of racial exclusion and disadvantage that persist today. This book demonstrates how systemic and structural inequalities woven into the fabric of metropolitan areas across the U.S. compares with similar practices that took place in Belfast, under the British Army in the 1970s and 1980s as part of an imperial boomerang that can be traced back to the sixteenth century. Consideration is also given to some recent efforts to address the spatial legacy of division in the U.S. through "restorative architecture". The final chapter examines what this means for how cities on both sides of the Atlantic, and further afield, ought to be planned and designed in the future in order to create more inclusive and sustainable urban spaces. It is first necessary however to consider the historical context within which coercive planning and architecture evolved.

1.3 COLONIAL URBANISM AND CORDON SANITAIRE

Walls and other physical structures have provided defensive cover for human settlements since antiquity. The earliest works of Western literature describe how the Greeks sought to breach the walled city of Troy, while the Book of Joshua recounts how the Israelites brought down the Walls of Jericho. Many European cities in the Middle Ages pioneered what we commonly understand today as "racial zoning," that is, the use of law to require designated racial or ethnic groups to live within specified territorial boundaries. Often, these boundaries were supported by walls and other physical barriers. Michael Radford's 2004 adaptation of Shakespeare's Merchant of Venice, starring Al Pacino as Shylock, vividly captured how the Jewish population of that city in the Middle Ages were corralled within walled ghettos locked at night. Venice was no means the only Italian city with a Jewish ghetto during this period. Pope Paul IV issued a decree requiring Jews in Rome to live in "a crowded and unsanitary area that regularly was flooded by the river Tiber."[5] The Roman ghetto served several functions, not least of which was visibly contrasting the splendor of the new city inhabited by enlightened Christians, from the squalor and deprivation visited upon those who denied the divinity of Christ.

The expressive role of architecture performed an equally important function in colonial settings by celebrating the power and grandeur of empire as well as helping to preserve the identity of European settler populations and police social and racial borders. Obert provides a valuable summary of the wide range of architectural and planning techniques deployed by the British, French, and German authorities across their empires that include intimidating monumentalization; the physical relocation of local populations to cities' outskirts, ensuring the current and future marginalization of colonized subjects; the mobilization of discourses of health and hygiene to justify racial segregation; the revision of place names and reorientation of memorial sites around colonial icons (often military heroes responsible for historic acts of violence against colored peoples); the projection of military might through menacing defensive fortification; and the celebration of metropolitan architectural styles. Obert notes that these similarities "point to the crucial, and often underexplored, significance of architectural and urban planning determinations in exercises of imperial domination…"[6] She argues that the patterns observed across colonial cities reflect the essential banality of imperial power that involves repetition of "the same predictable architectural strategies, over and over again across space and time."[7] The aim is to concretize power through architecture that is "steady, careful, legible," not just because it is planned and efficient, but because imperial power itself is relatively mundane and predicated upon violence and control – regardless of its claims to the contrary. In her study she identifies how colonial powers followed this approach of "planned violence," eschewing vernacular architecture, and difference, in favor of bland, "boring" designs and spaces that served to segregate and control native populations.[8]

Some of the techniques outlined by Obert are evident through the seizure of land and the erection of walled fortifications around several towns and cities in Ireland, including Kinsale, Coleraine, and Derry, to protect against attack from native insurrectionists in the late sixteenth and early seventeenth century. Faced with the threat of uprising in 1641, an earthen rampart was constructed around the town of Belfast with the native Irish required to live outside the wall. On the other side of the globe, Dutch colonialists forcibly established Batavia (Jakarta) in 1619 as the headquarters of their East India Company. By the end of the seventeenth century, non-Europeans were excluded from the city and required to settle in enclaves or kampongs scattered around the periphery. Following rebellion in 1740, populations were dispersed into ethnically defined settlements.

The experience of London merchants constructing walls around the city of Derry in Ireland was replicated by the British East India Corporation in Madras. Efforts to build walls around what was then Calcutta in the early years of the eighteenth century however were circumscribed by the local governor, or nawab, who had prohibited Europeans from fortifying their settlements. To circumvent this edict, London proposed taxing the native population to dig "a defensive ditch" around the English part of Calcutta and "to color over our reall design of making the place tenable against … the Moors [the Mughals] by alledging that the Ditch was to drain the Townes &c."[9] The British governor in Calcutta equivocated and the idea was dropped, but this proposal represents an early attempt at a practice that became commonplace in the twentieth century, and which features heavily in this book, namely using ostensibly legitimate civilian purposes (in this case irrigation) to camouflage military and defensive designs. This period also saw the pioneering of an authoritarian form of urban planning by Governor General Richard Wellesley in Calcutta that is usually credited to better known projects in Europe. Nightingale has pointed out that the clearing of what was regarded as degraded housing in "Black Town" predated the razing of London's St Giles slum in 1847 by a quarter of a century.

Colonial settings exemplified the relationship between the form of urban space and imperial authority. Scott has highlighted how "the relative illegibility to some urban neighborhoods ... provided a vital margin of political safety from control by outside elites." During the invasion of Algiers in the 1840s, Marshall Bugeaud surmounted this problem and gained control over the dense kasbahs by destroying entire neighborhoods and widening roads for military movements. Bugeaud returned to Paris in 1847 and proposed a similar radical reorganization of the city that involved new routes cut through the city and military units positioned within it. According to Israeli architect, Eyal Weizman, Bugeaud understood that there is a direct relationship between the organization of the terrain and the ability to control it militarily.[10] This logic, when extended to an urban context, "requires that military thinking guide urban design." This approach in turn provided the rationale for Louis Napoleon and Georges-Eugéne Haussmann's plans for Paris, with the redesigned city to be made safe against insurrection.[11] Military control of "insurrectionary spaces" that had not been well mapped was integral to Haussmann's plan through "simplification, legibility, straight lines, central management, and a synoptic grasp of the ensemble."[12] The densely packed working-class *quartiers* with illegible street plans, which Haussmann described as

> a dense belt of suburbs given over to twenty different administrations, built at random, covered by an inextricable network of narrow tortuous public ways, alleys, and dead ends, where a nomadic population without any real ties to the land [property] and without any effective surveillance, grows at prodigious speed.

His solution was a transportation network that would ensure direct rail and road links "between each district of the city and the military units responsible for ensuring order there." Where possible, "insurrectionary quartiers were demolished and broken up by new roads, public spaces and commercial development."[13]

As they rebuilt Moroccan cities, Lyautey and Prost developed "one of the most elaborate theories of racial segregation in the history of city splitting" namely, the complete separation of European and indigenous communities.[14] In Rabat, Morocco, Prost proposed "a building-free cordon sanitaire between the medina and the new city" to protect Europeans from "inevitable contagion" and "free up a corridor for commercial traffic." This approach preserved the medina as a tourist attraction, and avoided the bitterness that the destruction of the kasbah had generated from the local Muslim community. Renowned British architect Edwin Lutyens designed a plan that separated New Delhi from the old city with a cordon sanitaire, reinforced by a wall.

These examples illustrate Demissie's point that hygiene served a dual function in African colonies "as both preventative medicine, and cultural identity formation" with sanitary campaigns designed to restructure urban space to legitimate imperial rule. The geographic segregation of native populations was therefore deemed a sine qua non for safeguarding both the health and the identity of European settler populations. Swanson has noted how sanitation served to entrench residential segregation in the late nineteenth and early twentieth centuries through controlling the social and geographic mobility of Africans.[15] The outcome was that

> a system of boundaries (cordon sanitaire) and its supporting architecture was developed to reinforce the separation of the European towns from the 'native towns' ... walls, buffer zones and defensive architecture were incorporated into the planning of colonial cities as part of a wider spatial strategy to ensure surveillance and control over the African population.[16]

This model of "planned violence"[17] continued to be applied across colonial settings during the twentieth century. In Palestine, during the first Arab rebellion, mandatory British forces set about a "designed" destruction of Old Jaffa through Operation Anchor that cut a large anchor shaped "boulevard" through the old city, destroying between 300 and 700 homes. The boulevard allowed deep patrols into the heart of the city and put an end to resistance with a combination of design and military force. Serious concerns about the scale of the destruction were voiced in the UK Parliament, but the actions were defended on grounds of regeneration and public hygiene in an area lacking basic services. Further examples of the extensive reshaping of the built environment for counterinsurgency purposes took place in Algeria during the bloody and protracted armed conflict that took place in the decades following World War II. French civil and military authorities "profoundly reorganized Algeria's vast urban and rural territory, drastically transformed its built environments, and rapidly implanted new infrastructures and settlements across the country." The colonial regime issued laws requiring the evacuation of certain areas and the construction of developments to strictly control the Algerian population and defend the European population living in the country. The forced relocation and construction of settlements in both rural and urban areas was "a key factor in isolating the Algerian population from the influence of the revolutionary liberation fighters and in impeding the spread of the desire and the support for independence (or "contamination," to use the French army's technical term)."[18]

1.4 INTERNAL COLONIALISM AND THE IMPERIAL BOOMERANG

Lorenzo Veracini has argued that recent decades have witnessed a "dramatic resurgence" in the importation of historic colonial practices and techniques for the management of contemporary cities in Europe and North America as part of the "War on Terror." Steven Graham explains this reemergence through what Foucault characterized as the "imperial boomerang":

> ...while colonization, with its techniques and its political and juridical weapons, obviously transported European models to other continents, it also had a considerable boomerang effect on the mechanisms of power in the West, and on the apparatuses, institutions, and techniques of power. A whole series of colonial models was brought back to the West, and the result was that the West could practice something resembling colonization, or an internal colonialism, on itself.[19]

Demissie concedes that from its beginning, colonial architecture and urbanism in Africa was "intimately connected with the history of the architecture and urbanism in the metropolitan countries from which colonialism emerged." Weizman notes the symbiotic relationship between Paris and Algiers in the nineteenth century that ensured military planning practices developed in Africa helped shape the French capital producing "Haussmannian boulevard-building through neighborhoods of insurrection to domestic cities after first experimenting with them on colonized frontiers".[20] Police forces, fingerprinting and panoptic prison design were among the other measures designed to ensure the security of the state that boomeranged from African and Asian colonies to European theatres during the eighteenth and nineteenth centuries.[21]

Scholars have also highlighted how colonial planning and architecture shaped not just urban centers but also the suburbs surrounding them. The utopian promise that lay at the heart of American suburbia was always, at most, "a partial paradise."[22] From the outset, the pastoral vision of leisure, family life, and union with nature was predicated upon the principle of exclusion. Work was excluded from the family home, middle-class residences were segregated from their working-class counterparts, and the greenery of suburbia was separated from the drab smoke-filled urban environment. Fishman outlines how this quintessentially American vision owes much to the development of English suburbs going back to the eighteenth century. Archer however highlights how these "spaces of modernity" in turn were shaped and influenced by colonial practices in South Asia that "provided a perfect basis for the construction of a modern bourgeois identity."[23] Raffle's 1822 plan for Singapore is an "archetypal example of a city where particular sectors were designed for specific functions and population groups." This vision influenced Loudon's subsequent 1838 treatise *The Suburban Gardener, and Villa Companion,* which assumed the need for racial and social homogeneity. Loudon extolled the virtues of residential enclaves adjacent to metropolitan cities, in which "the houses and inhabitants are all, or chiefly, of the same description and class as the house we intend to inhabit, and as ourselves."

Graham provides several contemporary examples of "Foucault's Boomerang" including the construction of 'security zones' around the strategic financial centers of government districts of New York and London that "directly import the techniques used at overseas bases and green zones." Also included in his list is the cultural geographies which underpin the ideology of the political right and hawkish commentators within Western militaries that deem cities, per se, to be intrinsically problematic spaces that facilitate acts of subversion, resistance, dissent, and protest. From this perspective, rural or exurban spaces are viewed as bastions of traditional i.e. Christian values distinct from the urban metropolis that is cast in "the same Orientalist terms as the mega-cities of the Global South." Graham argues this is connected to the far right's conflation of terrorism and immigration that perceives power and national identity to be under existential threat – a view exemplified by military theorist William Lind who writes, "invasion by immigration can be at least as dangerous as invasion by a state army."

This book will fill in some of the gaps in the preceding and intervening period and show that rather than experiencing a resurgence, Foucault's Boomerang never lost its flight, or its power but rather was sharpened and perfected as it performed a north Atlantic oscillation during the twentieth century ultimately providing a template for city planning and design in the post-9/11 era. Of particular interest for this work is the way in which "Foucault's Boomerang" shaped planning and architecture on both side of the Atlantic from the early decades of the twentieth century. Perhaps the most influential architect and planner of that period was Charles-Edouard Jeanneret, better known as Le Corbusier. Widely regarded as the father of high modernist urbanism Le Corbusier eschewed traditional design and promoted a vision that was based on formal order that incorporated functional differentiation with geometric design. Like Haussmann, and Bugeaud before him, Le Corbusier was inherently suspicious of densely packed, illegible, vernacular urban spaces that provided fertile breeding ground for insurrection and criminality. Le Corbusier was also influenced by the large-scale military mobilizations that characterized WWI which led him to advocate extreme authoritarianism to deliver an urban vision of hygienic and standardized rectilinear shapes. Scott contends that the great majority of Le Corbusier's grand schemes were never built because they required a level of resolve and financial largesse "that few political authorities could muster." While this may be true in one sense, the racial and political dynamics of the U.S. ensured that Le Corbusier's modernist vision, merged with earlier colonial architecture and design practices to profoundly shape the dynamics of settlement on both sides of the Atlantic during the twentieth century. This is reflected for example in the area descriptions of Home Owners Loan Corporation (HOLC) officials in Chapter 2 that note the value of typography and infrastructure for halting the "invasion" of African Americans, Jews, and other immigrant groups into white territory. This is also reflected in the "Negro Resettlement Plan" devised by white elites in Miami in the 1930s that aimed to relocate the African American community of the city to the outskirts of Dade County. As Chapter 3 shows, lack of finance and legal authority prevented this from happening at that time. However, the aims of the resettlement plan were ultimately delivered in the 1960s when urban renewal and the highway program provided both the financial wherewithal and the legal authority to obliterate the African American district of Overtown. This book explores how this modernist/colonial vision was underpinned by strong political and economic imperatives, namely that segregation by design provided a way of circumventing court rulings that struck down laws promoting racial segregation and allowed for redevelopment of land previously occupied by non-whites.

This story will be told through examination of two documents, similar in language, and written almost 30 years apart, which illustrate how colonial urbanism shaped the urban fabric of cities from Atlanta, Georgia, to Belfast, Northern Ireland, as part of a transoceanic exchange of ideas and practices. In the U.S. the most significant document is the Federal Housing Association (FHA) Underwriting Manual, written in the early 1940s. This document was significant in the decades following World War II, providing an economic rationale for segregation by establishing decision-making rules for property values predicated upon maintaining the racial and social homogeneity of the surrounding neighborhood. The Underwriting Manual outlined a range of legal and architectural means by which racial segregation could be achieved, including the use of zoning ordinances, restrictive covenants, roads, walls, and other physical barriers. These ideas and practices had been in circulation among real estate professionals, municipal officials, and federal agencies for some decades. The FHA formalized them and crucially made adherence to racial segregation a requirement to secure Treasury-backed mortgage insurance. The outcome profoundly shaped racial patterns of settlement, concentrating poverty, ill-health, educational barriers, and social exclusion of hundreds of thousands of African Americans over the course of several decades.

The Belfast document, written in the early 1970s with the oversight of British army commanders who had arrived in the city to restore order following the outbreak of large-scale civil conflict, aimed to provide a template for reducing the immediate security threat posed by armed insurrectionists embedded within communities in the city. Reflecting the differing economic systems of the two jurisdictions, the Belfast document says nothing about housing markets or home loans, largely because at that time, residential accommodation in Belfast, especially in areas experiencing significant civil unrest, was dominated by public housing. Neither was there a need to resort to legal measures such as racial zoning or restrictive covenants to organize patterns of settlement since the military could approve or veto the location of public housing developments directly. Although the legal and economic conditions at the heart of both documents were different, the architectural and design aspects were remarkably similar and are of particular interest here. Both documents highlighted the importance of road design and physical barriers as instruments of segregation, for example. The design of the Belfast Urban Motorway, which was in the process of being constructed, was singled out as being of particular importance in securing a cordon sanitaire around areas of conflict.

1.5 SEPARATE AND UNEQUAL

The two documents examined here were applied in distinct ways that have important consequences for how patterns of settlement came to be viewed in each jurisdiction. The FHA Underwriting Manual outlined the conditions under which mortgages across the U.S. would be approved and as such needed to be circulated widely among real estate officials, lending institutions, developers, and anyone else involved in the home purchase process. The Belfast document was very different: it was marked secret and was uncovered, along with several similar documents from the public archive in Belfast after a protracted search by this author in 2015. As a result, within the U.S. there is a much greater awareness about the role of architecture and planning in promoting segregation and division than is the case in Belfast. In the U.S., legal challenges and protests at the erection of walls and barricades, as well as the highway routing process, gave rise to terminology such as "white men's roads through Black men's homes" and "Freeway Revolts." In recent years, the need to upgrade failing infrastructure and the impact of climate change has precipitated additional coverage in mainstream media about the racist legacy of the highway program. In Belfast the situation was quite different. The secret nature of the practices, and denial by the British government that they were actively involved in such activities, results in skepticism about claims that architecture and planning were manipulated for security objectives. The social, economic, and political consequences of these activities are therefore more easily overlooked, including the way in which patterns of inequality and deprivation were concentrated by architectural and design features.

In the U.S., despite greater awareness of the racist nature of architectural and planning activities, legal challenges, with one or two notable exceptions, have been overwhelmingly unsuccessful. Sarah Schindler and Deborah Archer separately identified the weaknesses that traditional forms of legalism and civil rights laws face when drawn upon to challenge issues like road design, walls, and other physical barriers designed for the purposes of racial exclusion. Judges typically fail to recognize the social impact of design features, including walls and street closures, as more than minor inconveniences justified by planning rationales affirmed by technocratic elites. Racial motives that give rise to these structures and design features are also often masked behind claims that exclusionary architecture promotes road safety or reduces crime. A similar phenomenon occurred in Belfast, where ostensibly benign motives were used to conceal security-led urban design features. Although these arguments were never tested in the Northern Irish courts, they did succeed in convincing many academic commentators and observers that decisions about street planning, road design, and the location of housing developments in Belfast emanated from mainstream civilian planning orthodoxy and not British Army HQ.

One lesson from this research is that societies around the world need a greater understanding of the link between spatial inequalities, conflict, and the planning and design of the built environment. This is especially relevant in a context in which "[p]lace has become a dimension of the new grievances; after a long period during which geographic economic inequalities narrowed, recently they have been widening rapidly."[24]

Some important findings emerge from this study regarding the various means by which planning and architecture can concentrate inequalities, segregation, and social exclusion. The first is the way in which planners can use their powers selectively to seize and raze property belonging to minority or suspect communities. In the U.S., the Fifth Amendment to the Constitution provides that private property shall not be taken for public use without "just compensation." This allows government and municipalities to legitimately claim property for the public good, such as building roads, government buildings, and public utilities, provided they compensate the property owner. This power received widespread use in the U.S. in the decades immediately following World War II as urban renewal and the interstate highway program took shape. A major problem was the highly racialized way in which eminent domain operated. Chapter 3 shows in some detail how African American communities were disproportionately targeted and typically denied just compensation. Part of the difficulty was that compensation was typically restricted to property owners – so renters, for example, received nothing if their homes were destroyed. In a context in which there was already a chronic shortage of available housing for African Americans, tens of thousands of people were rendered homeless. For property owners, the compensation provided did not reflect true market value, so a huge amount of African American wealth was obliterated without recompense. The land that was seized was then used for highway construction or redevelopment purposes to benefit white investors. The profits to be gained from this exercise were substantial but did not go to those communities who were displaced.

In addition to displacement of communities, highway design also played an important function in serving as racial barriers. Richard Daley, the "American Pharoah," understood this, which is why he ensured that the design of the Dan Ryan Expressway in Chicago presented one of the most formidable impediments "short of an actual wall" that the city could have built to separate the white South Side from

African American communities. This aspect of racial and social exclusion is perhaps one of the most effective but least understood techniques by which architects and planners can ensure some groups remain "stuck in place" and separated from the physical, social, economic, and cultural mainstream of city life. Chapter 3 explores how this process impacted cities on both sides of the Atlantic. To fully appreciate this phenomenon, it is worth considering the observations of Charles Wilson, who notes that architecture is of its nature assertive – it proposes

> a certain way of doing things, of bringing together or separating activities – and this will either create an order that affronts or one that enhances the quality of life … The form of our buildings is the form of our life ….[25]

Wilson draws his conclusions from personal observation and the work of Immanuel Kant. A fundamental tenet of the Critique of Pure Reason is that "all our awareness is grounded in spatial experience."[26] In a similar vein, and much more recently, Sarah Williams Goldhagen has noted that even when we pay no conscious attention to the built environment, it functions in our lived experience as a never-ending "concatenation of *primes* that can influence a person's thoughts, feelings, and responses by activating memories, emotions, and other kinds of cognitive associations."[27]

In some cases, these influences can be benign. For example, Thayler and Sunstein describe how the design of the road on Lake Shore Drive in Chicago slows traffic helping prevent collisions. At the beginning of the dangerous curve, a sign warns of the lower speed limit and a series of white painted stripes appear on the road. The stripes are not bumps but send a series of visual primes to drivers. When they first appear, the stripes are evenly spaced, but as drivers reach the most dangerous section of the curve, the stripes get closer together, creating the sensation that driving speed is increasing and the instinct for drivers is to slow down.[28] This is a classic example of architectural primes influencing thoughts and responses to achieve a desirable social outcome – in this case greater road safety.

The same authors point out that there is "no such thing as neutral design." Visual primes to slow traffic and improve road safety along Lake Shore Drive can be contrasted with the use of visual and material primes on the Dan Ryan Expressway to maximize the barrier effect for African Americans living in the Robert Taylor Homes (see Section 3.1). Danielle Wiggins (see also Section 3.1) describes this phenomenon in Atlanta, Georgia, where I-75/85 fundamentally altered the form of life on "Sweet Auburn Avenue." Prior to the arrival of the highway, Auburn Avenue had excellent connectivity with the rest of the city, which meant businesses and the local community benefited from economies of agglomeration as all manner of individuals walked to and from the main downtown area of the city. This flow of people included those who might wish to get their hair cut, their nails done, their shoes shined, buy a cup of coffee, or purchase food for lunch or dinner. Businesses offering these services and many more were able to benefit from this free flow of people who had money to spend. The pleasant ambulatory experience along this vibrant street was destroyed as I-75/85 split Auburn Avenue in half, and the remaining area was turned into a dead zone by a highway overpass that "served as a hotspot for petty thieves and drug dealers."[29] Businesses closed, customers were dispersed, and the area never recovered.

In addition to destruction of property and community life, one of the striking themes to emerge from this research is the capacity of coercive architecture to shape human behavior in ways that communities themselves often fail to appreciate. This research shows that when environmental stimuli consist of designs that promote isolation, anxiety, separation, and fear, the outcome is communities that are inward-looking, fearful, and defensive. The psychological impact of architecture and planning, and the capacity of both to concentrate poverty and social exclusion have been grossly underestimated by scholars and researchers. Too often, the form of the urban fabric is overlooked as a contributing factor in shaping patterns of area-based deprivation. This explains, partly, why legal challenges to divisive architecture have been so unsuccessful. Road designs and fences that impede movement are more than just encumbrances; they serve as technologies that ensure communities know their place. Literally and figuratively. Much more research is needed on this issue, and the hope is that this book will help prompt that.

Another notable finding of this research is the way in which government-funded programs on both sides of the Atlantic that were ostensibly established to address area-based deprivation were in fact used to fund structures that fenced communities in and inhibited access to jobs and resources. A perennial debate among social scientists, economists, politicians, and policy makers is how best to address area-based disadvantage. Some argue that the best method to address the problem is to invest in place-based initiatives, while others argue that the most effective interventions involve removing people from concentrated centers of poverty and exclusion to more affluent and successful communities. In an interesting variation of these approaches this research shows some area-based initiatives were used to support architecture that undermines social mobility and social inclusion. In Northern Ireland, initiatives like Belfast Areas of Need (BAN) and Targeting Social Need (TSN) were promoted as evidence of government commitment to addressing concentrated poverty and deprivation. The use of funds from these programs to fortify walls surrounding local communities and business parks sent subtle but powerful cues to local populations that they should remain within their own territory and had the effect of promoting exclusion, isolation and insularity. Similarly, the town of Hamden, Connecticut, erected a 16-foot-tall fence that cost almost $100,000, paid for with state Local Capital Improvement Program funds, to close its area off to residents from New Haven. As a result, poorer families in New Haven were significantly disadvantaged in accessing essential services. This book examines how walls, street design, and architectural features served to isolate communities on both sides of the Atlantic with similar consequences. The so-called Peace Walls of Belfast have been subject to a considerable degree of attention on the part of scholars. Much less attention has been paid to the network of similar structures across the U.S. This book aims to try and begin to redress that imbalance.

What is also notable from this research is that the capacity of architecture and planning to promote isolation, fear, and anxiety is by no means restricted to poorer communities. Chapter 5 highlights how many societal problems, including loneliness, and isolation can be explained, at least in part, by homes and communities that were specifically designed to promote separation and loneliness and are not all down to the algorithms of big tech. There is also some discussion of the environmental consequences of several decades of autocentric architecture and planning including the impacts on the health and economic well-being of communities who bore the brunt of environmental injustices.

1.6 THEORY AND METHODOLOGY

An important challenge for this research was establishing the extent to which architecture, planning, politics, and economics impacted upon structural and systemic inequalities. Part of the problem was that in many cases, the architectural and planning processes involved were carried out in secret and/or disguised with other motives. Another problem was that this researcher arrived at the work with a background in socio-legal studies and no training or background in architecture or planning. Given the range and type of disciplines involved in this study, a framework that is especially useful is that of autopoiesis, which is predicated upon the notion of closed, individual subsystems (e.g., architecture, law, politics, economics) operating within their own vertical chain of recursive meaning processes.[30] Schumacher argues that architecture can be most adequately understood if it is analyzed as an "autopoietic" system of communications,[31] while Luhmann has written extensively on law as a closed system.[32] Viewing law and architecture as one of a number of subsystems allows for a more nuanced and more comprehensive understanding of the way in which abuses of power can be shaped by political, economic, and other interests. Paterson and Teubner contend that at the most basic level, by being open to the possibility that more than one recursive chain will be in play across the entire range of the process, autopoiesis "can direct existing analytical tools to the most appropriate places."[33] Thus, instead of economic analysis seeking to account for the entire field – as attempted, for example, by the Chicago School – autopoiesis "can limit economic analysis to those recursive processes where the economic code and programs are in evidence." Similarly, instead of political analytical tools coming to a study with ready-made concepts into which the entire field must be squeezed, autopoiesis asks that political analysis more modestly restrict itself to those systems where the power code of politics is in evidence.[34] A systems approach is therefore useful for helping researchers to understand, and identify, interests and influences that might otherwise go unnoticed.[35] The value of considering the totality of influences on a field of study is especially important in this context, given the subject matter of this work, namely and the degree to which economic factors, and the architecture and planning of the built environment, as well as politics, are central to explaining spatial disadvantage.[36] A research methodology predicated upon identifying the operations of each subsystem also helps to illuminate the ways in which architecture and planning serve to marginalize and exclude certain communities in a context in which the real motives for street designs or highway routing are often concealed through benign motives.

Paterson and Teubner also argue that the identification of the specific autopoietic systems at play in any given context "is not a question that can be answered theoretically but only through empirical observation."[37] This author would agree. A mixed methods approach was adopted for this research that incorporated documentary analysis, qualitative interviews, walking ethnography, and photography to identify the autopoietic systems that shape urban space. A broad range of documentary resources were used to inform the wider theoretical and practical basis of the research, namely law, economics, politics, urban planning, transitional justice, and architecture. Documentary analysis included examination of the historic files contained within the Public Record Office in Belfast (PRONI). These files, which had been classified, revealed the true extent of security force involvement in civilian planning in Northern Ireland. The Belfast files were not digitized, so this research involved searching through materials on site at PRONI. Analysis of original records in the U.S. was assisted by the fact that Federal Housing Administration materials and other sources were digitized and could be examined online. In addition to documentary analysis, qualitative semi-structured interviews were conducted for the Belfast case study during three phases, the first of which was November–December 2011; the second of which was in January 2015; and the third and final set of interviews took place in July 2018.

The temporal aspects of autopoiesis are especially useful for urban research and have also informed the methodology for this work. Jane Jacobs considered the city ecosystem to be in the same state of flux as the natural world, so that observers of city life who thought they viewed static situations were actually seeing processes of beginning and processes of ending occurring simultaneously.[38] Similarly, autopoiesis is characterized by functionally differentiated subsystems located within their own vertical chain of recursive operations. This model cautions the researcher to locate observations within the context of social, economic, political, architectural, and planning processes that have taken place over an extended period. This research engaged documentary analysis, as well as other primary and secondary sources, to illuminate how current patterns of disadvantage were shaped by historical processes. This explains why initial interviews for the Belfast redevelopment case study were followed up to assess the extent to which the project evolved over an extended period. When the first phase of interviews was completed, there was political gridlock around the way forward for the Girdwood site, which was, at that time, still a cleared and derelict site. Subsequent interviews took place as this vacant site was transformed by the construction of a community hub, through the political discussions about the "Conceptual Framework" alternative to the original Masterplan. The point here is that city ecosystems and the patterns of inequality that are contained within them are not static, and therefore it was imperative as the research evolved that additional information was garnered to trace the development of the site. The U.S. research did not involve a single case study but included an overview of coercive architecture and planning in several cities across the country over an extended period. Semi-structured interviews took place in Chicago in 2011, with some additional interviews in Miami in 2025.

The urban fieldwork for this research combined photography with "walking ethnography." Rather than imagining walking ethnography as a unidirectional practice that brings us from one destination to another, it is a process of "orientations/reorientations and attractions/distractions," which can only be achieved through an open-ended approach that involves an element of "wandering" that brings our bodies "into conversation" with the environments we move through.[39] Photography and walking ethnography were used to help capture the "layers of subtlety" that are essential to understanding the "context forensics" of field urbanism.[40] There are limitations with respect to how well one can capture the essence of a city ecosystem without "visualization" of the context in order "to trace the agents and forces that have produced the image of the city."[41] The importance of rendering visibility on those fields that might otherwise be obscured was especially important in this context, given the "hidden" aspect of many of the architectural and planning practices. Regrettably, no measurement was taken of the step count required to complete this research, but it was considerable. Extensive walking ethnography and photography in Belfast was complemented by equally lengthy wanderings and observations in Chicago, New York, Washington, D.C., and Miami.

NOTES

1 Caro, R. *The Power Broker: Robert Moses and the Fall of New York.* (Vintage Books 1975), p. 278.
2 Ignatieff, M. *The Ordinary Virtues: Moral Order in Divided World.* (Harvard University Press 2017), p. 212.
3 Cassie, R. *Baltimore,* "Road to Ruin," February 2023.
4 Bayor, R. *Race and the Shaping of Twentieth-Century Atlanta.* (University of North Carolina Press 1997), p. 82.
5 Bayor, R. *Race and the Shaping of Twentieth-Century Atlanta.* (University of North Carolina Press 1997), p. 82.
6 Obert, Julia. (2023). *The Making and Unmaking of Colonial Cities: Urban Planning, Imperial Power, and the Improvisational Itineraries of the Poor.* OUP, p. 5.
7 Obert, Julia. (2023). *The Making and Unmaking of Colonial Cities: Urban Planning, Imperial Power, and the Improvisational Itineraries of the Poor.* OUP, p. 6.
8 Boehmer E and Davies D, eds. (2018) Planned Violence: Post/Colonial Urban Infrastructure, Literature and Culture. Palgrave Macmillan.
9 Nightingale, C. *Segregation: A Global History of Divided Cities.* (University of Chicago Press 2012), p. 79.
10 Misselwitz P and E Weizman, 'Military operations as urban planning', 28 August 2003. Mute. https://www.metamute.org/editorial/articles/military-operations-urban-planning.
11 Mumford, L. *The City in History.* (MJF Books 1961).
12 Scott, J. *Seeing Like a State.* (Yale University Press 2020), p. 39.
13 Scott, J. *Seeing Like a State.* (Yale University Press 2020), p. 61.
14 Nightingale, C. *Segregation: A Global History of Divided Cities.* (University of Chicago Press 2012).
15 Swanson, M. "The Sanitation Syndrome: Bubonic Plague and Urban Native Policy in the Cape Colony, 1900-1909", Journal of African History 18, 387–410.
16 Demissie F. "Colonial Architecture and Urbanism in Africa: An Introduction", in F Demissie (ed) *Colonial Architecture and Urbanism in Africa.* (Ashgate 2012), p. 3.
17 Boehmer E and Davies D, eds. (2018) Planned Violence: Post/Colonial Urban Infrastructure, Literature and Culture. Palgrave Macmillan.
18 Henni, S. *The Architecture of Counterrevolution: The French Army in Northern Algeria.* (GTA Publishers 2018).
19 Foucault, M. Society Must Be Defended: Lectures at the College de France, 1975-6, London: Allen Lane, 2003, 103 in Graham S. Cities Under Siege: The New Military Urbanism. (Verso, 2010). p. xvii.
20 Graham S. Cities Under Siege (2010).
21 Veracini L, "Colonialism Brought Home: On the colonization of the metropolitan space" (2005) University of Wollongong, Faculty of Law, Humanities and the Arts – Papers. 1338. https://ro.uow.edu.au/lhapapers/1338.
22 Fishman Robert, Bourgeois Utopias: The Rise and Fall of Suburbia (Basic Books, 1987), p. 4.

23 John Archer, "Colonial Suburbs in South Asia, 1700-1850, and Spaces of Modernity" in Visions of Suburbia, (ed.) Roger Silverstone (Routledge 1997), p. 29.
24 Collier, P. *The Future of Capitalism: Facing The New Anxieties* (Harper Collins 2018), p. 3.
25 Colin, W. *Architectural Reflections: Studies in the Philosophy and Practice of Architecture* (Manchester University Press 2000), p. 25.
26 Critique of Pure Reason, Transcendental Aesthetic, First Section, in Wilson, p. 24.
27 Goldhagen, S. *Welcome To Your World: How the Built Environment Shapes Our Lives.* (Harper 2017), p. 59.
28 Thayler, R. and Sunstein, C. *Nudge* (Yale University Press 2008), p. 38.
29 Danielle, W. Remembering sweet auburn before the expressway: what nostalgia reveals about the limits of postwar liberalism, *The Metropole,* https://themetropole.blog/2021/04/14/remembering-sweet-auburn-before-the-expressway-what-nostalgia-reveals-about-the-limits-of-postwar-liberalism/#_edn4.
30 Paterson, J. and Teubner, G. 'Changing maps: Empirical legal autopoiesis' (1998) 7 (4) *Social and Legal Studies* 457. Teubner contends that 'the economy for example, constructs its society through the language of prices. Law features in economic calculations not as a binding guide to conduct but as a cost factor (the severity of the sanction involved and the likelihood of it being applied. Politics constructs its "public" through the language of power, law its "legal reality" through the distinction between legal and illegal, and so on.' Teubner, G. *Law as An Autopoietic System.* (A. Bankowska and R. Adler tr, Blackwell 1993), pp. 102–103.
31 Schumacher, P. *The Autopoiesis of Architecture, Vol 1: A New Framework for Architecture.* (Wiley 2011) *(passim).*
32 Luhmann, N. *Law as a Social System.* (K. Ziegert, F. Kastner, R. Nobles, S. Schiff, and R. Ziegert tr OUP 2004) *(passim).*
33 Paterson, J. and Teubner, G. 'Changing maps: Empirical legal autopoiesis' (1998) 7 (4) *Social and Legal Studies* 457.
34 Paterson, J. and Teubner, G. 'Changing maps: Empirical legal autopoiesis' (1998) 7 (4) *Social and Legal Studies* 457.
35 At the same time, there is no suggestion that systems theory is the only framework that can be used here but rather the aim is to highlight the potential value of a systems approach in this context.
36 An example of the more typical approach characteristic of much transitional justice scholarship is evident in a recent report published by a group at Queen's University Belfast which looked at the impact of the conflict on housing and land in Northern Ireland. The report largely ignored the role of redevelopment, planning and architectural practices focusing exclusively on political violence. See Moffett, L. and others, *No Longer Neighbours – The Impact of Violence on Land, Housing and Redress in the Northern Ireland Conflict* (Reparations, Responsibility & Victimhood in Transitional Societies 2020), https://reparations.qub.ac.uk/assets/uploads/Land-Report-ENG-SP.pdf.
37 Paterson, J. and Teubner, G. 'Changing Maps: Empirical Legal Autopoiesis' (1998) 7 (4) *Social and Legal Studies* 451.
38 Jacobs, J. *The Death and Life of Great American Cities* (first published 1961, Modern Library 1993), p. xvii.
39 Yi'En, C. 'Telling stories of the city: Walking ethnography, affective materialities, and encounters' (2013) 17 (3) *Space and Culture* 3.
40 Pluntz, R. *City Riffs: Urbanism, Ecology, Place* (Columbia University GSAPP 2017), p. 14.
41 Alkhoven, P. 'The Reconstruction of the Past: the Application of New Techniques for Visualization and Research in Architectural History', (1991) *CAAD Digital Futures Proceedings*, 549.

2 The Origins of Urban Segregation in the U.S.

2.1 SEGREGATION BY LAW

2.1.1 Racial Zoning

The earliest example of racial zoning in the U.S. occurred in San Francisco in the late nineteenth century when the city enacted a law that required all members of the Chinese community, citizens as well as foreign nationals, to vacate their property and move to a designated district.[1] Failure to comply with the order would result in a prison sentence not exceeding six months; however, the courts ruled the ordinance unlawful under the equal protection amendment to the U.S. Constitution and several treaties with China. One might have expected legal segregation through racial zoning to end there, but the process was revived in the early decades of the twentieth century to target African Americans as the great migration changed the demographics of U.S. urban settlements.[2]

One of the notable features of antebellum southern cities was that they were quite racially integrated. Slaves were typically required to live near their masters, and fears of insurrection meant that whites were reluctant to allow African Americans to live together unsupervised in large districts. In the decades following emancipation, this approach gradually shifted, especially as rural African Americans flocked to cities in search of economic mobility and/or physical security. As the African American urban populations increased, they were increasingly forced into less-desirable areas on city peripheries, often near railroad tracks or dumps. Informal racial barriers, typically enforced by the threat of white violence, also limited African American residential mobility, especially for poorer sections of the community.

Some middle-class African Americans sought to escape overcrowding and substandard housing by moving into white areas, at which point city officials began to enact laws to prevent them. When a prominent African American physician in Baltimore purchased a home in a white neighborhood, the city passed a racial zoning ordinance that prohibited African Americans or whites from occupying houses on any street in which the larger number of houses was occupied by members of the opposite race. The measure was not without controversy. The *New York Evening Post* considered that the Baltimore ordinance was "utterly absurd in this day and generation, to return to the ghetto of the Middle Ages, abandoned by Europe long ago."[3] Notwithstanding these criticisms, other cities, notably Atlanta, Birmingham, Dade County (Miami), Charleston, Dallas, Louisville, New Orleans, Oklahoma City, Richmond (Virginia), and St Louis, followed suit.

When challenged in the courts, racial zoning ordinances were typically upheld on the grounds that they applied equally to both races, that is, African Americans were not permitted to live in white areas, but equally, the reasoning went, whites were also not allowed to live in African American communities. This rationale, based on a narrow

DOI: 10.1201/9781003208129-2

interpretation of "formal equality," ignored the fact that the purpose of these zoning laws was to keep African Americans confined within smaller, overcrowded, and less desirable districts often near industrial facilites. In 1917, the U.S. Supreme Court ruled in *Buchanan v Warley* that racial zoning ordinances were unlawful because they interfered with property rights protected under the Due Process Clause.[4] The right of a property owner to be able to dispose of their assets in a manner of their own choosing was deemed not to be a matter that ought to be constrained by government interference.

Notwithstanding the decision in *Buchanan*, many cities, states, and indeed the federal government ignored the ruling. One racial zoning ordinance, in West Palm Beach, Florida, remained in place until 1960. More commonly, however, direct racial zoning evolved from an outright ban on people of a certain race or ethnicity living in designated districts or streets to a more subtle form of land use zoning that would achieve the same objectives but survive challenge in the courts. One way in which this was achieved was by requiring the construction of single-family homes and/or by designating minimum lot sizes or square footage for new developments. These measures precluded low- and moderate-income residents of any color but were especially effective in excluding African Americans who disproportionately faced housing shortages and who were disproportionately of lower income in many cities. Rothstein highlights how this form of economic zoning was rare in the U.S. before World War I, but the *Buchanan* decision provoked urgent interest in economic and land use zoning to circumvent the ruling.[5]

Another way in which zoning worked to further racial segregation was to designate territory for commercial or industrial use to prevent it from being used for African American housing or by designating land for commercial or industrial use so that it could serve as a racial buffer zone. As outlined in Chapter 1, the idea of using buffers to segregate races had been developed by colonial powers, including Britain and France, in the nineteenth century. In St Louis, the city's more exclusive (and white) districts sought to curtail African American residential mobility by surrounding Black neighborhoods with commercial or industrial facilities.[6] In some cases, if African Americans moved into a particular district, it was subsequently re-zoned for industrial use. One of the difficulties with legal challenges to this form of exclusion was that zoning did ostensibly serve a legitimate purpose, namely, to help municipalities provide more healthy environments by regulating the location of certain undesirable and/or hazardous industries including slaughterhouses and stockyards.

One particularly obnoxious way in which this played out across the U.S. was the use of zoning to site toxic waste facilities within or adjacent to African American communities, which gave rise to the Environmental Justice movement in the latter decades of the twentieth century.[7] Courts generally are reluctant to intervene in such decisions, and in the absence of an explicit admission that siting of waste facilities had a racial intent, municipalities have been largely free to place industries and activities deemed detrimental to human health adjacent to African Americans and away from white communities. This has also allowed municipalities to locate schools serving African Americans next to quarries and other sites that contain health hazards.[8] Another way in which this form of zoning impacted adversely on African American neighborhoods was that the presence of industrial facilities, bars, workshops, etc., helped render these areas ineligible for federal government mortgage insurance.

Zoning remains a major impediment to African American residential mobility as many residential communities in the U.S. continue to permit only the construction of single-family homes for example.[9] The racial wealth gap means that the absence of duplexes and other types of more dense and more affordable housing, indirectly solidifies segregation.[10] In Norwalk, Connecticut, the local planning department recently issued a detailed response to concerns expressed by residents over proposed zoning changes in the city, specifically one that would allow for an increase in two-family homes that would boost density and diversity in the town.[11] Section 5.1 examines the racial, social, economic, and environmental impact of zoning in U.S. suburbs highlighting how this practice came to be mainstreamed into the design of housing and subdivisions with dire consequences for the country as a whole.

2.1.2 Racially Restrictive Covenants

From a segregationist perspective, the problem with zoning restrictions is that they are, by their nature, blunt instruments. The *Buchanan* ruling meant that even if municipalities used indirect means to exclude certain racial groups by limiting lot size and prohibiting muti-family units, or creating racial buffer zones, the problem remained that there was no means of limiting the purchase of "white" property by upwardly mobile affluent African Americans who had money to spend and were willing to leave their own neighborhoods. The solution to this problem came in the form of deed restrictions or, as they were more commonly known, restrictive covenants. These were private agreements that involved stipulations written into the title deed that prohibited the new owner from selling the property to people of a certain race. As with zoning, the use of restrictive covenants historically served a legitimate purpose, namely, to protect the rights of existing property owners in a community by limiting the kind of activities that the new owner of a property could engage in. These included turning the new property into certain types of undesirable commercial enterprises, for example, horse stables, glue factories, and forges. Following the rise in immigration from China at the end of the nineteenth century, California used restrictive covenants for the first time to bar certain types of people from purchasing properties.

In 1892, a federal court ruled that restrictive covenants were in violation of the U.S. Constitution on the ground that, although they were a private contract, they still had to be upheld by courts.[12] Following the ruling, clauses limiting the sale of property to certain types of people largely disappeared, but as with racial zoning, these were resurrected following the commencement of the great migration in the early twentieth century. Abrams notes that the adoption of restrictive covenants increased significantly following the *Buchanan* ruling, which had limited the scope of racial zoning. Restrictive covenants in the early decades of the twentieth century became fashionable and were regarded as a symbol of neighborhood exclusivity but had an especially pernicious effect on neighborhood integration.

> Once signed, there seemed no way of expunging it from the chain of title, which made it more mischievous than even segregation laws – which at least were responsive to the pressures of public opinion and subject to repeal.[13]

It is estimated that by the early decades of the twentieth century up to 80% of all the vacant land in Chicago and Los Angeles was proscribed against African Americans.[14] One study of the New York suburbs found that over half of all homes were forbidden to

African Americans.[15] The 'Mapping Prejudice' project at the University of Minnesota has uncovered thousands of restrictive covenants, forming a ring around downtown Minneapolis. Although the first wave of restrictive covenants excluded African Americans, the coverage quickly spread to other immigrant communities, notably Jewish, Mexican, Japanese, and Chinese communities. Organizations and institutions, as well as homeowners, began to covenant their property as a matter of policy, including the University of Southern California. It became "unethical" for builders and homeowners not to ban minorities. Ignoring the earlier ruling in *Grandolfo*, courts across the country, including the Supreme Court, ruled covenants lawful.[16]

In 1948, the Supreme Court finally agreed to review lower court decisions involving enforcement of covenants under pressure from the Truman administration.[17] A significant aspect of the case was the intervention of the Attorney General. Partly this was a result of pressure following the horrors of Nazi Germany and the responsibilities assumed by the nation under the UN Charter and the Report of the President's Committee on Civil Rights.[18] Supporting briefs calling for an ending of covenants came from, among others, the American Jewish Congress. Some institutions however, such as the National Association of Real Estate Boards, supported the covenants. The Shelley case gained national attention with legal arguments securing "front-page coverage and widespread comment on editorial pages."[19]

The Supreme Court ruled that the racial covenants were not unlawful between the respective parties but that enforcement by the courts was in violation of the fourteenth amendment. In other words, covenants were deemed to be unenforceable but not invalid. The *Shelley* ruling did not directly ban suits for damages against a signer of the covenants, and as a result, some courts continued to uphold lawsuits for damages against signatories who had sold their properties to African Americans or other proscribed groups. It was not until 1953 that the Supreme Court finally ruled against covenants, finding them "unworthy" and that award of damages for violating the covenant was "state action" and in breach of the fourteenth amendment. The Chief Justice of the Supreme Court dissented in this ruling, strongly defending the use of covenants as a means of safeguarding property values.

By prohibiting courts from enforcing covenants, more homes were opened to African Americans. The Federal Housing Administration (FHA), which had endorsed racial zoning and restrictive covenants in its Underwriting Manual (see Section 2.2), initially ignored *Shelley*, arguing that the ruling applied to court enforcement, not administrative enforcement. Following pressure from several rights groups and the Attorney General, the FHA eventually ruled that they would not insure properties under covenants after February 1950 but would continue to insure where covenants existed before that date.

As with zoning, the real estate industry used a range of means to get around *Shelley* by relying on the part of the judgment that held restrictions between parties were valid. This led to the development of a range of devices to uphold racial exclusion that did not require judicial enforcement. This included, for example, joint consent, where the approval of adjoining owners was required before land could be sold for building. This approach was used in the early 1950s to keep African Americans and Jews out of several developments. In other cases, real estate brokers and lenders agreed among themselves or through associations not to rent or sell property, or to provide loans to certain groups, notably African Americans. Some FHA directors refused to approve sites where there was a prospect of occupancy by minority communities. The Los Angeles Real Estate Board argued for a constitutional amendment to validate restrictive covenants and outlined several ways in which the industry could circumvent the *Shelley* ruling, highlighting how popular the measures were among some within the industry.

In the early decades of the twentieth century, legal instruments like racial zoning and racially restrictive covenants combined to establish a differentiated settlement pattern in the U.S with African Americans concentrated primarily in urban centers and increased white movement to suburban subdivisions. It is estimated that of more than 7 million homes constructed during the building boom in the 1920s, only a small fraction was eligible for African Americans. The consequence of this was that African Americans were prevented from gaining an economic foothold in the newly built suburban developments, and overcrowding in urban areas was increased when it might have been eased. From the outset, racial zoning ordinances and restrictive covenants were defended as necessary to safeguard property values and prevent conflict. This reasoning was taken up by the federal government and came to dominate urban planning, architecture, federal, state, and local government policymaking, and lending practices within the real estate industry for the rest of the twentieth century, and arguably, to the present day. Part of the reason this rationale came to such prominence is that it was enshrined in federal housing regulations and determined access to home loan insurance (see Section 2.2). What is also notable about this period is that although racial zoning and racially restrictive covenants together proved very effective for restricting African American residential mobility, the courts gradually limited the scope of both these measures. In response, other techniques were developed, drawn from architecture and planning that could achieve the same objective, notably road and street design, public housing site selection, and physical barriers, including walls and fences. The evolution of this approach to segregation by design is the focus of the next section as architecture and urban design came to accompany exclusionary laws to promote racial segregation.

2.2 SEGREGATION BY DESIGN

2.2.1 Redlining and the Economics of Segregation

Franklin Delano Roosevelt was elected U.S. President with a commitment to lift the country out of the Great Depression under a New Deal program predicated upon government intervention in many aspects of life that had hitherto been the preserve of private markets. One pillar of his New Deal was the establishment of a Home Owners Loan Corporation (HOLC) in 1933 to rescue homeowners facing foreclosure by refinancing existing loans on a long-term, low-interest basis with government support.[20] This measure was necessary to ensure the survival of the real estate and construction industries at a time when record numbers of U.S. homeowners were defaulting on their loans. The HOLC refinanced existing mortgages by extending the loan period and reducing repayments thereby significantly boosting the number of homeowners who were able to keep their property. Aside from the large number of mortgages that the HOLC helped to refinance, the organization played a crucial role in establishing a formal method of property appraisal that centered on assigning color codes to neighborhoods across the country that reflected property values within each area. The most desirable districts were given a first or "A" grade and were colored green. Second or "B" grade districts (colored blue) were regarded as still desirable areas that had somewhat reached their peak but were expected to remain stable for many years. Third-grade (yellow or "C") neighborhoods were generally considered to be past their peak and declining in value. Fourth-grade neighborhoods were the least desirable and were colored red, giving rise to the term "redlining."

Under the HOLC assessment process, real estate value was viewed as a package reflecting the physical condition of the property, local transportation networks, typography, landscaping, and, crucially, the racial composition of the area in which the property was located. The presence of African Americans, Jews, and immigrant groups in a particular neighborhood was deemed to be evidence that the area was in economic decline and therefore a poor risk for prospective lenders. Assessors also paid close attention to surrounding neighborhoods to anticipate the degree to which the racial composition of a given area might change in the future. By the same token, close attention was also paid to any barriers that would ensure the racial exclusivity of more desirable neighborhoods could be preserved. These barriers included legal measures such as racial zoning ordinances and restrictive covenants, as well as natural or man-made physical features, including rivers, hills, ravines, or roads that might serve to act as a buffer against "infiltration" of undesirable racial groups.

This research includes examination of original HOLC records, including maps and area descriptions contained within the digital archive "Mapping Inequality: Redlining in New Deal America," compiled by the University of Richmond.[21] These records illustrate how the HOLC assessment process operated in Bergen County, New Jersey, where legal measures like zoning ordinances and restrictive covenants were cited along with road and street design as means by which racial segregation could be preserved. In Hackensack, for example, appraisers expressed approval about the way in which road and street design served to maintain racial divisions and preserve the value of the neighborhood:

> [A]cross Passaic St to the south of the area, along Stanley Place in the adjacent area are negroes. However, the wide thoroughfare serves as a buffer between the two sections.[22]

In nearby Hillsdale,[23] which was occupied at the time by "white collar clerks and commuters," the HOLC appraisers found that the area had a "colored population concentrated on Lawrence Street at eastern end of area." Despite having a convenient shopping center, as well as "good transportation facilities, rail, and bus 50 minutes to downtown New York City," the verdict was "Mixed population, poor development and cheap construction in parts combine to make this a third-grade area." The records state that the African American population of the area was in fact 5%, highlighting the low threshold for an area to be deemed to have a "mixed population."

In the district of Ridgewood, assessors describe with approval the terrain "with oak and birch trees, and landscaping in evidence about the homes. Plots are large and houses well set back ... One of the best and most desirable sections in Bergen County."[24] In addition to the typography, landscaping, and architecture, assessors also noted the strict nativist zoning restrictions and restrictive covenants in the district. "Ridgewood is known as a community which excludes Jews from buying residences and this policy is strongly enforced." The area was deemed to be "unquestionably one of the best first grade areas." By contrast, in a less affluent area of Ridgewood[25] HOCL assessors concluded that

> no future is predicted for it … Highwood Ave. though predominantly "white" is threatened by negro invasion. Because of colored population and old poorly maintained structures, no better than 4th grade.

This verdict came despite the finding that "[s]chools and high school are accessible transportation is good and shops are convenient," illustrating how even the possibility of African Americans moving into an area trumped accessibility and services. In the Glen Rock area,[26] which at the time was occupied by white collar clerks, the value of the area was "affected by proximity to a fourth-grade neighborhood across the tracks, and the possibility of infiltration by colored population from that neighborhood." No African Americans lived in the Glen Rock area at the time although there was a "foreign families" presence of 10% that was listed as "Italian." This area received a third-grade listing.

Analysis of HOLC area descriptions in New Jersey, and in other cities across the country, indicates that the most desirable districts from the assessor's perspective were newly built developments in emerging suburbs outside town and city centers inhabited exclusively by white, Anglo-Saxon, and Protestant communities in single-family homes that reflected a nativist pastoral idyll. These neighborhoods were especially highly regarded if they had a combination of legal and *physical* measures in place to prevent racial integration. What is also notable about the area descriptions of this period is the militaristic tone of the language used by many of the assessors to describe how typography, roads, and other features could serve to protect white territory from "invasion" or "infiltration" by Jews, African Americans, or immigrant groups. Although written by U.S. civilians, many area descriptions could have been written by colonial officials in Africa or Asia and highlight the extent to which racial and immigrant groups were viewed as an existential threat to the success of suburban developments. Another notable feature of these area descriptions is that whether they were written by local assessors in Atlanta, Georgia, or in Bergen County, New Jersey, located directly across the Hudson River from Manhattan, the same nativist and racist sentiments were present.

Analysis of original records from this period also reveals that in addition to racial bias, there was a strong and inherently anti-urban bias at the heart of these assessments, which meant that older properties, mixed land use, density, and social or economic diversity, were deemed evidence of economic decline. In Hudson County, the smallest and most ethnically diverse county in New Jersey, for example, the communities of Bayonne, Hoboken, Secaucus, Kearny, Union City, Weehawken, Harrison, and Jersey City, taken together, did not have a single first-grade area and only two very small second-grade areas.[27] Several of these areas have undergone significant gentrification in recent decades and currently contain some of the most valuable and desirable real estate in the country.[28] Hoboken, for example, home to the prestigious Stevens Institute of Technology (then and now), is currently a thriving urban center with spectacular million-dollar views of the Manhattan skyline across the Hudson River. The prime location of Hoboken however cut little ice with HOLC appraisers – in fact this was perceived as the problem. Admittedly, the Hoboken shoreline was quite different and much more industrial in the 1930s than it is today. The area was immortalized by Elia Kazan and Marlon Brando in the 1954 movie "On the Waterfront," which focused on the links between organized crime and the stevedores who worked on the Hoboken docks. Nonetheless, the fact that private investors, existing homeowners, developers, and indeed everyone else was precluded from obtaining loans as entire counties were basically written off illustrates the blank and sweeping nature of the judgments made at that time about what parts of the country were considered worthy of investment. The attitudes of the HOLC assessors, evidenced in their area assessments, are somewhat bizarre when viewed from the perspective of contemporary urban theory and investment practices that regards run down or dilapidated real estate within the urban core as prime sites for redevelopment. The attitudes of the HOLC officials are much less puzzling however when viewed from the perspective of colonial cultural geographies that viewed urban centers as spaces of subversion and dissent that ought to be isolated physically and economically. The language of the HOLC assessments also reflects the contemporary disdain for cities that has seen fiery rhetoric from President Trump and the deployment of heavily armed I.C.E. agents, the National Guard and the U.S. Marines to quell protests in Los Angeles and other urban centers. As the next section will show, many of these contemporary anti-urban attitudes that view inner cities and the non-whites who live there as a contagion were formalized and implemented through the operations of the Federal Housing Administration in the middle decades of the twentieth century.

2.2.2 Formalizing Segregation: The FHA and the Economics of Exclusion

Another agency established during the New Deal era, and one that had a much greater influence over home ownership in the U.S. than the HOLC is the Federal Housing Administration (FHA). Established under the National Housing Act of 1934, the FHA had several objectives that included stabilizing the housing market, extending home ownership, improving housing standards and conditions, and, crucially, alleviating unemployment by boosting house construction. Taking their lead from the HOLC, the FHA also devised a formalized process for assessing the value of homes eligible for insurance predicated upon the condition of the property *and* the racial composition of the neighborhood in which it was located.

The operations of the FHA were enshrined in the Federal Housing Administration Underwriting Manual that contained detailed regulations stipulating the procedures to be followed by underwriting staff when determining whether mortgages were eligible for insurance under Title II of the National Housing Act. Issued in 1938, with some revisions and updates through the 1940a, this document shaped housing policy for decades to come. Following the model established by the HOLC, the FHA Manual adopted a strictly segregationist approach to insuring homes, linking home values to the identity of the communities in neighborhoods where the property was located. This was stated explicitly in the Manual:

> If a neighborhood is to retain stability it is necessary, that properties shall continue to be occupied by the same social and racial classes. A change in social or racial occupancy generally leads to instability and a reduction in values.[29]

This twin objective of ensuring both racial and social segregation was to have profound consequences for shaping patterns of settlement over subsequent decades. The message from the federal government to white homeowners, investors, developers, and mortgage lenders was that it was in their *economic* interests to keep their neighborhoods free from African Americans or other racial minorities, as well as occupants from different social classes. Also following the HOLC example, legal measures such as zoning ordinances and racially restrictive covenants, combined with architectural and design features like roads and other physical barriers, were viewed as the optimum means by which the economic integrity of an investment could be preserved.

The Underwriting Manual instructed appraisers to give higher ratings where "protection against some adverse influences is obtained" and that "important among adverse influences … are infiltration of inharmonious racial or nationality groups." Minority communities were explicitly equated with other unpleasant or undesirable land uses in the document.[30]

An entire section of the Underwriting Manual was dedicated to "Protection from Adverse Influences," with Underwriters instructed that this "is one of the most important features in the Rating of Location."[31] The Manual outlined a detailed framework for ensuring racial exclusion based on combining legal and physical barriers that covered not just the immediate property but the surrounding territory.[32]

The document stated explicitly that:

> Generally, a high rating should be given only where adequate and properly enforced zoning regulations exist or where effective restrictive covenants are recorded against the entire tract, since these provide the surest protection against undesirable encroachment and inharmonious use. To be most effective, deed restrictions should be imposed upon all land in the immediate environment of the subject location.[33]

The Manual conceded however that "some areas in which there is no zoning regulations or restrictions may be considered properly protected because of the favorable topography or geographical position of the land. The natural protection afforded in such instances might be sufficient to warrant a good rating."[34]

Further detail is provided on the role of physical barriers in preserving the territory of neighborhoods, with the FHA mirroring the militaristic and colonial tone of the HOLC and highlighting the protection afforded by hills and ravines.

> Natural or artificially established barriers will prove effective in protecting a neighborhood and the locations within it from adverse influences. Usually, the protection from adverse influences afforded by these means includes prevention of the infiltration of business and industrial uses, lower class occupancy, and inharmonious racial groups.[35]

Underwriters were also instructed to consider the role of infrastructure, particularly roads in maintaining racial exclusivity, again reflecting sentiments that had appeared in earlier HOLC appraisals. "A highspeed traffic artery or a wide street parkway may prevent the expansion of inharmonious uses to a location on the opposite side of the street."

While acknowledging the value of roads and highways as instruments of containment and segregation, the FHA Underwriting Manual also recognized the potential adverse impact of transportation infrastructure

> a highspeed traffic artery ... directly through a desirable neighborhood area with similar development on each side of the artery, the noise and attendant danger constitute an adverse influence, rather than a protection. The same holds true for the presence of railroads, elevated or surface line, and other means of transportation.[36]

These instructions mirror earlier area descriptions carried out by the HOLC and show that by the early 1940s, legal and architectural/design factors were both identified and used by federal agencies to maintain racial segregation. These sentiments were to have important consequences in subsequent decades as the Urban Renewal and Interstate Highway Programs were rolled out across the country, and roads and highways were typically routed through African American communities or used as physical barriers to support racial exclusion. As with earlier HOLC assessors, FHA underwriters were required not just to consider the area in which the property was located but also to examine the surrounding hinterland to safeguard against future population or demographic changes that might be seen to diminish property values. Once again, the threat of areas being "invaded" by other races was a prominent concern expressed in militaristic terminology that reflect the same thinking and practices of the British Governor General Richard Wellesley in Calcutta, or his French counterpart Marshall Bugeaud in Algiers a century earlier.

> Areas surrounding a location are investigated to determine whether incompatible racial and social groups are present, for the purpose of making a prediction regarding the probability of the location being invaded by such groups.

In 1947, the FHA removed words like "inharmonious racial groups" from the manual but "barely pretended" that this represented a policy change.[37] The revised 1952 Underwriting Manual continued to base property valuations in part on whether properties were in neighborhoods where there was "compatibility among the neighborhood occupants."[38] This is significant because this was a period when the Interstate Highway Program was getting underway, and road expansion was taking off across the U.S. As outlined in Chapter 3, the rolling out of the Highway Program shows how the FHA Underwriting Manual provided a template that federal, state, and municipal authorities could put into effect when designing highways and other transportation networks that often aligned with their own local racist agendas (see Section 3.1).

Over the coming decades, "natural or artificially established barriers" in the form of highways, roads, and street design would be erected across the country, with the approval of the FHA, to impede African American social, economic, and physical mobility. Where these features were unavailable, the FHA resorted to insisting that walls or other physical barriers were erected between white and African American communities to secure finance. Along the way, established centers of African American wealth would be obliterated amid the wholesale spatial realignment of cities that concentrated poverty and racial segregation within concrete boundaries under the guise of progress. The pattern that emerges in the U.S. in the latter decade of the twentieth century is one in which coercive architecture and planning practices, which were less visible, but arguably much more effective, emerged to ensure racial exclusion as rights activists and others persuaded courts to strike down some of the most egregious exclusionary laws and lending practices. Entire patterns of settlements were reshaped under the guise of slum clearance and modernist planning which provided a technocratic justification and rationale for the spatial realignments of urban populations envisaged by earlier racial zoning plans. This model would subsequently be exported across the Atlantic to reshape cities in the U.K. including Belfast, Northern Ireland. The impact of these actions, which are still felt today, will be considered in Chapter 3.

NOTES

1 In re Lee Sing, 43 F. 359 (1890).
2 Wilkerson, I. The Warmth of Other Suns: The Epic Story of America's Great Migration. (Vintage 2010).
3 Klarman, M. *From Jim Crow to Civil Rights: The Supreme Court and the Struggle for Racial Equality* (Oxford University Press 2004), p. 83.
4 Klarman, M. *From Jim Crow to Civil Rights: The Supreme Court and the Struggle for Racial Equality* (Oxford University Press 2004), p. 79.
5 Rothstein, R. The Color of Law (Liveright 2017).
6 Gordon, C. *Mapping Decline. St Louis and the Fate of the American City*. (University of Pennsylvania Press 2008), p. 124.
7 Bullard Robert. 2005. "Environmental Justice in the Twenty-First Century." pp. 19-38 in *The Quest for Environmental Justice: Human Rights and the Politics of Pollution*, (ed.) Robert Bullard. San Francisco: Sierra Club Books.
8 Benjamin Karen. 2012b "Suburbanizing Jim Crow: The Impact of School Site Selection, and the Rise of Residential Segregation in the Urban South, 1890–1930." Journal of Urban History 38 (2): 225–46.
9 Bronin S. Key to the City: How Zoning Shapes Our World. (Norton 2024).
10 Gray N. Arbitrary Lines: How Zoning Broke the American City and How to Fix It. (Island Press 2022).
11 Norwalk Planning and Zoning Commission. Memorandum. June 23, 2023. https://www.norwalkct.gov/DocumentCenter/View/30508/Regulation-Rewrite-Memo-6-23-2023-to-Commission-with-Prior-Planning-Docs
12 Grandolfo v Hartman, 49 F. 181.
13 Abrams C. Forbidden Neighbors: A Study of Prejudice in Housing. (Harper 1955). p. 218.
14 Robert, W. *The Negro Ghetto*. (Harcourt, Brace and Company 1948). p. 214.
15 Dean, J. P. 'No other than Caucasian: A study of race covenants' (November 1947) *Journal of Land and Public Utility* Economics: 428.
16 Corrigan V Buckley 271 U.S. 323 (1926).
17 Shelley v Kraemer 334 U.S. 1 (1948).
18 To Secure These Rights: The Report of the Presidents Committee on Civil Rights. U.S. Government Printing Office. (1947).
19 Abrams C. Forbidden Neighbors: A Study of Prejudice in Housing. (Harper 1955). p. 221.
20 In 1926, 68,000 homes were foreclosed across the U.S. By 1932 this figure had risen to 250,000, and by the spring of 1933, half of all home mortgages in the U.S. were technically in default, with foreclosures taking place at the rate of 1'000 per day. Jackson, K. T. *Crabgrass Frontier*. (Oxford University Press 1985). p. 193.

21 Nelson, Robert K., LaDale Winling, et al. "Mapping Inequality: Redlining in New Deal America." Edited by Robert K. Nelson. American Panorama: An Atlas of United States History, 2023. https://dsl.richmond.edu/panorama/redlining/about
22 Nelson, Robert K., LaDale Winling, et al. "Mapping Inequality: Redlining in New Deal America." Edited by Robert K. Nelson. American Panorama: An Atlas of United States History, 2023. https://dsl.richmond.edu/panorama/redlining/#loc=11/40.914/-74.227&city=bergen-co.-nj&area=B26&adview=full
23 Nelson, Robert K., LaDale Winling, et al. "Mapping Inequality: Redlining in New Deal America." Edited by Robert K. Nelson. American Panorama: An Atlas of United States History, 2023. https://dsl.richmond.edu/panorama/redlining/#loc=11/40.914/-74.227&city=bergen-co.-nj&area=C6
24 Nelson, Robert K., LaDale Winling, et al. "Mapping Inequality: Redlining in New Deal America." Edited by Robert K. Nelson. American Panorama: An Atlas of United States History, 2023. https://dsl.richmond.edu/panorama/redlining/#loc=11/40.914/-74.227&city=bergen-co.-nj&area=A5
25 Nelson, Robert K., LaDale Winling, et al. "Mapping Inequality: Redlining in New Deal America." Edited by Robert K. Nelson. American Panorama: An Atlas of United States History, 2023. https://dsl.richmond.edu/panorama/redlining/#loc=11/40.914/-74.227&city=bergen-co.-nj&area=D1
26 Nelson, Robert K., LaDale Winling, et al. "Mapping Inequality: Redlining in New Deal America." Edited by Robert K. Nelson. American Panorama: An Atlas of United States History, 2023. https://dsl.richmond.edu/panorama/redlining/#loc=12/40.945/-74.134&city=bergen-co.-nj&area=C19
27 Jackson, K. T. *Crabgrass Frontier*. (Oxford University Press 1985).
28 Currently in Hoboken, for example, a two-bedroom condominium in an older, redbrick property typically sells for about $1,000,000.
29 Federal Housing Administration. Underwriting Manual: Underwriting and Valuation Procedure Under Title II of the National Housing Act. U.S. Government Printing Office. (Washington D.C. 1938). para. 937.
30 Federal Housing Administration. Underwriting Manual: Underwriting and Valuation Procedure Under Title II of the National Housing Act. U.S. Government Printing Office. (Washington D.C. 1938). para. 935.
31 Federal Housing Administration. Underwriting Manual: Underwriting and Valuation Procedure Under Title II of the National Housing Act. U.S. Government Printing Office. (Washington D.C. 1938). para. 932.
32 Federal Housing Administration. Underwriting Manual: Underwriting and Valuation Procedure Under Title II of the National Housing Act. U.S. Government Printing Office. (Washington D.C. 1938). paras. 933, 934.
33 Federal Housing Administration. Underwriting Manual: Underwriting and Valuation Procedure Under Title II of the National Housing Act. U.S. Government Printing Office. (Washington D.C. 1938). para. 980 (1).
34 Federal Housing Administration. Underwriting Manual: Underwriting and Valuation Procedure Under Title II of the National Housing Act. U.S. Government Printing Office. (Washington D.C. 1938). para. 980 (5).
35 CFederal Housing Administration. Underwriting Manual: Underwriting and Valuation Procedure Under Title II of the National Housing Act. U.S. Government Printing Office. (Washington D.C. 1938). para. 935.
36 Federal Housing Administration. Underwriting Manual: Underwriting and Valuation Procedure Under Title II of the National Housing Act. U.S. Government Printing Office. (Washington D.C. 1938). para. 935.
37 Abrahms C. Forbidden Neighbors: A Study Of Prejudice in Housing (Harper 1955)
38 Rothstein R. The Color of Law. (Liveright 2017). p. 66.

3 Modernist Planning
A Road to Wreck and Ruin

3.1 HARD CHOICES: RACIAL DIVISION IN CONCRETE

3.1.1 Introduction

There was an important but unstated defensive aspect of the U.S. highway program that was central to the subsequent planning and design of the new system. Following the playbook outlined in the FHA Underwriting Manual, highways, roads, and street design across the U.S. were used as barriers to afford "protection against adverse influences" and prevent the expansion of "inharmonious racial groups" into white territory at a time when the Supreme Court and civil rights activists were championing racial integration. The racial aspects of road design in the U.S. took many forms and had "a profound impact" on the composition of cities and patterns of racial segregation across the country.[1] In some cases, highways were deliberately designed as barriers to segregate African American communities within discrete territorial boundaries. In other instances, highways were used to bring about the forced relocation of Black and Brown citizens to more distant areas outside the urban core.[2] The result was the reconfiguration of patterns of settlement that heightened racial segregation and racial exclusion, driving African American communities into more spatially isolated and concentrated "second ghettos" that were disconnected from the social, economic, and cultural centers of city ecosystems.

As outlined in the previous section, Home Owners Loan Corporation (HOLC) area descriptions show that real estate appraisers considered roads and highways to be valuable "buffer zones" to support racial segregation in the 1930s and included reference to this in their files. Predating HOLC, there were instances of roads being considered as tools of racial segregation as far back as 1917 when officials in Atlanta proposed building a 150-foot-wide parkway that would serve as a racial dividing line in the city.[3] The highway was never built but provides an example of how road design and racial politics were linked from the earliest decades of the twentieth century. In Atlanta in the 1940s, some of the same city planning officials who had worked on earlier racial zoning measures in the city proposed building a parkway that would purportedly prevent African American incursion into traditionally white neighborhoods on Atlanta's west side. The parkway design of the 1940s aimed to separate Black and white communities by establishing two parallel highways with a park in the land between. The parkway "was to run in an "L" shaped manner, with the white community located on the land west of the road and African Americans on the territory to the east." Lest there be any prospect of racial intermingling, the Atlanta Planning Commission proposed the installation of a fence "to prevent any integrated use of the parkland."[4] This project was abandoned following the outbreak of World

DOI: 10.1201/9781003208129-3

War II but again illustrates how the design skills of architects and urban planners were used as instruments for supporting racial segregation in the pre-war years.

3.1.2 Planned Violence and Land Grabs

Key to understanding the importance of highways for reconfiguring the racial composition of U.S. cities in the postwar period is a recognition that at the highest levels of the federal government, expressway planning in the 1950s and 1960s was as much about urban development and racial politics as it was about transportation.[5] Several decades of redlining and the anti-urban bias at the heart of financial appraisals by the Federal Housing Association (FHA) left many cities starved of investment. This was augmented by a declining tax base fueled by the exodus of middle-class white families who had left for the emerging suburbs. As outlined in Chapter 2, African Americans were largely denied this opportunity for social and economic mobility by racial zoning and racially restrictive covenants.

Traffic congestion was also a problem with an increase in automobile use impacting older narrow streets in Central Business Districts. Cold War anxieties about the threat posed to U.S. cities by nuclear weapons also drove political imperatives predicated upon dispersal of people and businesses to the suburbs and the need for new transportation infrastructure that would allow for rapid evacuation of civilians and the deployment of troops in the event of conflict. The Great Depression of the 1930s, and the lack of building during the war years meant that city elites were "hungry for investments to stop the downward spiral"[6] and the fact that the federal government agreed to pick up 90% of construction costs through the newly created Highway Trust Fund offered an unprecedented and welcome opportunity to address the economic doom loop.

Although the federal government provided 90% of the funding for Interstate Highway construction, state highway departments working with local officials selected the actual routes – a framework that persists to the present day. This allowed urban expressway design and routing to be used to pursue local race, housing, and residential segregation agendas with some particularly egregious consequences. In Montgomery, Alabama, Sam Englehart, the leader of the white hate group Alabama White Citizens Council, and Director of the Alabama Highway Department, personally intervened to have I-65 rerouted through the neighborhood where Rosa Parks, Martin Luther King, and other civil rights leaders lived in an attack on the movement that had led the Montgomery bus boycott and other protests.[7] Also in Alabama, I-85 was routed to physically separate the historically Black Alabama State University (ASU) from its nearby residential neighborhood of doctors, faculty, and civil rights intelligentsia. Englehart even went so far as to intentionally target the home of civil rights leader Ralph David Abernathy for destruction.[8] Abernathy protested, writing to President Kennedy complaining that I-85 would destroy "one of the best black neighborhoods in the South and aid segregationists in their attempts to undermine the black vote in the city." Abernathy's residence was ultimately saved due to the relocation of a proposed highway interchange, but many other prominent African Americans lost their homes, thousands of small businesses were destroyed, families were displaced, and property de-valued. I-85 also displaced many middle-class African American registered voters, highlighting how highway planning served to further white political interests by suppressing Black electoral representation at a time when advancing voting rights was a key objective of the civil rights movement.[9] This example illustrates how highway design impacted civil, political, economic, social, and cultural rights in a more pernicious way than the "Jim Crow" laws that are typically the focus of historians of this period.

Further south, the experience of New Orleans illustrates how the dynamics of highway planning and racial politics played out to produce very different results between neighborhoods in the same city, one predominantly white, the other, African American. Proposals for a waterfront freeway through the French Quarter were halted following "intense opposition from historic preservationists."[10] There was to be no such reprieve, however, for the Tremé district of the city, home to one of the oldest and most successful Black neighborhoods in the country.[11] Tremé was devastated by the Claiborne Expressway section of I-10, which displaced residents, over 300 Black-owned businesses, as well as approximately 400 mature oak trees that provided shaded walkways and gathering spots.[12] The Claiborne Avenue Alliance is a local non-profit that was established to try and revitalize the community and aims to highlight the social, political, economic, environmental, and health impact of the expressway on local communities. The Claiborne Avenue History Project has also sought to highlight the community that was lost[13] through multimedia and oral history.[14] Film footage and other accounts from the 1940s and 1950s show that like many African American neighborhoods, Tremé was far from a slum but was in fact a vibrant center of life where many business owners lived above premises that lined Claiborne Avenue.

The activities of Robert Moses in New York City have been well documented by others and are covered in Section 5.1, which examines how African American and ethnic communities were devastated by his road and urban renewal programs. Outside the city, and across New York State, there were several other notable examples of African American communities that were segregated and had property destroyed under the sucker punches of highway planning and urban renewal. In Syracuse I-81 divided the city and destroyed countless African American homes and businesses which led to long-term exclusion and segregation.[15] In Buffalo, the Kensington Expressway created a physical barrier isolating residents on the city's primarily Black east side and severing east-west roads, plunging the Humboldt Parkway Community into several decades of economic and social decline.[16]

In Pittsburgh, Pennsylvania, the city's Hill District, known as "Little Harlem," was devastated by I-579, which displaced thousands of residents and severed the neighborhood's connection to downtown.[17] Anthony Foxx, Secretary for Transportation in the Obama administration, described how the "connective tissue" of his own neighborhood in Charlotte, North Carolina, was shattered by a series of highway construction projects that began in the 1950s.[18] Between 1960 and 1967, some 1,480 homes were demolished in the city, with 1,007 families displaced, and over two hundred Black businesses, which relied upon Black communities to thrive, were forced to close.[19] Foxx also recounted the history of the Brooklyn neighborhood in Charlotte, a thriving African American community that was founded by freed slaves after the Civil War and was home to churches, schools, doctors, lawyers, residents, and many businesses. A document from 1912 describes how the area was considered by white elites to be too valuable to be left for African Americans. A new "independence boulevard" cut a gash through the community, destroying historic Brooklyn with no trace of the area remaining today. The Brooklyn example is part of a pattern that emerges in which white supremacist opprobrium toward African American settlements in the early decades of the twentieth century was subsequently pursued through highway routing in the 1950s and 1960s.

In Nashville, Tennessee, the State Highway Department "went out of their way to put a kink in the urban link of Interstate 40 as it passed through the city," gouging a "concrete swath through the North Nashville Black community, destroying hundreds of homes and businesses and dividing what was left of the neighborhood."[20] Outraged residents formed the Nashville I-40 Steering Committee, taking a legal challenge to the federal court, alleging racial discrimination. The challenge failed, and I-40 proceeded as planned, obliterating the North Nashville Black community.

In Baltimore, Maryland, roughly 3,000 residents were displaced, and hundreds of homes and businesses, including bakeries, pharmacies, and grocery, hardware, and clothing stores, were destroyed for an expressway that was never completed because residents of more affluent areas of the city blocked the other stages of the route.[21] A decade later, Baltimore was left with "a multilane trench carving a mile-and-a-half gash across several neighborhoods." Known officially as the US 40/Franklin-Mulberry Expressway and unofficially as the "Highway to Nowhere" or "The Ditch," the "stub of sunken freeway has been a corrosive presence in the West Side ever since, and a conundrum for city leaders and planners."[22] Viewed in isolation, the Highway to Nowhere appears to be something of an aberration, or a quirk. Viewed through the lens of what was happening in other cities across the country at that time, with FHA Guidance extolling the virtues of "natural and artificial barriers" as means of preventing "infiltration by inharmonious racial groups" the true purpose of the "Highway to Nowhere" is revealed.

In the upper Midwest, Akron, Ohio, has a mile-long section of the city's Innerbelt freeway that separates "a mostly Black neighborhood from the commercial and cultural life of downtown."[23] Also known locally as the "road to nowhere," this stretch of freeway was built through a thriving Black community, destroying hundreds of homes and businesses, including the Hotel Matthews, which hosted stars such as Duke Ellington and Louis Armstrong. Following the standard playbook, nearby streets were cut off to make way for the freeway ensuring that economic life was choked from the neighborhood as natural linkages between homes, local businesses, and the commercial center of the city were ruptured. Also in Ohio, Ryan Crane has described how Cincinnati could be "considered a poster child for the devastation wrought by the highway builders of the past century" with the predominantly African American community in the Kenyon Barr neighborhood on the West Side displaced and properties razed to facilitate I-75.[24]

In nearby Michigan, the African American neighborhoods of Black Bottom and Paradise Valley in Detroit were both destroyed and 100,000 residents displaced by I-375.[25] Similar patterns emerged under the leadership of Richard Daley in Chicago. Adam Cohen and Elizabeth Taylor outlined how the "American Pharaoh" shaped the city on the "unstated foundation" of a commitment to racial segregation with urban renewal and the highway program central to his efforts. Daley "preserved the city's white neighborhoods and business district by building racial separation into the very concrete of the city."[26] New developments, including housing, highways, and schools, were located where they would serve as a barrier between white neighborhoods and "the black ghetto." Daley also worked with powerful business leaders to revitalize downtown by removing poor African Americans and replacing them with middle-class whites. African Americans, who might otherwise have moved into white neighborhoods, were instead shunted into densely packed high-rise public housing developments, the largest of which was the State Street Corridor on the city's South Side that became the largest concentration of public housing and poverty in the nation. His coup de grace was routing the Dan Ryan Expressway to follow the neighborhood's traditional racial boundary with the fourteen-lane highway separating "the State Street Corridor from the white, working-class neighborhoods of the South Side – including Daley's own neighborhood of Bridgeport." The Illinois state DOT had wished to build highways in the city for years, but it was only with the advent of the Interstate Highway Act that their dream became a reality.

Cohen and Taylor highlighted how the original design of the Dan Ryan Expressway was altered to maximize its impact as an instrument of segregation. The Expressway had been planned to cross the Chicago River "almost directly north of Lowe Avenue, Daley's own street, and then to jag east several blocks, at which point it would turn again and proceed south." When the final plans were announced, the road had been "realigned" several blocks to the east so that it would head south along Wentworth Avenue. This was a less direct route and required the Expressway to make two sharp curves in a short space, but the alternative route turned it into a formidable racial barrier separating the Black and white South Sides.[27]

In its new location, "the Dan Ryan Expressway reinforced the South Side's oldest racial dividing line along Wentworth Avenue." This was the line Daley himself had defended in his youth but like many well-established and historic racial dividing lines, Wentworth Avenue was not a "hard" boundary meaning that there was always a likelihood that African American residents would eventually cross over and move into white neighborhoods like Bridgeport. These fears were allayed by the architects and planners, who ensured that the Dan Ryan Expressway would provide a barrier between the races. Cohen and Taylor provide a devastating critique of the way in which the design of the Expressway was as much about racial politics as it was about traffic flow:

> one of the widest highways in the world, with a "dual-dual" design consisting of seven lanes in each direction, four high-speed through traffic and three slower-moving lanes. It was the most formidable impediment short of an actual wall that the city could have built to separate the white South Side from the Black Belt.

Daley's actions delivered an economic and social "spatial mismatch" in Chicago, with potential workers living in the Robert Taylor homes and other developments separated from jobs, which in turn fed long-term unemployment and welfare dependency. As Cohen and Taylor observed, "the distance between the Black Belt and the world of work in downtown Chicago was psychological as well as physical." They noted that researchers who subsequently interviewed residents of the State Street

Corridor found that a surprisingly large percentage of them had never been to the Loop, and many had never journeyed outside their own neighborhood. The result was that development like the Robert Taylor Homes, served, in the words of one longtime tenant, as a "public aid penitentiary."[28]

In Minnesota, the historic Rondo neighborhood of St Paul was devastated by I-94, which split the community in half destroying 700 homes and 300 businesses and plunging the remaining community into sharp decline as they were cut off from their social, economic, and cultural hinterland. In this case, a less destructive route through underutilized industrial land to the north was ignored as I-94 targeted a thriving and historic community with an established Black middle class.[29] Most of those displaced, who failed to receive adequate compensation, were African Americans who had been concentrated into the Rondo area by discriminatory lending practices, racial zoning, and racially restrictive covenants. One study has estimated that the destruction of home equity brought about by the clearing of homes and properties for I-94 led to an intergenerational wealth gap of almost $160 million.[30] Ben Goldfarb calculated that amount was sufficient to put every Black child in the area through college.[31] In addition to the financial loss, the destruction of Rondo also brought about the demise of a neighborhood that was an important cultural and civic center for African Americans. According to former resident Marvin Anderson, Rondo was the only neighborhood in St Paul "where Black people could walk around and not feel like a stranger or oddity."[32]

Over on the west coast, the city of Los Angeles provides another prime example of the way in which architects, planners, white elites, and private developers followed the template set out by the FHA to segregate their city. In the early decades of the twentieth century the L.A. Red Car transit system was one of the best in the U.S. Integrated and racially diverse neighborhoods like Watts and Boyle Heights emerged and thrived along transit corridors that boosted homeownership and prosperity. As the population of the city expanded from 320,000 in 1910, to more than 1.2 million in 1930, including tens of thousands of African Americans moving from the South, white LA followed the template set out by the FHA with racial zoning, racially restrictive covenants, and other legal instruments deployed to uphold segregation in the newly emerging suburbs. These measures combined with racial violence, including bombings, cross burnings, and drive-by shootings, to contain the growth of the African American presence in the city. As elsewhere, infrastructure planning and design were arguably the most effective and long-lasting means by which the racial and spatial demographics of the city were reconfigured. Freeways in the city were deliberately routed – often at considerable additional expense – to destroy thousands of homes in racially diverse communities. Gilbert Estrada described how mixed-race Boyle Heights was gutted by freeways.[33] Despite a mandate to avoid parks at all costs, planners put lanes through the middle of Hollenbeck Park in Boyle Heights while spending millions to reroute around a park in the white suburb of San Dimas. Dozens of Boyle Heights homes were also destroyed "to provide white suburban shoppers easier freeway access to a Sears department store."[34]

> Officials justified these actions as slum clearance – intended to upgrade the city's supposedly crumbling housing stock. But their racially malign intent was

> obvious, laid bare when officials moved the Santa Monica Freeway so that it ran directly through the stately African American middle-class neighborhood of Sugar Hill – wiping it off the map.[35]

Adam Paul Susanack, a New York architect, has highlighted images of the stately mansions that were destroyed in Sugar Hill, illustrating the substantial amount of Black wealth that was obliterated in the process.[36] Protests from African Americans were ignored, but white areas like Beverly Hills and South Pasadena successfully fought off freeways planned through their neighborhoods. As freeways enabled L.A.'s car-dependent suburbs to expand outwards, they did so "under the guidance of federal and local policies that were explicitly designed to keep those neighborhoods whites only." Developers were barred from selling to non-whites by the lending policies of the FHA, while the Underwriting Manual provided the template for segregationist road and highway design that concentrated African Americans within discrete boundaries frozen in concrete. Of "the 125,000 Federal Authority Housing units built in L.A. County between 1950 and 1954, less than 3% were open to people of color."[37] Unable to move into white suburbs, those displaced by highway construction and African Americans fleeing the South were forced to find housing in the already crowded and segregated areas of South and East Los Angeles, away from established and emerging job centers but in the middle of concentrated freeway pollution corridors.

Of all the cases of African American destruction in the twentieth century the experience of the Greenwood district in Tulsa, Oklahoma stands out as especially heinous. African American wealth in that state was boosted by the fact that after the Civil War the U.S. Government imposed stricter terms for reconstruction on slave-owning Native American nations that had allied with the Confederacy than it had on Southern states. Slave-owning tribes were required to share tribal land and other resources and rights with freed African Americans.[38] This land, and the wealth that accrued from it, provided the seed money that helped turn the Greenwood area into the "Black Wall Street." Spurred by the oil boom, many African Americans seeking to escape the Jim Crow South subsequently moved to Tulsa in the early 1900s to create a "haven for the burgeoning Black middle class."[39] The Black Wall Street was a thriving community but was obliterated by white violence that erupted following an accusation that a Black teenager, Dick Rowland, had assaulted a white woman named Sarah Page, who worked as an attendant in an elevator. Amid a moral panic in Tulsa about inter-racial mingling, threats were made by white members of the city, to have Rowland lynched. After several other lynchings of Black men by white mobs a show of force from Greenwood men to protect Rowland escalated into an all-out attack on Black Tulsans by white vigilantes, some of whom were deputized and handed weapons by the police.[40] Over the course of two days Greenwood lost 1,256 homes to burning and another 215 to looting. Martial law was declared, and National Guardsmen imprisoned "all Black Tulsans, over 6,000 people, in the Convention Hall and the fairgrounds."[41] There were eyewitness reports that planes were flown over Greenwood dropping kerosene bombs. A report in 2001 found that "Tulsa was likely the first city [in the United States] to be bombed from the air."[42] The entire business district of Greenwood was obliterated, including the high-end Stradford

Hotel and Dreamland Theatre. One estimate put the number of businesses destroyed at 150. In addition, ten churches were burned, including the newly opened Mt Zion.

In the immediate aftermath, community leaders tried to rebuild Greenwood in the face of intense resistance from white elites and civic authorities. Insurance claims were rejected, property theft took place, and there were numerous arrests of African American victims of the massacre, many of whom were charged with inciting the violence. An ordinance was introduced by the city banning new wooden structures that made rebuilding "prohibitively expensive" to frustrate reconstruction efforts by the local community. Civic leaders refused financial assistance from other parts of the country to help victims rebuild the area, even returning a check from the Chicago Tribune for $1,000, while state authorities also refused to provide financial assistance to help rebuild Greenwood. The speed with which white elites in Tulsa responded to the massacre with plans to redevelop the area as an industrial site fed speculation that the destruction was part of a pre-planned land grab by city officials. Support for the destruction of Greenwood within white Tulsa is evidenced by an editorial published shortly after the massacre: "Such a district as the old 'Niggertown' must never be allowed in Tulsa again."[43] Francis Burke, a social service director, wrote in a 1936 thesis, "Prominent businessmen of Tulsa today stated that there was a definite organized effort to displace the Negroes from the south portion of Greenwood because of developmental possibilities."[44] The idea that the massacre was planned is also supported by the work of Charles Moreno and others who point out that the attack involved planes dropping firebombs on Greenwood and residents being rounded up into three detention centers, all of which would have taken some time to organize. The fact that white businessmen had unsuccessfully tried to buy up parts of Greenwood in the years prior to the massacre leads several historians to the conclusion that the massacre was part of an orchestrated land grab.[45]

Despite actions on the part of white elites to frustrate the efforts of the Greenwood community, the area was successfully rebuilt relatively quickly. W.E.B. Du Bois wrote in 1926,

> Black Tulsa is a happy city. ... Five years ago, fire and blood and robbery leveled it to the ground. Scars are there, but the city is impudent and noisy. It believes in itself. Thank God for the grit of Black Tulsa.[46]

By the 1950s, the area was once again "an oasis of Black economic self-sufficiency, inside an Oklahoma city flush with oil industry wealth."[47] Despite this rebirth, Greenwood was destroyed in a second wave of violence the form of highways and urban renewal. I-244, which runs directly through Tulsa's Greenwood district, "did what the Klan and white racists couldn't do: demolish and depopulate the place."[48] Carlos Moreno has pointed out that what often gets erased is the "45 years of prosperity after the massacre and the events that led to Greenwood's second destruction: The Federal-Aid Highway Acts of 1965 and 1968."[49] Moreno found that "Greenwood looks the way it does today, not because of the massacre ... [but] because of the federal highway project, and because of urban renewal."[50] This view is supported by

Freeman Culver, president and CEO of the Greenwood Chamber of Commerce, who summarized the impact of I-244 on the African American community in Tulsa:

> Imagine going through a massacre and rebuilding your district bigger than it ever was before. And then imagine, less than 40 years later, the federal government got involved and built a highway right through your commercial district.[51]

Victor Luckerson described how "waves of re-destruction" in the form of urban renewal and the highway program did as much, if not more damage to the African American community in Tulsa than the original conflagration in 1921. In an interview with Bloomberg, he describes how "After the race massacre, we have this really powerful story about resiliency and Black people rebuilding. … I would say the massacre was more devastating in the short term, and urban renewal more devastating in the long term."[52] The highway removed the business district and the core area, with urban renewal projects eventually taking out much of the surrounding residential district. The Model Cities program, which ostensibly aimed to promote community involvement in renewal efforts lasted just long enough to raze the area but funding was ended by the Nixon administration before any of the vacant land could be rebuilt, leaving a plethora of parks and empty spaces that still dominate the area.

Atlanta, Georgia, was another city profoundly shaped by racially motivated road planning during the twentieth century. As the city embarked on an extensive postwar road building program, "the racial functions of highway design were never far from the minds of city planning officials."[53] Mayor Bill Hartsfield described how I-20 served as "the boundary between the white and Negro communities on the west side of town."[54] The highway's awkward, circuitous path was deliberately plotted to contain Black neighborhoods on one side while "protecting" White areas on the other. Though racial residential patterns have since shifted, I-20's circuitous route illustrates the spatial legacy of racist planning practices. The construction of the Atlanta expressway devastated the thriving Black business district of Auburn Avenue, known as "Sweet Auburn."[55] This area was home to some of the country's leading Black-owned businesses and cultural institutions, including Citizens Trust Bank, Atlanta Life Insurance Company, and the city's first Black-owned radio station, daily newspaper, and office building.[56] Unlike many African American business districts, Sweet Auburn was advantageously located adjacent to the White business district on Peachtree Street, allowing for physical, economic and social connectivity that boosted the wealth of both areas.[57] The interstate succeeded in displacing or cutting off many patrons of Sweet Auburn's businesses and the area never recovered.[58] Danielle Wiggins describes the impact:

> In splitting the avenue in half, the expressway fundamentally altered Sweet Auburn's atmosphere. The looming and loud highway disrupted the ambulatory experience of Auburn Avenue. No longer a pleasant experience, Atlantans no longer walked the avenue. The underpass it created below separated the eastern and western sections with a dark stretch that served as a hotspot for petty thieves and drug dealers.[59]

The experience of Sweet Auburn illustrates again how urban renewal and the highway program combined to target the African American presence in cities per se, not just low-income "slums." In what has been described as a "two birds with one stone" tactic i.e. commercial downtown expansion was facilitated by reclaiming land while removing the African American presence from the urban core. Black housing in downtown Atlanta that was destroyed was replaced with expressways, a civic center, and a sports stadium serving as a new "racial buffer" between the remaining African American communities and downtown. Atlanta city officials had agreed to grant additional land for Black housing in previously agreed "expansion areas" in exchange for the destruction of Black homes and businesses downtown but far more housing was destroyed than was replaced and much of the new housing development was deliberately placed outside city limits. In addition to removing African Americans from proximity to jobs, retail, and economic vibrancy, Black political power was also undermined by dispersing voters to the outskirts of the metropolitan area at a time when Black politicians were advocating for civil rights reform.

Ronald Bayor has highlighted how urban renewal and public housing site selections that characterized Atlanta during the 1950s and 1960s largely followed the framework set out in a 1922 racial zoning plan for the city, illustrating how urban renewal and highway planning allowed white elites to carry out the very actions that the courts had begun to prohibit.[60] The refusal to locate any of the public housing developments in the more economically vibrant northeast of the city heightened the economic and social isolation of African Americans. Henderson describes how this ensured that "low-skill, low-wage jobs in the retail and service sector are located vast distances from where available low-wage, low-skilled workers live (central city and inner suburbs)."[61]

Similar patterns emerged in Miami, Florida, where state highway planners and local officials deliberately routed Interstate-95 directly through the thriving African American community of Overtown in the center of the city. As with many other urban centers, especially in the South, segregation laws enacted in the decades following the Civil War severely limited housing options for African Americans in Miami. Overtown was the only central location in the city not officially "whites only" and despite overcrowding the neighborhood maintained a strong Black middle class. Known as the "Harlem of the South," owing to its strong local economy and entertainment scene, its streets were lined with hundreds of Black-owned businesses including "dentists, law offices, restaurants, laundries, beauty salons, and drugstores."[62] Between businesses were "wide, large brick and stucco homes, some with two stories, large front porches, and occasionally, two and three bathrooms to spare" belonging to Miami's African American professional class.

When the downtown section of the expressway was completed in the mid-1960s it obliterated huge amounts of housing as well as Overtown's main business district, the commercial and cultural heart of Black Miami. The huge I-95/I-395 interstate interchange "took up twenty square blocks of densely settled land and destroyed the homes of some 10,000 people." *The Miami Times* observed that no corner of Overtown "seemed to have escaped the angered wrath of the bulldozers and wrecking cranes that have been busy at work demolishing homes, churches, apartment houses, and business places."[63] The eight-lane highway interchanges destroyed 87 acres of housing and commercial property in the area. In addition to the highway, large parcels of land in Overtown were taken for parking lots, garages, and city and county office buildings. Almost 40,000 Black people lived in Overtown before the highway expansion, but only about 8,000 remained after the highway was built.

Among the properties destroyed was The Sir John Hotel. This was originally built as the Lord Calvert Hotel and was home to the popular nightclub Knight Beat which hosted some of the country's most famous Black entertainers. Overtown was an essential stop on the "Chitlin Circuit," the parallel music world that many Black artists traveled in the days of segregation.[64] But it was also close to Miami Beach, where local musicians and national stars like Lena Horne, Count Basie, Nat King Cole, Duke Ellington, B.B. King, Ella Fitzgerald, James Brown, and Aretha Franklin entertained at white hotels. In the Jim Crow world of 1950s Miami, African Americans, even if they were nationally renowned celebrities and stars were prohibited from staying in the hotels in which they were performing. After their shows to all-white audiences, many of these stars would travel, often accompanied by white fans, back across the railroad tracks to Overtown to the popular hot spots that included Knight Beat, Harlem Square, the Island Club, and Mary Elizabeth's Hotel Fiesta Club. These clubs provided an opportunity for integrated entertainment and socializing in a wider world of segregation. The swimming pool at the Lord Calvert Hotel, which was desegregated, was a popular hangout for visiting artists and prominent Black leaders. The pool provided the location for one of the most iconic images of the twentieth century when Flip Schulke photographed a young boxer named Cassius Clay holding a fighting stance underwater wearing his Everlast boxing trunks.[65] Much like the rest of Overtown – which city authorities had officially deemed a "blighted slum" – the hotel was demolished in the 1960s and replaced with a parking lot for a U.S. Postal Service distribution facility. Adam Paul Susaneck captured before and after images of the location for his website Segregation by Design, which illustrates the extent of the destruction.[66]

Authorities in Miami, as elsewhere, compensated building owners with amounts well below market rate and offered renters – who made up most of the community – nothing at all. The outcome was that vast amounts of Black wealth in Miami were obliterated, and the highway significantly widened Miami's existing "color line." The extent of the devastation is evident in Figures 3.1 and 3.2.

Wandering around the huge swathes of negative space and derelict wasteland that constitute contemporary Overtown it is hard to believe that this was once a densely populated and thriving community. Multistorey car parks, empty spaces, concrete overpasses, and an absence of any form of street life give the area an eerie and desolate feel. The only evidence that remains of the "Harlem of the South" is a visitors' center with photographic exhibits reminding visitors that "Overtown was not a slum" and a handful of historic preservation sites such as the Overtown Performing Arts Center, and the Historic Black Police Precinct and Courthouse Museum. These sites, while commendable for their efforts at historic preservation, are scattered amid empty tracts of land and are insufficient to enhance an unpleasant and depressing walking experience.

As with Atlanta, the destruction of Overtown was the realization of a decades-long campaign by white business leaders to remove Black residents from the urban

FIGURE 3.1 Northwest 11th Street in Overtown, Miami. The Metrorail line is visible on the right-hand side of the image. This street is typical of many others in the area dominated by vast swathes of negative space and urban voids. Walking along this route is an unpleasant experience with no street life or frontal development. The area is full of what Jane Jacobs characterized as "shatter zones" i.e. areas where the urban fabric has been shattered by road development.

Photograph by Tim Cunningham, October 2024.

core and reclaim the land to expand Miami's central business district. Raymond Mohl identified how plans to remove African Americans from central Miami can be traced to the Great Depression when white elites in the city used undeveloped land five miles northwest of the central business district to build a New Deal public housing project for Black residents called Liberty Square. Opened in 1937, this development was intended to replace Overtown as the center of the Black community. In 1936 the Dade County Planning Board proposed a "negro resettlement plan" that called for the city to assist "in removing the entire Central Negro town [Overtown] to three model negro towns located on the distant agricultural fringes west of Miami."[67] In 1937 George Merrick, a prominent local developer, proposed a "complete slum clearance effectively removing every negro family from the present city limits."[68] These objectives continued through the 1940s and 1950s with the involvement of federal housing officials in plans for the development of public and private housing for African Americans on the metropolitan fringe. Although the earlier proposals never made it off the drawing board they are important in providing a context for the motives that drove urban renewal and the routing of I-95 in subsequent decades. As with the Rondo community in St Paul, an alternative and much less disruptive route for I-95

FIGURE 3.2 Northwest 1st Ct and 6th Street, Overtown, Miami, a few blocks from I-95. This area is currently dominated by parking garages, storage and postal facilities, and highway overpasses. The walking experience here is unpleasant with blank walls and an absence of eyes on the street creating a hostile environment in daylight. During the hours of darkness, feelings of anxiety and isolation created by this hostile environment would be significantly enhanced. Photograph by Tim Cunningham, October 2024.

along an abandoned Florida East Coast Railroad corridor was rejected to provide "ample room for the future expansion of the central business district in a westerly direction." In his impressive study of the remaking of South Florida, N. D. B. Connolly summarizes the racist intentionality behind the planning and urban design decisions that characterized slum clearance, urban renewal, and the highway program in Miami:

> The resulting disruption and pain many of these projects wrought was not, as some have argued, the result of some political accident or bureaucratic misstep on the part of otherwise earnest housing reformers. Displacements were intentional. They represented, for growth minded elites, successful attempts to contain Black people and to subsidize regional economies with millions in federal spending.

Local race agendas during this period need to be considered within the wider context of an enthusiastic endorsement of racial segregation by the federal government and the value of urban renewal and expressway design to achieve this objective. In other words, local officials, city planners, and architects were not exactly "going rogue" as they set about reshaping racial patterns of settlement in cities across the U.S. that enshrined older and more permeable racial barriers in concrete.

Urban renewal and the highway program presented white elites with the legal and financial means to bring about the spatial realignments that they had long been looking for, and which the courts had circumvented by proscribing overt racial zoning practices. In cities across the U.S., new civic centers, hotels, sports stadiums, office buildings, universities, and hospitals were all constructed on land that had been cleared by displacing African Americans under the benevolent guise of slum clearance. Another goal was keeping Black communities "away from downtown business districts so that white commuters, shoppers, and business elites would not be exposed to Black people."[69] Urban renewal frequently "eliminated poor and working-class neighborhoods and replaced them with high-rise, luxury structures for the more affluent while concentrating public housing sites within black neighborhoods."[70] Rather than alleviating the African American housing crisis, urban renewal significantly exacerbated the problem with much more housing destroyed than was replaced putting extra strain on an already chronic shortage, and rendering huge numbers of people homeless.

Given the amount of land that expressway building required, it was clear from the outset that highway construction was inextricably linked with housing policy. Notwithstanding concerns, President Truman rejected co-ordination of highway and housing programs, citing high costs and the difficulty of securing Congressional approval.[71] The outcome was that entire neighborhoods across the country were displaced with little regard for those who lost their homes, businesses, churches, schools, or community organizations.

> According to the 1969 report of the National Commission on Urban Problems, at least 330,000 urban housing units were destroyed as a direct result of federal highway building between 1957 and 1968.[72]

In the early 1960s, the House Committee on Public Works found that "federal highway construction dislocated an average of 32,400 families each year." According to estimates from the U.S. Department of Transportation, more than 475,000 households and more than 1 million people were displaced nationwide because of the federal highway roadway construction.[73]

In some cases, displaced African American communities were relocated to new large-scale, high-rise public housing projects such as the Robert Taylor Homes in Chicago, which were to become synonymous with urban decay, disinvestment, deprivation, and crime. Many other African Americans "were forced to move into other inner-city neighborhoods that were transitioning from white to Black which further concentrated poverty and disadvantage." The AARP (American Association of Retired Persons) compiled a series of stories from those who were displaced by the interstate highway system that provides detailed accounts of the history of the program and its devastating impact on local communities torn apart and isolated as hulking concrete structures replaced the tight-knit and walkable streets containing homes, businesses, and churches that characterized many communities.[74]

The long-term impact of concentrating entire communities into spatially isolated housing projects has been well documented. As Massey and Denton concluded

> That social pathology would follow was inevitable. Concentrating poverty concentrates things that correlate with poverty ... People adapt to hostile and violent environments by becoming hostile and violent.[75]

The impact of these policies is evident in cities like Chicago where nine of the nation's ten poorest census tracts in the 1990s were in its housing projects; and most of these were in the South Street Corridor.[76] Warnings about the destructive consequences of concentrating poverty in this way predated construction of these developments but the warning was ignored because the social, economic, and physical isolation of African American communities was not an unfortunate consequence of misguided practices but was in fact one of the primary objectives. These objectives were set out the federal government in the FHA Underwriting Manual in 1938, and were pursued with gusto across the country in the subsequent decades.

3.1.3 Redlining and Highway Robbery: The Rhetoric and the Reality

This section analyzes the impact of disciminatory lending, the highway program, and urban renewal to show how a highly racialized series of policies spanning several decades came to shape racial contemporary inequalities. This section challenges some accounts that attribute current patterns of racial disadvantage solely to the HOLC and overlooks how many urban communities engaged in efforts to circumvent redlining practices. This section argues that more research is needed on the role that urban renewal and the highway program played in contributing to the current racial wealth gap.

One consequence of several decades of racial zoning and racially restrictive covenants, supported in many cases by the threat and reality of racial violence was that most U.S. cities in the early decades of the twentieth century had heavy concentrations of African Americans in areas with some of the oldest and most dilapidated housing. This allowed state highway engineers and city officials to claim technocratic justification for routing roads through African American communities on the grounds that right-of-way expenses would be lowest and revitalization most needed. The reality however is that these technocratic explanations masked the deeply political reality of highway planning and design, which was predicated upon highly racialized notions of urban space. The land valuations and route designs that took place during the highway program were conducted by the same local appraisers who had been inculcated over several decades with bogus and pseudo-scientific theories that African American and minority ethnic presence within a given area was synonymous with decay and blight. While there were undoubtedly decaying properties in many African American communities, not least given that homeowners in these areas faced significant obstacles to accessing credit, there were also valuable and thriving businesses that white elites ignored or in some cases deliberately targeted for destruction as part of a racial land grab.

In some cases, white elites were quite open about the racial dimensions of expressway planning, as evidenced by observations from the former executive director of the

American Association of State Highway Officials, Alfred Johnson, the lobbyist most deeply involved with the congressional committee that formulated the 1956 Highway Act. He subsequently recalled that some city officials expressed the view in the mid-1950s that the urban Interstates would give them "a good opportunity to get rid of the local niggertown."[77] More commonly, however, the fate of African American communities was not mentioned explicitly in discussions but "… when the highwaymen talked about clearing out central-city blight in the postwar era, everyone knew what they meant."[78]

As the examples in the last section illustrate, in some instances highway design was explicitly linked to local and even personal racist agendas. Another pattern emerges however that challenges the notion that highway planners were simply following the path of least resistance where land values and political opposition were lowest. The routing of I-85 to separate the historically Black Alabama State University from its nearby residential neighborhood of doctors, faculty, and civil rights intelligentsia is part of a wider trend that includes the intentional destruction of thriving communities like Sugar Hill in L.A., Rondo in St. Paul, and Overtown in Miami, where African American middle- and in some cases upper-middle-class communities were deliberately targeted. In the cases of Overtown and Rondo, alternative and less expensive routes for the highways were available in the form of disused railroad tracks (Overtown) and vacant industrial land (Rondo). Cheaper, less disruptive, and certainly less controversial routes were discounted to maximize the destruction inflicted upon African American communities and wealth. In other instances, such as in Nashville, Atlanta, and Chicago, highway routes were deliberately altered, with a "kink" added to I-40 and the design of the Dan Ryan Expressway changed with sharp curves inserted to maximize its impact as a racial barrier around the Robert Taylor Homes. In Chicago, Richard Daley's insistence on only siting public housing within existing African American territorial boundaries also meant that land costs for public housing projects were far higher than in most other cities in the country.[79] In L.A., highway routes were altered, sometimes at considerable expense, to target minority communities, including the African American mansions of Sugar Hill. The notorious "Highway to Nowhere" in Baltimore, Maryland, has been described as a planning conundrum and appears to be the result of incompetent transportation planning, only if one discounts the other "Highway to Nowhere" in Akron, Ohio, and all the other assorted kinks, curves, and dead-ends that characterize other interstates, from Miami, Florida, to Syracuse, New York, and from Charlotte, North Carolina, to Los Angeles, California.

The reality is that transportation planning in the U.S., epitomized by the interstate highway program, is often not driven by efficiency and cost-effectiveness. Raymond Mohl, the preeminent historian of U.S. highways, has observed that "very few Blacks lived in Minnesota, but the road builders found them."[80] The politics of this period determined that although highways were ostensibly designed to facilitate "slum clearance," African American and other urban ethnic neighborhoods were targeted for destruction – regardless of the level of wealth that they contained.

In this respect, the routes and designs that shaped the interstate highway program followed inexorably the reasoning and guidance of the FHA Underwriting Manual that the "infiltration of inharmonious racial groups will produce the same effects as those which follow the introduction of incongruous land uses, when the latter tend to lower the level of land values and lessen the desirability of residential areas."[81] Under this framework the only way to preserve land values and economic value was to deploy natural or artificially established barriers to protect a neighborhood and the locations within it "from adverse influences … [including] the infiltration of … inharmonious racial groups."[82] Chief among these was the use of a "highspeed traffic artery or a wide street parkway" that "may prevent the expansion of inharmonious uses to a location on the opposite side of the street."

Adherence to the rationale of the Underwriting Manual led to the widespread inability of African Americans to access long-term, low-interest, federally guaranteed mortgages, which had huge consequences for the racial composition of suburbia (see Section 5.1). The role of the federal government in championing policies that promoted segregation through denial of home loans has been the focus of a significant amount of study, not least given its impact in perpetuating racial wealth inequality and racial segregation. Writing about the FHA, Charles Abrahms concluded,

> The publication of anti-racial doctrine by a federal agency established official justification for prejudice and gave the semblance of right to lawlessness. The evil that FHA did was of a peculiarly enduring character. Thousands of racially segregated neighborhoods were built, millions of people re-assorted on the basis of race, color, or class, the differences built in, in neighborhoods from coast-to-coast.[83]

The role of redlining and the FHA in denying African Americans access to home loans and entry into suburbia has been the focus of much recent attention, but much less has been written about the way in which the highway program and urban renewal destroyed huge amounts of pre-existing Black wealth.

Interest in redlining has been fueled in the past decade following the wave of killings of African American men at the hands of police officers including Michael Brown in Ferguson, Missouri, in 2014; Freddie Gray in West Baltimore in 2015; and George Floyd in Minneapolis in 2020.[84] This in turn has led to a welcome focus on racial disadvantage including the racial wealth gap.[85] The average African American family currently "has around one-tenth the wealth of the average white family."[86] Much of the contemporary analysis however tends to focus on the role of redlining, FHA lending practices, and other historic barriers to homeownership.

Redlining in particular has become something of a bête noir for a new generation of scholars, activists, and commentators, who have sought to link the color-coded maps with contemporary patterns of inequality. Richard Rothstein's book *The Color of Law* for example has a redlining map on the cover, even though his book says comparatively little about the actual practice of redlining – and in fact is much more critical of other federal agencies like the FHA. Numerous other articles and journalistic pieces have also featured the color-coded HOLC maps which appear to provide a compelling and accessible visual portrait of contemporary racial inequality. In the digital age, colored maps can be reposted, shared, or retweeted providing a snapshot that captures the imagination and is prime material for clickbait. The mapping inequalities database established by the University of Richmond in Virginia, which provided many sources highlighted above, has also served to focus attention on redlining practices, as has award-winning essays like "The Case for Reparations" by Ta-Nehisi Coates in The Atlantic.[87] While renewed attention on original HOLC maps and redlining is very welcome, observers should be careful about the visual appeal of color-coded images that provide an overly simplified explanation for complex patterns of racial disadvantage that have been engineered over the past century or more. HOLC assessments, while questionable as an instrument for predicting *current* levels of racial disadvantage,[88] are a hugely valuable resource for providing a snapshot of the prevailing zeitgeist of the period that was predicated upon the view that racial and land use diversity reflected economic decline and financial risk.

There are several reasons to exercise caution when determining the long-term causal impact of the HOLC practices on tracts of land, especially when drawing a direct link between redlining maps and current patterns of deprivation. First, there is a paradox about the work of the HOLC, in that notwithstanding the racist nature of the redlining maps, Kenneth Jackson has shown that the HOLC did in fact issue mortgages impartially and make most of its obligations in "definitely declining" hazardous neighborhoods. The "lasting damage caused by the HOLC came therefore not through its own actions, but through the influence of the appraisal system on the financial decisions of other institutions."[89]

Chief among these institutions was the FHA, which enthusiastically embraced HOLC appraisal methods, but unlike the HOLC ruthlessly denied financing to minority homeowners and urban centers. Entire areas of cities were deemed ineligible for loan guarantees. By 1966, the FHA did not insure a mortgage on a single home in Camden or Paterson, both declining industrial cities. In any market, the inability to purchase is inexorably bound up with the inability to sell. As a result, homes in neighborhoods that were ineligible for federally insured loans often remained empty, which in turn precipitated a steep decline in value.[90] The main beneficiary of the $119 billion in FHA mortgage insurance issued in the first four decades of its operations was suburbia.[91] The corollary was a doom loop in which entire areas of cities were starved of finance, and middle-class occupancy was reduced which in turn led to further decline. This had devastating consequences for those living in cities who were not able to avail of an FHA-financed mortgage and purchase a home in the suburbs cheaper than they could rent i.e. African Americans. In 1967, Nolan summed up the indictment against the FHA by asserting, "The imbalance against poor people and in favor of middle-income homeowners is so staggering that it makes all inquiries into the pathology of slums seem redundant."

The Urban Institute and several other leading scholars including Rob Gioelli and Scott Markley have also sought to expand understanding of redlining by pointing out the importance of charting the range of processes that took place in the decades between the drawing up of the original HOLC maps and the emergence of the Black Lives Matter movement eight decades later. This includes assessing the role of the FHA, other lenders, the real estate industry, suburban developers, and the range of stakeholders who sought to limit African American residential mobility during the twentieth century. Included in this list are law enforcement officials who turned a blind eye to, or in many cases actively assisted in, acts of racist violence to impede African American residential mobility.

Another difficulty with drawing a direct parallel between redlining maps and current patterns of deprivation is that this approach overlooks the agency of the communities that successfully fought and resisted redlining. Much of the contemporary commentary implies that communities that were "redlined" passively accepted the process. This feeds into the conventional narrative that redlining precipitated a process of urban decline that brought designated areas to their knees which in turn made them ripe for urban renewal and targeting by highway officials. From this perspective, redlined communities were not targeted because of their racial composition per se but because they had gone into significant social and economic decline that could only be alleviated by the intervention of the wrecking ball. There is some element of truth to this argument. But this analysis is limited.

The main problem with the Redlining = Deprivation = Urban Renewal = Displacement thesis is that it obscures the significant efforts of those who managed to circumvent the racist policies and practices of the federal government at that time. Examination of writing from the period highlights the various measures that urban communities deployed to challenge the widespread problem of being denied access to finance. Jane Jacobs used "credit blacklisting" rather than "redlining" to describe how "dooming" a neighborhood was both common and destructive.[92] In 1959 Jacobs visited the North End in Boston, which at that time

> was cut off from building, expansion, or rehabilitation credit from the American lending system almost as effectively as if it were a community in Tasmania.[93]

Notwithstanding credit blacklisting practices, she discovered that the North End was one of "the most vibrant communities in the city" with property improvements in the area financed by earnings generated within the neighborhood with local craftsmen (electricians, plumbers, stonemasons, etc.) engaged in "primitive methods of barter." The inability to borrow money for property improvements was a serious concern for residents who were aware that this improvised form of repair and refurbishment could not last forever and were wary that the only investment their areas could secure was through urban renewal which would wipe out local businesses – a fate that had befallen the nearby West End district of the city. As it happens, the North End of Boston was spared the worst excesses of urban renewal in the 1960s, and the contrast between the two districts is apparent today.[94] The North End remains a historic, walkable, and thriving community, while the West End is synonymous with many of the bland and sterile urban renewal developments pioneered during that era.

Another community Jacobs identifies that successfully overcame credit blacklisting in the 1950s was the Back-of-the-Yards neighborhood in Chicago.[95] In this instance, the local neighborhood council (BYNC)[96] which included a Bishop, prominent sociologist Saul Alinsky, and community organizer Joseph B. Meegan, met with representatives of several local banks and savings and loans institutes.[97] The lending institutions were informed that local depositors – businesses as well as individuals – would withdraw their money if credit blacklisting continued in the district. The banks relented, and between 1953 and 1963, the BYNC fostered the rehabilitation of 90%

of the community's housing stock. Jacobs also describes how her own neighborhood, Greenwich Village, resisted credit blacklisting for 12 years. She also recounts how in East Harlem, John J. Merli, city councilman and leading figure in the New York County Committee of the Democratic Party advanced money for buying necessary materials and arranged for labor barter and labor donation in the fashion of Boston's North End. He secured bank loans for the property owners so that they could repay him for materials. African Americans in Atlanta found another more straightforward way of circumventing the refusal of white lending institutions to provide mortgages. Ronald Bayor highlighted how the ability of African Americans to "migrate quickly across Atlanta's west side" in the 1940s and 1950s surprised the city's white elites, who were unaware of the economic power that they wielded through the Black-owned Atlanta Life Insurance Company, Citizens Trust Bank, and Mutual Federal Savings and Loan Association of Atlanta.[98]

It is impossible to determine how many communities across the country followed the examples of those in the North End in Boston, the Back-of-the-Yards in Chicago, or Atlanta's west side however accounts of the impact of the highway program and urban renewal on African American neighborhoods highlight the success and wealth that existed in neighborhoods such as Overtown in Miami, Sweet Auburn in Atlanta, Tremé in New Orleans, Black Bottom and Paradise Valley in Detroit, Rondo in St Paul, Brooklyn, North Carolina, the Hill district in Pittsburgh, and Greenwood Avenue in Tulsa. Each of these communities, and indeed many others across the country were variously labeled the "Harlem of the West," "Harlem of the South," "Little Harlem," or "Black Wall Street." These monikers reflected the affluence and vibrant culture that existed within neighborhoods that had been subject to official redlining for decades. The question for contemporary scholars therefore is how exactly did these areas become so vibrant and successful despite being "redlined" for decades? And what does this say about the impact of "redlining"?

Some communities undoubtedly thrived against the odds by engaging in some of the informal trade and barter practices like the residents of East Harlem, or borrowed money from local civic leaders or businesspeople, undoubtedly at a premium. Despite racist employment practices, many African Americans were economically successful through running their own businesses or by virtue of their own ingenuity and forbearance. The original HOLC maps and area descriptions are useful in identifying how some successful African American neighborhoods emerged. The description of the Rondo district in St Paul (designated D4) for example states:

> The class of colored people in this area are somewhat better than other districts, many of them, due to the city of St Paul being the General Headquarters for the Northern Pacific and Great Northern Railroads, are pullman porters who do acquire ownership of their homes.[99]

Pullman Porters were originally former slaves employed by the Pullman railroad company to work as stewards in their private dining carriages and were among the more prestigious positions available to ordinary African Americans in the early decades of the twentieth century.[100] Despite the area consisting of homeowners of "a better class," it was still listed as a "D" grade; it highlighting the racist judgments that were rendered by appraisers. Rondo was also a vibrant center of African American culture, providing a haven to live with dignity and respect.

Some scholars have claimed that the racial wealth gap "is largely explained" by the discriminatory lending practices of the FHA. Undoubtedly, discriminatory lending practiced by the FHA severely restricted African American homeownership in the postwar years, which in turn resulted in less property wealth to be passed on to successive generations. Redlining and FHA lending practices were, however, fundamentally about refusing access to credit which impeded the ability of African Americans to invest in property in the first place. The highway program and urban renewal were about obliterating wealthy and thriving businesses that were already well established. Smart Growth America carried out a study to estimate the amount of wealth that was destroyed by two interstates, one in Atlanta and one in Washington, D.C., from the perspective of property tax revenue lost to both cities. Reconnect Rondo has also sought to estimate the value of Black property that was destroyed in St. Paul. This author is not aware of a body of work that has estimated the total amount of Black wealth that was destroyed by the highway program across the country but such a study ought to be a priority.

Any assessment of the African American communities that were devastated during the 1950s must also acknowledge the chronic levels of overcrowding and decay with many properties subject to vermin infestation and unsafe conditions that led to deaths from arson (sometimes by slumlords) and gas poisoning. Many of these areas contained successful entrepreneurs, some of whom made fortunes by exploiting members of their own community. White businesspeople were very happy to exploit racialized housing markets by charging African Americans extortionate rents or by engaging in unscrupulous mortgage lending practices such as "contract sales."[101] These practices were only possible because white elites were so keen to uphold "color lines" limiting the territory into which African Americans could move and because the FHA denied African Americans the kind of long-term low-interest U.S. Treasury-backed loans available to whites fleeing to the suburbs. In "The Case for Reparations" Coates includes redlining practices and the refusal of the FHA to grant home loans to African Americans as part of his argument for reparations. Coates however makes no mention of the amount of Black wealth that was obliterated and seized under urban renewal projects and the highway program. In fact, he is quite dismissive of inner-city African American neighborhoods, claiming,

> It is common today to become misty-eyed about the old black ghetto, where doctors and lawyers lived next door to meatpackers and steelworkers, who themselves lived next door to prostitutes and the unemployed. This segregationist nostalgia ignores the actual conditions endured and ignores the actual conditions endured by the people living there – vermin and arson, for instance – and ignores the fact that the old ghetto was premised on denying black people privileges enjoyed by white Americans.[102]

One does not need to engage in "segregationist nostalgia" to appreciate the wealth and success that existed in African American neighborhoods prior to the highway program and urban renewal. To ignore the level of wealth that was accumulated despite the obstacles that the federal government put in place denies the agency, determination, and resilience of African Americans during this period in the nation's history. Perhaps the best example of African American ingenuity and fortitude is the case of Greenwood in Tulsa (see previous section). This "Black Wall Street" was obliterated by white violence in 1921, but despite resistance from white civic leaders it was rebuilt and thrived until it was devastated by the waves of additional violence in the form of urban renewal and I-244 in the 1960s. Again, this was an area that experienced redlining and other petty roadblocks to redevelopment in the form of municipal ordinances but was nonetheless rebuilt. In an interview with Bloomberg, Victor Luckerson argues that the plethora of wooden shacks used in the hasty reconstruction of Greenwood made the area vulnerable to destruction by the planners and architects of I-244. But Luckerson's own book also highlights how many whites were vehemently opposed to rebuilding Greenwood after the massacre arriving on the scene in the immediate aftermath to buy up land with some sources indicating the destruction may have been a pre-planned racial land grab. Regardless of the quality of the structures in Greenwood, whether they were timber-framed shacks or made from brick and steel, arguably the area was destined to be destroyed in the same way that Overtown, Rondo, Sweet Auburn, and other Black centers were also targeted. The stately mansions of Sugar Hill were among the finest properties in the country – but that did not prevent them being obliterated.

Undoubtedly, there were limits to what the informal methods of neighborhood regeneration could achieve. Jane Jacobs was scathing about the practice of credit blacklisting describing it as "outrageous and destructive" and arguing that it led inexorably to social and economic decline in affected areas. But her analysis also points to a symbiotic relationship between slum clearance practices and refusal of credit that in some cases was not contingent on the condition of the actual area:

> Credit-blacklist maps are identical, both in conception and in most results, with municipal slum clearance maps ... Sometimes the planners anticipate the lenders; sometimes the lenders anticipate the planners ... Credit blacklisting maps are accurate prophecies because they are self-fulfilling prophecies ... There is no telling how many city districts have been destroyed by blacklisting.[103]

In her arguments with city officials in Boston Jacobs described a conversation with one municipal official who agreed with her that the North End of Boston was not a slum and that he liked to go to the area himself to wander around. The official conceded that despite its vibrancy the area was scheduled for destruction under urban renewal because that was what the map he had been given indicated. Credit blacklisting, while "outrageous and destructive," was not responsible for the wholesale obliteration of urban communities but rather was part of an exercise that ensured certain neighborhoods would be identified, labelled, and targeted for destruction by urban renewal regardless of the level of deprivation or level of wealth they contained.

Evidence here shows that HOLC maps were not the main driving force behind shaping current patterns of concentrated poverty and deprivation but rather part of a package of measures that included the FHA, local real estate professionals, planners, architects, road engineers, and municipal authorities. What all these actors had in common was that they pursued, with ruthless efficiency, the containment and isolation of African American communities that followed a playbook developed and honed by British, French, German, and other colonial administrators across Africa and Asia in the eighteenth and nineteenth centuries. As outlined in Chapter 1, Haussmann's solution to the insurrectionary densely packed working-class *quartiers* of Paris, with illegible street plans, was a transportation network that would ensure direct rail and road links "between each district of the city and the military units responsible for ensuring order there." Where possible, "insurrectionary quartiers were demolished and broken up by new roads, public spaces and commercial development."[104] This thinking is apparent in cities from Minneapolis to Miami as African American neighborhoods were subject to the same rationale. Chapter 1 also includes Obert's list of architectural and planning techniques deployed by the British, French, and Germans across their empires that include the physical relocation of local populations to cities' outskirts, ensuring the current and future marginalization of colonized subjects and the mobilization of discourses of health and hygiene to justify racial segregation.

Analysis of the operations of the urban renewal and the highway programs also reveals that African American communities were, in the words of Coates, "viewed as a contagion." Even middle-class residential districts in central cities were considered suspect, since there was always the prospect that they too might change their residential composition as African Americans or poor whites continued to pour into cities, and as middle- and upper-middle-income whites continued to move out.[105] With military efficiency, and in accordance with FHA guidance, physical and legal barriers sought to ensure African Americans were segregated and isolated by concrete cordon sanitaire developed and designed by architecture and planning professionals across the U.S. This was a process that spanned several decades and went way beyond the color coding of maps in the 1930s as technocratic elites sought to reimagine the spatial arrangements of urban areas using highway design and urban renewal programs to marginalize and exclude minority communities. What is also significant about this process is the way in which it followed the path of "Foucault's Boomerang" and was exported across the Atlantic in the mid-twentieth century, with equally devastating consequences. This will be explored in the next section.

3.2 BUILDING CONFLICT IN NORTHERN IRELAND

This section examines the role that planning and architecture played in helping to precipitate political violence in Northern Ireland during the 1960s. Included in the analysis is examination of the ways in which modernist planning orthodoxy of the 1960s, imported from the U.S. and characterized by large-scale urban redevelopment programs and displacement of inner-city communities, accentuated pre-existing tensions within a political system that was already under pressure from a growing and discontented Catholic population. The violence that erupted in many U.S. cities in the late 1960s mirrors similar experiences in Belfast, with minority communities on both sides of the Atlantic falling victim to planning and architecture practices implemented by ruling elites who were keen to use technocratic arguments to enhance their social, political, and economic advantage.

3.2.1 Spatial Inequalities and the Legacy of Conflict

Shortly after resigning from the British delegation at the Paris Peace Conference in 1919, John Maynard Keynes observed,

> The great events of history are often due to secular changes in the growth of population and other fundamental economic causes, which, escaping by their gradual character the notice of contemporary observers, are attributed to the follies of statesmen or the fanaticism of atheists.[106]

Keynes' maxim about the importance of systemic changes over time fits with Jane Jacobs' temporal vision of a city ecosystem in which observers, who often thought that they were viewing static situations, were in fact seeing "processes of beginning and processes of ending occurring simultaneously."[107] In several respects, urban planning and architecture in Belfast over the past century has been characterized by successive failures to recognize and manage processes of beginning and processes of ending which ultimately helped precipitate the political violence that would engulf the city in the late 1960s. The most important secular "process of ending" in Belfast during the twentieth century has been deindustrialization. In 1920, Belfast was the industrial heartland of Ireland with an economic base centered on shipbuilding, linen, engineering, and a ropeworks that was the largest in the world.[108] The Victorian grandeur of the city was reflected in a population growth level unequaled by any other urban center in the British Isles during that period – the population of the city rose from under 20,000 at the beginning of the nineteenth century to almost 350,000 in 1901.[109] Significant growth levels in population and economic fortunes during the nineteenth century were mirrored by sharp differences in the political, social, and economic status of Catholic and Protestant communities. Most notably, prosperity and wealth generated during the nineteenth century was largely concentrated within the hands of the Protestant community – the owners and management of most industrial and commercial concerns were Protestants as were the bulk of the workforce and the skilled trades.[110] In 1901, Protestants comprised 76% of the population but constituted 93% of Belfast's skilled working class; Catholics, constituting 24% of the city's population as a whole, provided the remaining 7%.[111]

Another important characteristic of Belfast during the nineteenth century was the presence of significant levels of sectarian violence which in many instances focused on attempts by the growing Catholic population to access the city's housing stock and collective attempts by Protestants to contain them.[112] The work of Hepburn, Boyd, and others has highlighted how sectarian street violence arising from Catholic attempts to move into traditional Protestant territory was a major factor in precipitating riots in 1872.[113] These riots culminated in the Catholic victory at the Battle of Leeson Street which secured the expansion of the main Catholic district in the city, the Falls, into the south-west of the city.[114]

The growth levels in the population of the city during the nineteenth century continued through the first half of the twentieth century so that by the late 1940s the population of Belfast had reached its high point.[115] The 1951 census figure put the population of the city at 443,671, 32% of the entire population of the region. By 1971 the population was down to 362,082, little more than it had been at the start of the century. This change was due to the "steady movement of people from the core of the city to more suburban areas in the period 1951–1971, with the greater urban area reaching around 600,000 in 1971."[116] This pattern continued through the 1970s and 1980s, which saw

> a continued decline in population and the number of households in the inner core of Belfast city so that from 1971 to 1991 there was a 55% loss of population. At the same time the outer regional area of the city grew by 39%.[117]

These figures illustrate a general pattern of decline in the population of the city since World War II following the massive growth of the nineteenth century. This pattern is common to older industrial cities across the U.K. and the U.S., where the hollowing out of the industrial base, and the decline of traditional industries allied to suburbanization, saw similar falls in population in the urban core.[118] What is especially notable about the phenomenon in Belfast however is the way in which these population trends differ across the two communities. Most of those who left Belfast in the 1970s and 1980s were Protestants with the result that Belfast is now a more "Catholic" city than at any time in its history.[119] The extent to which these demographic changes have impacted upon political representation in the city is evidenced by the fact that only representatives of the Unionist community held the position of Lord Mayor of the city from 1842 until 1996. In the years since, the role of Belfast Lord Mayor has alternated on a yearly basis between Unionist, Nationalist, and Alliance party representatives.[120]

In terms of the spatial distribution of the current population, increases in the Catholic percentage of the population in the city have been accompanied by high levels of residential segregation.[121] Segregation levels between the two communities can be seen starkly in the division of the city between the two sides of its main river, the Lagan.[122] Whereas the core city to the west of the Lagan is 55% Catholic, to the east it is only 12%. In terms of overall levels of segregation within the city, Murtagh and Carmichael have pointed to the fact that 68% of electoral wards in Northern Ireland have a population in which more than 70% of the residents belong to one community or the other.[123] In the late 1960s, six in ten of public-sector households in the urban area lived in streets that were segregated, but a decade later, it was nine in ten households.[124]

In the north and west of the city in particular, segregation by community background is heightened by the presence of concrete or steel barriers euphemistically referred to as "peace lines" – see Figure 3.3.[125]

The demographic characteristics of Protestants and Catholics living at these peace lines or interface areas as they are commonly known differ significantly, with Catholic communities characterized by higher family sizes, higher fertility rates, and a more youthful demographic structure. On the other side of these barriers, Protestant

FIGURE 3.3 This security barrier aka "Peace Wall" is located at Cupar Way, West Belfast. The Catholic Falls Road is located to the left of the picture behind the barrier, the photo is taken from the "Protestant" Shankill side. Photograph by Tim Cunningham, 2015.

populations are older, with smaller household sizes and lower-than-average fertility rates. When these demographics are played out in highly segregated space, very different housing need profiles are reflected in terms of the number, size, type, and location of accommodation required.[126] This in turn has serious and important consequences for the post-conflict regeneration of these areas which experience chronic levels of social and economic deprivation. Another striking feature about contemporary Belfast is that segregation by community background is augmented by socio-spatial segregation, creating an "acute intersection of socio-economic and identity-based inequality."[127] The uneven distribution of spatial deprivation is evidenced by the fact that the north and west Belfast parliamentary constituencies account for nine out of the top ten most deprived areas within the entire region of Northern Ireland.[128]

In summary, the Catholic community is more likely to live within the most deprived areas, deprived spaces are more likely to be segregated than non-deprived spaces, and there is a significant association between deprivation and political violence in Belfast. If the electoral ward death rates per 1,000 of the population are correlated with the Belfast deprivation scores that were generated in 1991 (Robson et al., 1994), the coefficient is 0.7 – in short, "the level of deprivation in a local area predicts about half its experience of political violence."[129] The areas that experience the highest levels of politically motivated violence during the conflict have experienced the least in terms of a peace dividend and are becoming relatively worse off compared to the rest of the city.

One of the striking features about these demographic changes is that they occurred despite a range of government-sponsored equality and anti-poverty measures spanning several decades.[130] As this research will show, many of these demographic imbalances and socio-economic and identity-based inequalities are the direct result of high-level planning and architectural processes that were established over several decades and wove sectarian divisions into the urban fabric of the city. In some cases, funding from government programs to address area-based deprivation was deployed for coercive architectural and planning technologies, instigated by a security-based approach to urban development, which produced physical, social, psychological, and economic barriers to residential mobility. This is an important finding and highlights how place-based initiatives to address socio-economic inequalities can be undermined by architectures of exclusion.

3.2.2 The Belfast Ecosystem and the Economics of Decline

To understand how the Belfast ecosystem came to be fractured in the late 1960s it is first necessary to consider the systemic weaknesses that lay at the heart of the wider Northern Ireland economy going back over a century. The beginning of the end for many established communities in Belfast, particularly Protestant ones, began at the turn of the twentieth century when the decline of traditional industries got underway. The difficulty for Belfast, as with many other industrial cities entering the twentieth century, was that the city was heavily weighted toward sectors with poor long-term prospects; however, it was not until after World War II that the scale of decline in traditional industries became apparent.[131]

In the late 1950s and early 1960s, Northern Ireland experienced a dramatic alteration in its manufacturing mainstays unequaled by many regional economies in the rest of the United Kingdom. The most concentrated job losses were in shipbuilding, where competition from a range of new competitors, and changing demand in shipping, "which favoured the new purpose-built yards in Europe and the Far East, meant that between 1960 and 1964 no fewer than 11,500 jobs, some 40 per cent, were lost in shipbuilding, repairing and marine engineering."[132] Similarly, the linen industry, another mainstay of the Belfast economy, fell victim to a range of secular changes notably competition from synthetic as well as natural fibers, the challenge of new foreign producers, and a change in consumer tastes away from linen fabric. The net result was that almost one-third of linen plants in Northern Ireland closed and "27,000 employees, over 45 per cent of the labour force, lost their jobs during the period 1958-64." Employment in agriculture, another important sector of the Northern Ireland economy, continued to contract as a consequence of modernization.[133]

The impact of these systemic weaknesses within the economy of the region led the government of Northern Ireland in the early 1960s to commission several reports exploring how to best achieve economic modernization.[134] At this time there emerged under the premiership of the Northern Ireland Prime Minister, Terence O'Neill, a "new technocratic unionism" which championed the twin objectives of a more accommodating relationship with the Catholic minority community while at the same time embarking on a program of economic modernization.[135] The basis of this "new technocratic unionism"[136] were two reports commissioned before

O'Neill became premier – the Hall Report and Professor Sir Robert Matthew's Belfast Regional Survey and Plan (the Matthew Plan) – as well as an economic plan drawn up by Professor Thomas Wilson at O'Neill's request.[137]

These reports represented an attempt to reconfigure the economy of Northern Ireland toward newer growth industries such as artificial fibers, and away from traditional industries notably, shipbuilding and linen. Belfast was the heartland of these traditional industries which provided not just the linen mills in and around north and west Belfast but also the Harland and Wolfe shipyard and numerous ancillary businessessuch as rope makers and engineering works representing the "economies of agglomeration" necessary to maximize production and reduce costs which characterize any urban center.[138] The production and transportation processes associated with the new growth industries, notably artificial fibers, required large brownfield sites on the city's periphery which militated against the location of new factories in the urban center, where space and land were at a premium. Successful reconfiguration of the Northern Ireland economy therefore required the relocation of large sections of the urban population of Belfast out of the city to provide a workforce for these new industries in outlying suburban brownfield sites. This objective was in line with modernist planning theory and practice of the time emanating from the U.S. and predicated upon large-scale demolition and redevelopment of urban centers, decanting of urban populations to surrounding new towns, and a massive expansion of road building.

The "Matthew Plan," which examined a number of these issues, including the need for new housing, an expanding population, and the problem of the substantial growth of the population of Belfast, was submitted to the Northern Ireland government in October 1962.[139] Matthew proposed a radical solution to fence the city of Belfast with a "stop line" beyond which the growth of Belfast would be halted while the people living in substandard housing within the city would be housed in seven existing towns or "growth centers" that would be greatly expanded, and one entirely new city "Craigavon" to be positioned in the countryside between Portadown and Lurgan.[140] Many of those who relocated to the new growth centers would be required to commute into Belfast and in line with conventional wisdom at the time, this was to be achieved by car with a series of motorways connecting these towns to Belfast. Such was the influence of Matthew's proposals that Belfast remains a commuter city, and with a few exceptions, "stop line" remains largely intact.[141] The model applied by Matthew fit within planning orthodoxy of the period and originated far beyond the streets of Belfast. The problem is that the failings of this orthodoxy, especially in the U.S., were ignored which would lead to profound and unforeseen consequences for sectarian politics and the ever-present tensions around land and identity in the city.

3.2.3 Authoritarian Modernism and Planning for Conflict

The new spatial vision for Northern Ireland that emerged in the 1960s reflected what became known as the dominant Anglo-modernist planning paradigm of that period.[142] These practices included large-scale demolition of substandard housing known as "slum clearance" and the decanting of inner-city populations to new tower blocks on the periphery of city centers, or other growth centers (i.e., "new towns"), with the cleared land in city centers used for new highways and commercial development.[143] This paradigm was presented as a technocratic solution characterized by tower blocks and new roads to replace the Victorian slums that blighted the urban landscape of the U.K. at that time. The problem was that the politics of the region meant that these planning decisions took on an ethno-national salience.[144]

One of the most problematic aspects of the new, Anglo-modernist vision for Belfast concerned housing provision. A key element of the Matthew Plan was the need to launch an urgent and concerted attack on dereliction which culminated in the 1969 Belfast Urban Plan. As with U.S. cities, there was nothing inherently problematic about a city plan that aimed to address significant levels of housing need. Several decades of under-investment in housing meant that many traditional working-class communities in Belfast experienced severe overcrowding and poor quality dwellings.[145] Apart from new housing, land in the inner city was to be designated for the new Belfast Urban Motorway (BUM) and new commercial/industrial development. The outcome of the Matthew Plan was to place the new road system at the heart of the development of the Belfast region with little enthusiasm for public transport and to accept that Belfast would become a city of commuters – a pattern still reflected in the current transport network. In order to create space for this development the Matthew Plan required that large sections of the inner-city population were to be encouraged to disperse to the suburbs and new towns, the locations for anticipated multinational investment.

Reflecting the command-and-control nature of the planning system at that period, there was scant evidence of public participation in these large-scale population movements.[146] As with U.S. cities, not only did many people in working-class areas of Belfast wish to remain where they were but the plans themselves increased housing need through the spreading of "blight." Once an area was listed for redevelopment, property values in the area slumped and existing landlords lost any incentive to modernize properties or carry out repairs.[147] Feelings ran particularly high in the Shankill area of west Belfast, where blight had 14 years to spread in the period between the time when the plans were first mooted and in January 1974 when the area was finally vested.[148]

Housing need was compounded by the effect of the Matthew "stop line," which halted most of the building that was already taking place in suburban areas and slowed down Belfast Corporation's already inadequate redevelopment program.[149] Maguire has highlighted how municipal leaders in Belfast Corporation, faced with the problem of finding houses for large numbers of people from inner-city areas that were in need of redevelopment, "were very reluctant to accept the stop line, and were by no means convinced that the growth of Belfast was as problematic as the planners thought."[150] The outcome of the Matthew Plan and the BUM project therefore was demolition of an already inadequate housing stock at a significantly faster pace than replacement dwellings were constructed further exacerbating the pre-existing housing crisis.[151] Housing need was compounded by the hubristic high modernist orthodoxy that embodied urban planning during this period, notably the belief that inner-city communities should be broken up and relocated regardless of whether they actually wanted to go. This view was epitomized by figures like Robert Moses in the U.S. and exported across the Atlantic, characterizing attitudes to redevelopment throughout the U.K. during this period.

The planning system in Northern Ireland remained oblivious to the difficulties facing urban communities, not least given the enthusiasm with which many politicians in other regions of the U.K., and beyond, embraced the modernist vision that had been presented to them by planners and architects.[152] In Britain, by the autumn of 1960, the pace of change in the urban built environment was becoming "as rapid as any time since the Industrial Revolution."[153] In fact, "revolution" was the term adopted by junior housing minister, Sir Keith Joseph, who told the Town Planning Institute's annual dinner in March 1961

> We are on the edge of a great job of urban renewal and a vast revolution in our industrial areas is taking place … heavy responsibility falls on the members of the planning profession. They are confronted with the challenge of not only filling in the void but of creating something out of existing disorder and creating ordered beauty out of empty space.[154]

Although there was a desperate need for old and substandard housing to be replaced in many British industrial cities not everyone was enamored with what was being proposed. In Glasgow for example, one local elected official declared that what disturbed him most about that city's radical Development Plan was that they were treating people like cattle, and "the proposals were being imposed from the planners' point of view without regard to other considerations."[155] In Britain, such concerns tended to be limited to a small number of local, usually independent elected officials and church leaders in addition to one or two dissenting voices within the planning and architecture professions, most notably Nairn:

> Does any planner in Birmingham, Liverpool or Manchester know how the Corporation dustmen live, what they enjoy in the city centre and what they abhor? The problem is at least as important as plot ratios, daylight angles and sight lines. To take one concrete example, most Cockney families like to live in the kitchen. The architect's job is not to ask why, nor to persuade Cockney's to live like young married architects, but to build big kitchens in the specific proportion that they are needed, as far as can be assessed. But do they? Not on your life …[156]

A key factor behind the pace of change that took place during that period was that radical redevelopment programs were supported across the political spectrum.[157] Most of the inner-city redevelopment took place with the active and vocal support of local Labour party officials and elected councilors in cities such as Glasgow, Sheffield, Liverpool, Newcastle, and Hull, although the urban renewal program itself had originated with the Conservative government under Harold Macmillan. Kynaston concludes that the push for radical urban renewal during this period became part of the prevailing "zeitgeist … a certain something in the air."[158] Among elected officials, the momentum was firmly behind redevelopment, reconstruction, and "slum clearance" to deliver a modernist utopia that was predicated on the needs of cars, not people

> with much greater vigour, we must rebuild our whole environment of working and living in terms of the motor car.[159]

From an economic perspective, the construction industry and those who made a living, both directly and indirectly, from that industry had a strong and vested interest in pursuing "revolutionary" changes in the built environment. The importance of economic stimulus has also been highlighted by Plunz, who argues that the motor industry played an important role in shaping postwar urban development in U.S. cities in favor of the car and away from public transport.[160] In autopoietic terms therefore, it can be said that during this period in the U.S. and the U.K., a structural coupling took place in which modernist planning and architecture became aligned with political and economic interests in order to deliver a modernist utopia that required large-scale destruction of major urban centers with scant regard for the needs of the local communities. The Anglo-modernist planning model that shaped British and U.S. cities in the 1950s and 1960s was directly replicated in Belfast through the Matthew Plan and other projects, notably the Urban Motorway. In some respects, this was hardly a surprise, given that these projects were designed by the same technocratic elites who brought not just their expertise but also the flaws and controversies that their projects had caused in British cities. Unlike the situation in Britain however, a lack of political legitimacy for the "revolutionary" changes taking place was compounded by the sectarian territorial claims within the region.

One of the main difficulties with the redevelopment of Belfast during the 1960s was that the political system in Northern Ireland, which excluded representatives of the Catholic and Nationalist community from both central and local government in Belfast, ensured that important decisions that required large-scale demolition of housing, the breaking up of communities, and forced relocation, lacked legitimacy.

This lack of legitimacy was further compounded by the way in which sectarian politics impacted upon systemic inequalities experienced by the Catholic community. The Protestant and Unionist-controlled Belfast Corporation had little interest in rehousing Catholics who were displaced within its own boundaries. Moreover, outside Belfast, other Unionist-controlled councils used their planning powers to limit the development of much-needed housing for Catholics.[161] The dominance of Unionist political representation across Northern Ireland from the 1920s until the late 1960s had led to increased housing supply in predominantly Protestant towns such as Craigavon, Antrim, Bangor, Newtownards, and Carrickfergus, and suburban developments on the outskirts of Belfast such as Monkstown, Glengormley, and Whiteabbey.[162]

Additionally, several green-field mixed housing developments (i.e., those occupied by both Catholics and Protestants) were constructed beyond the Belfast urban fringe notably Rathcoole, New Barnsley, Springmartin, Lenadoon, and Twinbrook. This left a lack of specific house building programs directly targeted at the Catholic population to address their chronic levels of need. This problem was not brought under public scrutiny until the advent of the Northern Ireland Civil Rights Association in 1968. In fact, the reluctance on the part of Unionist councils to house Catholics was identified in the inquiry into the first wave of disturbances that took place concluding that

> among the immediate and precipitating causes of the disorders which broke out, was a rising sense of continuing injustice and grievance among large sections of the Catholic population in Northern Ireland … in respect of inadequacy of housing provision by certain local authorities.[163]

The reality for those experiencing chronic housing need in Belfast therefore was that the demolitions that took place to clear land for the BUM and the decanting of populations outside the city significantly increased the housing crisis by exacerbating the imbalances that had built up during the 1940s and 1950s. What this period shows clearly is the relationship between urban planning and civil unrest, and the capacity of the former to precipitate the latter. The former Sinn Féin President Gerry Adams has described how his involvement as a political activist began with opposition to the plans for redevelopment of the Pound Loney area of the Lower Falls in west Belfast which was the site of the Divis Tower blocks that opened in May 1968. In many ways, the objections of Adams and the protestors on the Lower Falls mirrored those in cities across the Irish Sea and the Atlantic:

> While we were all for the replacement of old, substandard housing, we were opposed to the plans which saw system-building in multi-story units as the answer to the demands of the housing crisis, and which structured grants to local authorities to favour the new methods. The old housing was desperately poor, but the Loney was a good area with its own sense of community, in the which the same families had lived in the same streets for generations. We were convinced that high-rise flats weren't the answer. How could you live up there? How would you get the pram down? Where would the children play? In a balcony fifty feet up in the air? …. there had been no consultation at all with local people.[164]

The Divis tower block complex, which replaced the Pound Loney housing and opened in May 1968, like many similar developments of that period in Britain and the U.S., was mostly demolished less than 20 years later and came to epitomize the worst example of the system-built developments of that era.[165] Adams describes in his memoirs how the public protests that he was involved in around opposition to this redevelopment led to confrontations with the police and disintegrated into rioting and civil unrest, a pattern that was to be repeated on a much larger scale in 1968–69 as other civil rights protesters, campaigning on similar issues such as housing, employment and electoral reform met the full force of the armed wing of the Unionist government.[166]

The comments by Adams regarding confrontations between protestors and the police highlight another difficulty with the Anglo-modernist planning framework imported from Britain and the U.S., namely, that opposition within the Catholic community, coupled with lack of political representation, led protesters against urban redevelopment onto the streets of Belfast who were brought into direct confrontation with an almost exclusively Protestant police force. The Royal Ulster Constabulary

(RUC) and the Ulster Special Constabulary ("B Specials" or "specials") showed themselves more than willing to flex their muscle when faced with protestors they viewed as "fifth columnists" intent on undermining the Union which in turn precipitated further disorder and street violence.[167] This dynamic is "widely regarded as having been a significant factor in the events precipitating the conflict."[168] Much like the Freeway Revolts that took place during the same period in the U.S., those opposed to the demolition of housing in Belfast or the construction of the urban motorway were protesting about political choices ostensibly promoted as technocratic planning decisions.

3.2.4 Urban Planning, Populations, and Political Violence

Morison and Livingstone contend that it was exactly at the point at which the Northern Ireland government began to act like a modern welfare state that issues of legitimacy of the state began to be posed.[169] This is a variation of De Tocqueville's thesis that the most perilous moment for a bad government is one when it seeks to mend its ways.[170] Close examination of the impact of the urban renewal programs that took place in working-class communities in Belfast during this period, and indeed, the way in which the economic modernization program of the 1960s exacerbated sectarian disadvantages, would indicate that the situation in Northern Ireland was somewhat more complex.[171]

There is little doubt that the technocrats within the planning profession adopting an Anglo-modernist vision could objectively justify several of the controversial decisions made during the 1960s. The decline of traditional industries required new inward investment which in turn required new infrastructure. Moreover, there were at the time sound technical reasons for believing that older, dilapidated and overcroweded dwellings ought to be replaced by new tower blocks. What the planners and architects in the U.K. failed to appreciate however is that planning is fundamentally predicated upon the control and manipulation of space, and in a divided society where territory and identity are inextricably linked "there is no such thing as a neutral design."[172]

The Unionist Government of Northern Ireland from the early 1960s onward was content to adopt the reports and recommendations that coincided with its own narrow sectarian interests. Examples of this include the redevelopment of "Protestant" towns while at the same time continuing to pursue sectarian policies that failed to provide adequate housing for Catholics who were displaced by the large-scale urban renewal programs and the construction of the new BUM.

At the same time, evidence from the period also illustrates how the Matthew Plan and the BUM project played a not insignificant and important role in destabilizing Protestant communities. For Belfast's Protestant community, the outcome of the modernization program that began in the late 1950s and early 1960s was an exodus of the more educated, skilled, and upwardly mobile members of that community from traditional Protestant heartlands.

This in turn would have disastrous consequences in terms of both social capital and, crucially, electoral support for political Unionism that has lasted until the present day. A study of labor mobility in Northern Ireland has shown that those who moved to new towns were young married couples whose parents were better educated, had better aspirations, and had higher incomes than those who did not move.[173] A further study of newcomers to one of the Craigavon "new town" developments found that the intake population tended "to consist of young married couples with children of school age, of whom 65% were semi-skilled and where 49% owned a car."[174] In fact, concerns about the adverse consequences for social capital and political representation within Protestant communities were brought to the attention of the Unionist government at the time by local political leaders. In January 1965, the Unionist Association of St Anne's Ward in West Belfast expressed concern at the fact that the plan for the BUM involved the loss of a significant number of homes in the area, all of which were "good residential properties," occupied by the "backbone of the community."

> It is appalling to consider the cold blooded inhumanity that regards human beings as no better than cattle to be moved around at official edict ... there is no possible building ground inside the City boundary where these extra 5000 replacement houses could be constructed ... presumably these 5000 families just vanish ... The more responsible, sensible people who are the backbone of any community, are going ... property and business values have dropped to slum clearance levels ... and there is the potential loss of a Unionist seat ....[175]

This letter also illustrates a further problem that the Matthew Plan and Urban Motorway created for Protestant communities namely, political representation. This fact is especially relevant in a context in which populations "are voters as well as occupiers of land."[176] Part of the difficulty for the Unionist community was that the sectarian nature of the political system in Northern Ireland dictated that Unionist hegemony was contingent upon maintaining Protestant electoral majorities. This meant that not only were Unionist-controlled councils outside of Belfast reluctant to provide housing for new Catholics leaving the city but also that Unionist councillors within Belfast Corporation were also fearful of measures that would disturb or threaten their political control within the city, especially anything that would reduce the population of Protestant residents.[177] The latter concern is highlighted above by the reference to the "potential loss of a Unionist seat."[178] This issue was raised as far back as 1958 when a delegation of Unionist councillors from west Belfast raised concerns with the then Prime Minister of Northern Ireland that house demolition for the urban motorway would commence a "drift away" from the constituency and that the Unionist hold on the U.K. parliamentary seat would be "seriously imperilled."[179]

This judgment proved prescient given that just over a decade later West Belfast ceased to have a Unionist representative in the U.K. parliament. Records show that between 1958 and 1965 successive delegations of Unionist councillors got short shrift from their own government regarding concerns about the adverse consequences of the redevelopment of the city for Unionism. For the Stormont government, concerns about the loss of Unionist votes in West Belfast were the price to be paid for the economic modernization program that promised to deliver wider prosperity for the region. The reference to Unionist communities in Belfast being "treated like cattle" in the letter from the St Anne's Unionist Association mirrors the terminology used by the councilor in Glasgow referring to the development plan in that city (see above) and some of the commentary from the U.S. during this period about the impact of programs led by Robert Moses and others. What was different between Northern Ireland and Britain during this period was the way in which political representation impacted upon overtly sectarian planning decisions and the response of the government of the region to dissent.

In this respect, the Belfast experience with urban renewal and modernist planning is more akin to that of the U.S. where African American communities were targeted for displacement, many of whom faced similar problems regarding limited housing options in the emerging suburbs. As outlined in the previous section, racially restrictive covenants, racial zoning, discriminatory lending practices and racial violence all combined to prevent African Americans from moving outside the urban core and limited their options for relocation. In Northern Ireland, where the private housing market played a much lesser role, the threat of sectarian violence and discriminatory housing allocation and planning policies on the part of municipal authorities were the key factors that prevented Catholics from moving to many of the new towns and suburbs surrounding Belfast. The pattern however was similar in that minority communities on both sides of the Atlantic bore the brunt of utopian visions formulated by ruling elites and their technocratic fellow travelers in the architecture and planning professions.

Part of the difficulty in Northern Ireland in the 1960s was that, in so far as the Catholic community was concerned, no effective mechanism existed within the political or legal systems for airing grievances about the way in which Anglo-modernist planning was impacting on the ground. At that time, the political system of Northern Ireland ensured that authority over urban regeneration programs in Belfast was exclusively in the hands of the Unionist Party so concerns about the impact of the wholesale destruction of working-class Catholic communities and their forced relocation had no adequate political outlet. Catholic elites in Northern Ireland often sought through "gentlemen's agreements" to negotiate informally with Unionist political leaders about house building programs in the same way that African American elites also bargained on behalf of their constituents to address chronic housing shortages in cities like Atlanta and Miami. These were ad hoc arrangements were undermined by the fact that there was never any doubt who held the upper hand in the negotiations, and there was little or no recourse when, as was often the case, bargains were not kept. Within the Protestant community, fears that dissension would be used by political opponents to undermine the Unionist cause ensured that public opposition to measures like the Matthew Plan and the Urban Motorway was somewhat muted. It was one thing for Unionist Associations to send a delegation to the meet with government officials to express concerns; it was quite another, and generally beyond the pale, to publicly voice opposition to government policy in the form of street protests or demonstrations.

For the Catholic community, the situation was quite different. Opposition to the impact of large-scale urban redevelopment programs in Belfast was compounded by wider grievances about lack of political representation, electoral gerrymandering, discrimination in relation to the allocation of jobs and housing, and in particular, the way in which housing was allocated by Unionist controlled councils to protect electoral majorities.[180] These tensions came to the fore in the early 1960s in towns such as Dungannon, Derry, and Newry when a younger and educated Catholic middle class emerged, giving rise to a burgeoning civil rights movement that challenged existing Protestant and Unionist hegemony.[181].

The Anglo-modernist planning of the 1960s, which ostensibly aimed to facilitate a program of economic modernization had the effect of further alienating the wider Catholic community through a series of actions that built upon and compounded pre-existing grievances. For example, the major new population growth center in Northern Ireland – "Craigavon" was named after the first Prime Minister of the region, James Craig, who famously claimed that he wanted a "Protestant parliament for a Protestant people." The Lockwood Report in 1965 controversially recommended that the region's new university be located near the predominantly Protestant town of Coleraine rather than the second city of Derry which had a large Catholic majority. The modernization of the transportation network also precipitated discontent within the Catholic community with the new motorways linking Belfast to the major unionist towns of Ballymena, Portadown, and Enniskillen, leaving most of the Derry-Belfast link single-carriageway. In other words, new road designs linked predominantly Protestant areas and largely neglected Catholic ones. Other controversial decisions included the Benson Report on railways in 1963, leading to the removal of the only railway line to the west, cutting off Derry from Strabane, Omagh, and Dungannon, again, predominantly Catholic centers of population. In 1966, the closure of Derry's naval base was another blow to Northern Ireland's second and overwhelmingly Catholic city.

There is no doubt that objectively many of the plans that Robert Matthew, John Lockwood, and others formulated for Northern Ireland fit with the wider planning orthodoxy and the zeitgeist of the time and could be justified on grounds of making the Northern Ireland economy more competitive by improving transportation networks and providing a ready source of labor outside Belfast to service the new artificial fiber industries that were being attracted to the region.[182] Many of the decisions taken during this period – economic development concentrated in new growth towns and contraction of the rail network paralleled contemporary developments in Britain and the U.S.[183] Each of these decisions could be justified on purely "technocratic" planning grounds.[184] For example, the naval base closures were part of a general cutback in British naval expenditure from which Northern Ireland could not expect to be immune.[185] Railways in Northern Ireland were losing money and, as was the case in Britain and indeed the Republic of Ireland, lines were closed in order to limit losses – the Benson Report had in fact recommended the closure of both lines to Derry. The Lockwood Committee produced reasoned arguments for concluding that Coleraine best fulfilled the requirements of a site for a new university, among them the availability of an area large enough to allow for expansion; a criterion that the existing university college in Derry could not meet.[186] In Britain, the prevailing orthodoxy was that new university campuses ought to be located in smaller towns, away from large urban centers.[187] The Unionist government could also point to sound practical reasons why entrepreneurs preferred to settle in the eastern areas of Northern Ireland since that was where skilled labor, docks, and communications were most heavily concentrated.[188] Furthermore, many of the planners and architects who formulated these proposals were not even from Northern Ireland and had no connection to the Unionist establishment. Sir John Lockwood for example was Master of Birkbeck College in London and his committee included three other academics from England.[189] Sir Robert Matthew, before coming to Belfast, had played a key role in reshaping several British cities, notably Glasgow, Birmingham, and Newcastle.[190]

Crucially however, much of the Anglo-modernist planning orthodoxy of the period originated in the U.S., where, as outlined in the previous section, ostensibly neutral practices involving housing demolition and highway routing were designed to advance white interests and perpetuate historic patterns of exclusion. Similar dynamics played out in Belfast where sectarian politics led to the selective implementation of a process that was already fundamentally disastrous for poorer urban communities. Just as historically discriminatory practices meant that African Americans were especially vulnerable to modernist planning and architecture practices in the U.S. and political exclusion meant that they lacked a political voice to champion their cause, so too Catholic communities in Belfast and across Northern Ireland found themselves falling victim to political decision-making that was given a seal of approval by ostensibly neutral architects and planners.

The political system in Northern Ireland, which excluded representatives of the Catholic and Nationalist community from both central government and political power within Belfast Corporation ensured that important decisions that required large-scale demolition of housing, breaking up communities, and forced relocation lacked political legitimacy. Moreover, a political system that was also predicated upon maintaining Unionist control of electoral districts and retaining a sectarian majority mitigated against adequate new build housing for Catholics who were displaced because of the large-scale redevelopment taking place in Belfast. In the context of Northern Ireland, opposition to these planning policies was taken up by the burgeoning civil rights movement, becoming a focus for public protest which in turn led to violent confrontation between protesters and the overwhelmingly Protestant police force.[191] This problem became especially apparent as the largely peaceful civil rights movement was met with a violent backlash from the Northern Ireland police force, as well as counter-protests organized by a number of radical and emerging figures within the Protestant community which in turn precipitated serious public disorder.[192] The violent reaction from the RUC and the B Specials toward the civil rights protestors precipitated a violent response from within the Catholic community. The disturbances that erupted in smaller town centers such as Dungannon and Newry were a prelude to the much larger conflagrations in Derry that led to the Battle of the Bogside and the arrival of the British Army on the streets to restore order. In Belfast, with its pre-existing sectarian divisions and strict demarcation of territory in tightly packed urban spaces, the 1960s vision of a modernist utopia disappeared in the flames that spread across the city in 1969.

Ultimately, the redevelopment and modernization program envisaged by Sir Robert Matthew and technocratic unionist leaders would go no small way to precipitating the violent street confrontations that would overshadow the entire economic modernization program. This would in turn bring Northern Ireland into a conflict lasting several decades claiming over 3,000 lives. Writing about the redevelopment schemes of the early 1960s, the former chairman of the Northern Ireland Housing Executive and leading figure in the Northern Ireland Labour Party explains,

> The authors of these schemes failed to learn from the bitter experiences of redevelopment in Britain; and it is arguable that the widespread demolition which preceded these sweeping and ill-thought-out schemes contributed largely to the violence of the sectarian strife which marked the Troubles.[193]

In Britain, the "revolutionary" changes that took place across the urban landscape from the late 1950s onward did not immediately precipitate civil unrest – that was to come later as high-rise tower blocks became synonymous with concentrated deprivation, crime, and a lack of social capital.[194] Paul Collier traces the origins of the political backlash which has characterized much white working-class discontent in recent years to the planning orthodoxy of the 1950s and 1960s.

> The growing number of cars needed flyovers and the growing number of people needed housing. In response, entire streets and neighbourhoods were bulldozed, to be replaced by modernist flyovers and high-rise towers. Yet to the bewilderment of the Utilitarian vanguard, what followed was a backlash. Bulldozing communities made sense if all that mattered was to raise the material housing standards of poor individuals. But it jeopardized the communities that actually gave meaning to people's lives.[195]

Several of the technical experts responsible for designing the Belfast Urban Motorway had visited a number of U.S. cities in order to gain expertise in road design but had singularly failed to learn lessons about the impact that these projects had on precipitating confrontation and the adverse consequences that motorway construction had for displacing communities.[196] During the early 1960s, urbanists such as Jane Jacobs were quite vocal in directing their ire at the way in which the substantial redevelopment that was taking place in the U.S. at that time was eviscerating entire cities and causing social discontent. Figures like James Baldwin declared that urban renewal was in fact "negro removal." Unionist elites, and the planners and architects who advised them failed to take into account these concerns as they were caught up in a modernist zeitgeist that ultimately disappeared in flames.

3.2.5 Shared Operating Systems and Divided Cities

In contrasting the experiences of Northern Ireland with those of Britain and the U.S. during the 1960s, it is worth considering Ignatieff's argument about the need for passably fair public institutions, decent policing, and a "subliminal operating system" characterized by "horizontal governance" when cities are ethnically diverse.[197]

As the evidence from the period considered above shows, Belfast during the 1960s was characterized by a total lack of political coalition building, or "horizontal governance." In fact, the redevelopment of the city was overseen by a top-down command-and-control model, with one single party, namely, the Ulster Unionist Party, maintaining control over contracts and patronage. Moreover, the fact that public institutions, especially policing, remained within the control of the Protestant community ensured that these basic requirements necessary for allowing ethnically diverse communities to live side by side were absent. The situation in Belfast in the 1960s can be contrasted with that in other cities in Britain during the same period, where the "revolutionary" changes that shaped cities enjoyed support across the political spectrum.[198]

In the U.S., where much larger urban redevelopment programs took place without the "buy-in" of the African American community, the Anglo-modernist utopias envisioned by postwar planners also disappeared in flames amid the race riots and freeway revolts of the late 1960s. In many respects, the findings of the Kerner Commission, established to investigate the causes of disturbances in US cities, parallel some of the key findings of the Cameron Commission in Northern Ireland.[199] Specifically, the Kerner Report found that population movements and discriminatory practices combined to concentrate poverty in racial ghettos and help precipitate violence.[200] In Northern Ireland, a major concern among Unionist elected representatives was to preserve electoral majorities and control of territory by limiting the spatial expansion of the Catholic community. Similar experiences took place in the U.S. to prevent the growth and expansion of African American communities beyond specified and discrete territorial boundaries using restrictive covenants that prohibited houses from being sold to non-whites, and FHA lending practices that prohibited non-whites from securing loans in certain districts.[201]

The Unionist politicians governing Northern Ireland can be seen, with the benefit of hindsight, to be at best naive with their application of Anglo-modernist planning regarding their failure to recognize the adverse social consequences of their actions. As outlined above, much early opposition to these measures came from within the Unionist community itself.[202] This opposition was managed however through political representation and the ability of elected officials to confer legitimacy upon the spatial realignments that took place. For the Catholic community, the situation was quite different and illustrates the complex relationship between systemic inequalities, urban planning, and conflict.

The evidence here would indicate that several factors, including lack of political representation, a sectarian police force, large-scale demolition of housing which had the effect of exacerbating existing levels of housing need, and the forced relocation of entire communities, all contributed to igniting violence in Northern Ireland in the late 1960s. That huge numbers of houses were demolished, and communities broken up and forcibly relocated illustrates how the uneasy balance that had held the city ecosystem together for several decades came to be shattered. Moreover, the adoption of the Matthew stop line, which ended the already limited housing construction that had been taking place under the auspices of Belfast Corporation, and which was opposed by its leadership, further illustrates the ways in which planning policy at the time inadvertently served to turn serious levels of housing need into a housing crisis. The experience of Northern Ireland during this period and the conclusions reached here resonate with findings that poverty and inequalities do not generate political conflict per se but rather that violent conflict arises when systemic inequalities are combined with other factors, notably, a lack of political legitimacy impartial institutions, and fair policing.[203]

The role of planning and urban development in precipitating violence in Northern Ireland, like the highway revolts in the U.S., illustrates the need for adequate consideration of how planning choices will impact an ethnically divided city where space and identity are inextricably linked and where conflict is fought out

> on a narrower ground than even the most impatient observer might imagine, a ground every inch of which has its own associations and special meanings.[204]

The evidence considered in this chapter from the 1960s illustrates the consequences of a failure to take account of the political impact of the "territorial imperative" on economic development and urban planning in divided cities. The role of planning and urban development in precipitating conflict in the 1960s lends support for the more nuanced approach that is currently being taken regarding redevelopment in Belfast, which is the focus of Chapter 4. It also highlights the fundamentally political nature of planning and architecture practices that are often presented as neutral judgements packaged in technocratic terminology. The Belfast experience represents the end game for authoritarian modernism. Le Corbusier famously proclaimed "the death of the street" as part of his modernist vision. These words were to prove remarkably prescient, but in ways that he probably never imagined as rows of Victorian terraces in Belfast were burnt to the ground in the sectarian conflagration that his modernist vision had helped bring about.

3.3 MILITARY URBANISM AND THE ARCHITECTURE OF SEGREGATION

This section explores the role that the Northern Ireland conflict played in shaping the contemporary urban fabric of Belfast. By examining hitherto undiscovered documents from the public archive, this section shows how planning and architecture in Belfast were reconfigured by the security forces in the early 1970s as part of a conflict management strategy with road networks, housing developments, and financial resources targeted at techniques that prioritized segregation and separation of the two communities. One important legacy of the defensive planning strategy established in the 1970s was the dismembering of some communities from the wider city ecosystem which in turn accentuated patterns of spatial inequality that persist to the present day. As with the previous section, important lessons will be drawn about the important role that planning and architecture can play not just in shaping the urban fabric of a city but also in shaping systemic and structural patterns of disadvantage.

3.3.1 The Planning System and the Politics of Containment

Evidence of the extent to which the planning system in Belfast was reconfigured by the security forces in the early 1970s as part of a conflict management strategy is provided by a series of documents uncovered from the Public Records Office for Northern Ireland (PRONI) during this research. These documents show that in October 1970 the Northern Ireland government established a Joint Working Party under the Chairmanship of the then Minister of State for Home Affairs, John Taylor MP,[205] to examine "areas of confrontation and public disorder" and provide recommendations on these issues toward future government policy.[206] The findings of the Joint Working Party (hereafter the "JWP Report") were marked "secret" and submitted to the then Prime Minister of Northern Ireland, Brian Faulkner, in April 1971 with a series of recommendations regarding the future planning and design of the city. This report provided a blueprint for the reconfiguration of the planning system that was to have significant consequences for the urban fabric of the city of Belfast for subsequent decades. Specifically, the JWP Report recommended that the substantial redevelopment of Belfast, which had begun several years previously under the auspices of the Belfast Urban Plan, should provide for the "maximum natural separation" between opposing areas.[207] The Belfast Urban Motorway, construction of which had also begun several years previously, was used as a physical "cordon sanitaire" by creating a "cleared belt" up to 100 yards wide to the west side of the City Centre.[208] Beyond the Urban Motorway, the report concluded that across Northern Ireland generally "opportunities to create natural divisions between difficult areas by means of road re-alignment should not be overlooked."[209] The JWP Report also stated that the Ministry of Home Affairs, as the department responsible for security and law and order, ought to be given an opportunity to consider

> any plans for substantial development or re-development (both for housing and for industry and commerce) at an early stage, consulting the security forces and other departments where necessary.[210]

These recommendations outline how the planning and design of the city was to serve as an instrument of counter-insurgency by dividing and segregating specific areas of Belfast and the communities living within them. The legacy of the JWP Report can be traced by examining the planning and design of the contemporary road network in and around Belfast, especially the "Westlink" motorway, which separates north and west Belfast from the rest of the city.[211] Specific examples of the use of the Belfast motorway as a "cordon sanitaire" include the barrier effect created by the canyoning of the road at the junction of Clifton Street (Figure 3.4), where the road was depressed by military-inspired design, in some places up to 30 meters.

Gehl contends that there are five different means with which architects and planners can promote isolation and prevent contact namely, walls, long distances, high speeds, multiple levels, and orientation away from others.[212] As illustrated in the images of the Westlink (Figures 3.4–3.6), all these elements were present in the architecture that shaped this section of the road network in the city.

The cordon sanitaire effect generated by the Westlink was especially pronounced at the Broadway roundabout, which became a border zone that split the old Donegall

FIGURE 3.4 The junction of Clifton Street and the Westlink. This image shows the cordon sanitaire effect created by the depressed road structure, which replaced the earlier elevated road design to maximize severance. The "seven towers" of the New Lodge area are visible in the background and retain evidence of the British Army observation posts that were located on the roof. The tower nearest is undergoing refurbishment with new roofing and cladding added in place of the former army posts. Photograph by Tim Cunningham, July 2015.

FIGURE 3.5 Grosvenor Road police station, to the east of the Falls Road. The Westlink motorway is depressed and runs perpendicular to the Grosvenor Road, beyond the police station. The bridge over the Westlink is visible in the distance.

Photograph by Tim Cunningham, July 2015.

FIGURE 3.6 Evidence of the cordon sanitaire provided by the canyoning of the Westlink (A12) at Millfield. Continuous panels provide excellent sightlines for the security forces that were based in observation towers along the road network and inhibit lateral movement across the city. Belfast Metropolitan College campus is on the right-hand side of the photograph.

Photograph by Tim Cunningham, July 2015.

Road in half and created a permanent boundary between the Protestant "Village" and Catholic Falls Road areas. Again, several of the elements outlined by Gehl feature prominently in this section of the road, which is characterized by long distances, high speeds, and orientation away from others. In terms of the social and economic legacy of these planning and architectural practices, Johnston has pointed out that the Westlink was subsequently to have

> a profound and permanent effect on the areas through which it passed, restricting access to the central area of the city to just five streets (Grosvenor Road, Divis Street, Peter's Hill, Clifton Street and North Queen Street) and one footbridge (serving Divis Flats).[213]

The Westlink significantly reduced connectivity between north and west Belfast by cutting off the natural linkages between homes, shops, leisure facilities, and workplaces that had previously characterized this area of the city.[214] The Forum for Alternative Belfast has highlighted the "fracture and disconnection" caused by the Westlink and established several projects, notably "Connect West," to address the legacy of the design practices of the 1970s and 1980s.[215] Faligot argues that a key objective of both the Westlink and the M2 motorways was to ensure the "fragmentation and isolation" of communities as well as provide the security forces with the means by which to respond rapidly when needed in these areas.[216]

Documentary evidence uncovered during this research confirms that the lack of connectivity and high degree of severance and isolation caused by the Westlink did not take place by accident, nor was it an unfortunate consequence of poor planning. Rather, as Hillyard has pointed out, the aim was to contain "dissident populations" as part of a wider conflict containment strategy.[217] This objective is confirmed in a letter written by Colonel K. J. Carter, Deputy Commander, 39 Airportable Brigade, at British Army HQ in Lisburn.[218]

> The reduction of latitudinal roads across the city will assist the security forces in checking movement. From the security aspect alone it would be desirable for the motorway to be constructed with a continuous central wall or panel to prevent any form of lateral movement underneath in the areas between through routes .... It would be appreciated if the current plans could be made available so that they can be studied from the long-term security viewpoint for the construction of the motorway would seem to be an important, integral part of the overall plan for the development of the city.[219]

This letter is important in two respects. First, it highlights clearly that the aim of the strategy was to prevent and restrict movement – the very antithesis of conventional planning which aims to achieve connectivity through facilitating the flow of people and goods throughout a city. The second point to note is that this letter indicates the degree to which plans were to be studied from a long-term perspective regarding the redevelopment of the city.

In addition to the design of the Westlink, there are numerous other examples that further illustrate how the framework laid down in the JWP Report shaped the city ecosystem. For example, the original Belfast Transportation Plan that preceded the outbreak of the conflict and which was published in 1969, had, in line with many other cities at that time, proposed a ground-level ring road encircling the city center within the urban motorway.[220] The intention was that vehicles arriving into the central area on the motorway would leave at the nearest exit and then join the inner ring road to guide them to their final destination.[221] Significantly, the 1978 Transportation Plan recommended completion of the ring road structure in an amended form. While the amended ring road functioned reasonably well from a traffic perspective, at least in the early years, it helped destroy many working-class communities around the city. On the eastern side of the city center, Thompson notes that "the Dunbar Link was appalling with no serious attempt made to reconstruct the truncated streetscape around it, almost no frontal development," with pedestrian facilities "sparse" rendering the area "an urban wasteland devoid of life for the next 30 years."[222] Figure 3.7 illustrates how the Dunbar Link served to act as a barrier around the eastern side of the urban core in line with security imperatives.

This approach to urban planning and design ensured local communities were physically, socially, economically, and psychologically separated and isolated from the city ecosystem. These adverse social consequences were not, as some commentators

FIGURE 3.7 The Dunbar link, one of the stretches of the inner ring road that runs along the northeastern side of the Belfast urban core. The lack of frontal development and limited pedestrian access significantly impeded movement and urban life around this area of the city and was devastating for local communities. This was an intentional objective that worked well from a security perspective. The heavily fortified Musgrave Street police station is located just to the south of this image.

Photograph by Tim Cunningham, July 2015.

have suggested over the years, an unfortunate by-product of incompetent or unimaginative planners and architects in the city but rather were the outcome of a deliberate strategy that sought to enhance legibility of local communities and impede movement and access.

Severance at the Divis Street junction (see Figure 3.8) was especially acute, given that pedestrians were required to navigate both the Westlink motorway and the inner ring road, which served to provide an additional barrier between the west of the city and its natural hinterland. The Divis Street junction was also provided with a British Army observation post located on the roof of Divis Tower to augment the defensive design of the road network. The Divis Street junction is a key route linking the Lower Falls and, indeed, Greater West Belfast to the rest of the city. In the early 1970s, these areas were the heartlands of Irish Republican Army (IRA) activity, which explains why securing this road junction was, from a military point of view, a key strategic objective. The Forum for Alternative Belfast has described the Divis Street junction as an "empty and inhibiting" place to walk and cycle and established

FIGURE 3.8 Divis Street junction, west Belfast. The depressed Westlink is perpendicular to and runs underneath this junction. From the late 1960s until the late 1990s, the British Army had an active observation post located at the top of the tower which is still visible, though now empty The wide interchange and lack of frontal development ensure clear sightlines for soldiers based in the observation post on top of the tower, enhancing legibility of local population. The unusually wide intersection also allows for the hasty deployment of military vehicles.

Photograph by Tim Cunningham, July 2015.

the "Divis Link" project to attempt to mitigate some of the worst aspects of the design of this area of the city.[223]

The redesigning of the road network to achieve security objectives was especially extreme but not solely restricted to north and west Belfast and the urban core. The Bridge End Flyover in the east of the city had the effect of converting the streetscape of the Ballymacarrett area to "an expansive urban wasteland that persists to the present day, acting as a significant psychological barrier between the Newtownards Road/Short Strand area and the city centre."[224] Dawson's detailed study of the area illustrates how security considerations shaped the road layout of the Catholic Short Strand.[225] The portion of the Lower Albertbridge Road that was widened to dual carriageway standard was the same portion of the road that acted as a sectarian interface, thereby ensuring that the road-widening created an additional security barrier or cordon sanitaire between the two communities. In a similar vein, Bryan has pointed out that the eight lanes of the M2 motorway, which run northwards out of the city, "cut right through the traditional docks or Sailortown district," rendering the York Street area "almost unrecognisable and effectively destroyed it as a residential district."[226]

FIGURE 3.9 The M2 motorway heading north out of the city at York Street. The M2 obliterated the Sailortown community and the urban voids that remain provided valuable sightlines for the military observation posts in the nearby New Lodge as well as impeding movement between the York Street/Lower North Belfast area and the city center.

Photograph by Tim Cunningham, July 2015.

Evidence of the impact of the conflict management strategy applied during the conflict is further illustrated by the fact that the working-class Protestant areas of the Shore Road and the working-class Catholic areas of the New Lodge and lower Antrim Road had major road junctions placed between them and the city center, thereby ensuring severance of these areas from the urban core (see Figure 3.9). These images provide a visual representation of the spatial legacy of the Northern Ireland conflict, illustrating how architecture and planning can deliver subtle forms of coercion or "spatial violence" over a period of several decades.[227]

3.3.2 The Design Legacy: Shatter Zones and Desert Planning

The Belfast Metropolitan Area Plan 2015 (BMAP 2015) has acknowledged the social impact of many of the planning and architectural practices of the 1970s and 1980s highlighting how the Inner Ring Road created both a physical and psychological barrier for surrounding communities:

> The Inner Ring Road which currently surrounds much of the city centre has left a scar on the urban fabric with large expanses of roadway inappropriate to the scale of the city, a fractured townscape and a physical and psychological barrier separating city centre from surrounding residential communities.[228]

In a similar vein, Grahame Fraser, former Director of Network Services at Road Service Northern Ireland, reflecting back on the impact of the road design policies of the 1970s onward on inner-city Belfast has concluded that "the York Street area was basically decimated, and it lay that way for years."[229] Figures 3.10–3.13 illustrate how the road network around York Street and the adjacent New Lodge area was used to inhibit connectivity and promote isolation.

Jane Jacobs directed much of her ire at the capacity of the road network to eviscerate cities, likening the communities around them to "amputated areas typically developing galloping gangrene."[230] The autocentric aspects of modernist planning favored in the postwar urban renewal programs in the U.S. were exported to the U.K. and Northern Ireland in the 1960s through the original plans for the BUM and the inner ring road. The adverse social and economic consequences of these design practices were further heightened by the impact of military imperatives which sought to use the road network to separate and segregate areas of confrontation.

In addition to the ongoing conflict secular developments also impacted how roads and highways were planned and designed in Belfast during this period. The oil shock of the early 1970s and the recession that it precipitated circumvented many of the more elaborate plans and designs of the early 1960s, as did growing citizen activism as local communities recognized the adverse social and economic consequences of urban expressways. In Belfast, as with other cities in the U.K., plans for an urban motorway encircling the entire city were abandoned in the 1970s.[231] This

FIGURE 3.10 The inner ring road at York Street to the north of the city center. The M2 flyover is visible on the right-hand side of the image in the background, the Yorkgate shopping complex is on the left, just out of the picture.

Photograph by Tim Cunningham, July 2015.

FIGURE 3.11 Henry Street, adjacent to the Catholic New Lodge area. The Yorkgate shopping complex is on the left fortified by a wall and metal barrier. The blank walls and orientation of the shopping complex away from the local community highlight the defensive design style developed in the early 1970s and implemented in subsequent decades. The road barrier, visible in the background of the image shows how direct access from Henry Street to the York Road was blocked. The blank walls also provided valuable sightlines for the military located in the observation towers in the New Lodge, immediately to the rear.

Photograph by Tim Cunningham, July 2015.

decision had the effect of preserving much of the urban and social fabric of south and east Belfast, notably in the Queen's University area where the southern section of the urban motorway would have required the demolition of large sections of Victorian residences and the communities living in and around them. Again, this was a pattern that emerged in other cities as historic preservationists and local activists were able to redirect or cancel highways as disillusionment with the modernist planning agenda evolved.

Crucially, however, the security imperative behind the JWP Report ensured that the lessons that had been learned about the adverse social consequences of road design were not applied to "difficult areas," especially those in the north and west of the city. In fact, for "areas of confrontation" the opposite was the case in that the original design for the western section of the urban motorway was altered to *increase* severance, segregation, and divisions. This is evidenced by the fact that the original plans for the Belfast Urban Motorway in the 1960s specifically ruled out canyoning of the road on grounds of cost, and the adverse impact that depressed road structures would have on connectivity for the city. Following the outbreak of conflict and the JWP Report, the canyoning of the road show how the desire to create a "cordon

FIGURE 3.12 Henry Street from the pavement at York Street showing how vehicular access via the east side of the greater New Lodge area was blocked by road severance.

Photograph by Tim Cunningham, July 2015.

sanitaire" and "maximum separation" replaced the earlier aim to minimize costs and disruption to connectivity. As outlined above, the net effect of this revised approach was to increase the social, economic, and psychological isolation of communities in certain areas of the city, while ensuring that other districts, notably around Queen's University in the south of the city remained reasonably well connected to the wider ecosystem.

Bryan has outlined how the north and west of the city, and the working-class residential areas in the east, (i.e., the predominantly Catholic Short Strand and Protestant Ballymacarrett), already separated from the city center by the river, are now further dislocated by large road junctions so that the pedestrian walking to the city center from most working-class residential areas is confronted by a desolate and intimidating combination of road junctions and empty wasteland. He contrasts this appearance with the commercially vibrant corridor linking predominantly middle-class residential areas of south Belfast directly to the city center.[232] Significantly, these "working-class residential areas" correspond to the areas of confrontation identified in the original Joint Working Party Report of the early 1970s illustrating the social consequences of the "cordon sanitaire" approach. The evidence examined here shows how the planning and design of the city of Belfast from the 1970s represents the apotheosis of desert architecture and urban design as entire areas of the city were dismembered by physical and psychological barriers of tarmac and hulking concrete.

FIGURE 3.13 The southern end of the greater New Lodge area with housing around Henry Street on the right and the "cordon sanitaire" provided by the Westlink on the left. The image shows the proximity of the houses to the busy road which closes the Catholic community off to the south. Vehicle noise and air pollution amplify the discomfort and adverse physical and mental health consequences of living next to this volume of traffic.

Photograph by Tim Cunnningham, July 2015.

There is another important legacy of the way in which the design of the urban environment can have adverse consequences for the city ecosystem and it is this. During her research into the impact of the large-scale redevelopments that characterized U.S. cities in the years following World War II Jane Jacobs identified how planners could themselves heighten racial tensions and the sense of ownership of "turf" through the very kind of urban design practices that characterized the security-led approach to planning in Belfast from the 1970s onward. By using the design of the built environment to restrict movement and demarcate territory identity-based tensions and competition around ownership of urban space were increased. Kadar Asmal has described how this phenomenon played out under the apartheid regime in South Africa

> a country with forced cleavages … brought about by a system based exclusively on separating people from one another on the basis of the colour of their skin … resulting in rabbit-warrens of housing separated on a racial basis.[233]

The security-led approach to planning that characterized Belfast from the early 1970s onward did the same thing. Prime examples of this include the "forced cleavages" achieved by the design of the Westlink motorway, the inner ring road, and the

M2. This was an issue that the Churches Central Committee for Community Work (comprising the Presbyterian, Church of Ireland, Methodist, and Catholic churches) and the Northern Ireland Council of Social Services picked up on in the early 1970s pointing out the various ways in which planners were in fact increasing sectarian polarization and territoriality

> putting into fixed moulds, and in physical moulds, and in physical form, segregated areas of the city … for the next one hundred years at least.[234]

The practices that characterized the planning and design of the urban environment during the years of the conflict served to exacerbate pre-existing tensions around ownership of territory, not alleviate them. In fact, as the documentary evidence uncovered during this research shows these consequences were known to public officials at the time. In addition to the main JWP Report a dissenting note written by Anthony Hewins from the Office of the United Kingdom Representative in Northern Ireland expressed concern that the proposals contained in the main report would mean accepting divisions in the community as a feature of life "for a hundred years or more."[235] Hewins considered the approach taken by the Joint Working Party for the future management of areas of confrontation to be "a counsel of despair … expressed in terms of bricks and mortar." Specifically, Hewins questioned whether the Protestant or Catholic populations would themselves be content to remain in "straight-jacketed communities," which he feared might well "become sources of major discontent."[236] He concluded his minority report with the observation that "the word ghetto has been lightly and loosely used in the past but that 'these proposals would give the name substance and would attract criticism from all over the world.'"[237] Hewins and the Church leaders of the 1970s correctly identified the way in which the framework outlined in the JWP Report would have long-term consequences for relationships between communities in the city. This point is especially evident regarding the way in which the securitization of the urban environment impacted upon housing development, which is examined in the next section.

The original JWP Report had specifically concluded that "plans for housing adjacent to the main lines of confrontation between the two communities, where not already approved, should be revised to provide more open space, particularly on either side of the new natural barrier" and that the number of "open routes" between the Falls and the Shankill areas should be substantially reduced.[238] In addition, the report considered it "to be of great importance that there should be a further reduction in the overall percentage of residents going back into the area after re-development"[239] concluding that "there may be no alternative to increasing rather than discouraging segregation of the two communities through the creation of natural barriers."[240] The JWP Report also recommended that as part of the overall redevelopment program for the city, various factories and warehouses would have to be situated and that "consideration should be given to the possibility of locating such accommodation along the line of confrontation, with high walls forming natural barriers and special provision made against fire risks etc" with "a special subsidy" made available to make this possible.[241] In addition, the report concluded that it was necessary in the future "for careful planning of the layout of new housing areas to avoid physical features leading to the creation of areas of confrontation."[242]

The subsequent implementation of these recommendations can be traced by examination of the minutes of another hitherto secret entity called the *"Standing Committee on Security Implications of Housing Problems in Belfast"* (SCH). The SCH comprised several government agencies including the Belfast Development Office (BDO), which was the main agency within government responsible for urban development, and the Department of the Environment, the government department responsible for planning, housing, and urban affairs. As with the original Joint Working Party, the SCH also contained representatives from the Northern Ireland Office (NIO), the government department responsible for law, order, and security as well as representatives from the security forces, namely the RUC and the British Army.[243] The minutes of a series of meetings that took place in the early 1980s and subsequent correspondence between the security agencies and the planning service indicate the degree to which the security forces were involved in the planning of housing in the city at that time.[244] There are several themes to emerge from examination of these documentary sources not least of which is the efforts made to ensure that the north and west areas of Belfast were separated from the main body of the city. The second theme is that the provision of additional and much-needed housing for the Catholic community ought to either be delayed or halted altogether to contain the risk of violence.[245] Another theme relates to the emphasis on "reducing the percentage of residents going back into the area after re-development" which was to have significant consequences for the Protestant community in the city.

These themes are especially apparent regarding the way in which commercial development was planned and designed to act as a buffer between contested areas. During the 1990s urban regeneration policy in Northern Ireland led to the construction of new business development or "enterprise" parks, subsidized by the government, which offered incentives such as low rent for new investors with the aim of revitalizing stagnant communities. This was part of a wider program of measures aimed at helping regenerate economically disadvantaged areas by redeveloping vacant brownfield sites into commercially viable land. One example was the North City Business Park in Duncairn Gardens in the north of the city which transformed a vacant brownfield site into a business development park. The original JWP Report had recommended that as part of the redevelopment of the city factories and warehouses ought to be located "along the line of confrontation" – in the case of the North City Business Park this was between the "Catholic" New Lodge and "Protestant" Duncairn Gardens areas (see below).[246] This approach is confirmed by Richard Needham the Minister with responsibility for planning and economic development in Belfast during this period:

> We did not advertise it but wherever we could we tried to find neutral uses for the no-go land. Factories, offices, warehouses, old people's bungalows, schools, leisure centres – all were used to make the divisions less obvious ….[247]

The difficulty with this approach was that in a divided city, where territory and identity are inextricably linked, there is no such thing as "neutral" uses for no-go land.[248] The construction of the Duncairn Business Park is of particular significance given

FIGURE 3.14 North City Business Park is located on the right, photographed from the "Catholic" end at Duncairn Gardens at Edingham St. The Business Park provides a "buffer" separating the Catholic area of Duncairn Gardens from "Protestant" Tigers Bay. The Tigers Bay housing is visible in the distance. The heavily fortified walls around the business park illustrate how defensive architecture continues to shape the urban fabric of the city. Photograph by Tim Cunningham, July 2015.

that this development ensured that no additional housing could be provided on the site, something that residents from the overcrowded Catholic New Lodge area were demanding at that time.[249] The net effect of the Duncairn Business Park therefore, illustrated in Figures 3.14 and 3.15, was to halt the growth and expansion of the Catholic community of the New Lodge northwards toward traditionally "Protestant" territory creating a buffer zone between the two communities.

Further examples of the way in which a security-based approach to urban planning impacted upon the spatial politics of the city are evidenced by the way in which government programs ostensibly aimed at benefiting deprived areas in the 1980s were also used to contain and control local communities.[250] In this case, funding from the Belfast Areas of Need (BAN) program supported the construction of a series of security barriers, euphemistically named "environmental improvement" walls.[251] One interviewee encountered this phenomenon in the Catholic New Lodge area of North Belfast, where he was an elected representative at the time

> At one stage I discovered in the early 1980s that what was supposed to an environmental improvement grant for these kind of areas … they were known then as the Belfast Areas of Need (BAN) … and I discovered that in many of these areas, particularly the New Lodge, the environmental improvement grant was actually being used to build walls around the place, and they were calling these

FIGURE 3.15 One of the current entrances to the North City Business Park which is accessible by a buzzer/intercom system for pedestrians on Duncairn Gardens. The gate closes automatically when staff have passed through concealing the access point. The defensive architecture which dominates this area of the city is evident with fortified walls, shrubs, and railings, used to disguise the entrances sending a message to locals that they are unwelcome and should keep away.

Photograph by Tim Cunningham, July 2015.

> environmental barriers, and trying to create them in such a way that they didn't look like fences, they would be nice walls … and it would be £45K here, and £88K there and so on … I can remember at the time issuing statements about it and trying to complain and so on, and getting absolutely nowhere.[252]

As illustrated in Figures 3.16 and 3.17, many of these "environmental improvement" walls remain in place illustrating how defensive architecture became mainstreamed into the fabric of the city. The "straight-jacketing" of the Catholic community in the New Lodge is evident from the way that the walls block off access to and from Duncairn Gardens and illustrates Hewins' warning that the proposals in the JWP Report would give the word "ghetto" substance.[253]

In some cases these barriers surrounding the new Lodge area served to block off vehicle and pedestrian access; in other instances, they were specifically targeted at pedestrians, as illustrated by the example in Figure 3.18, which shows how "anti-climb" paint complemented the metal fencing.

These images show how the strategy outlined in the JWP Report "to provide for the maximum natural separation between opposing areas"[254] and to ensure that

FIGURE 3.16 One of the "environmental improvement" barriers blocking vehicular and pedestrian access from the northern side of the Catholic New Lodge onto Duncairn Gardens at Lepper Street. The pedestrian gates are typically closed at night.

Photograph by Tim Cunningham, July 2015.

"opportunities to create natural divisions between difficult areas by means of road re-alignment should not be overlooked"[255] played out in practice. Taking the Greater New Lodge area as an example, the use of environmental improvement grants to build walls and close off streets, along with the construction of the North City Business Park, halted the expansion of the Catholic population of the area northwards, creating a buffer between the New Lodge and Protestant Tigers Bay. Vehicular access northwards was also eliminated by the closure of the streets fronting onto Duncairn Gardens (see Figures 3.16 and 3.17) which permitted some limited and controlled pedestrian access to and from the area. Access to, and expansion of the New Lodge area to the south was circumscribed by the Westlink motorway which served as a "cordon sanitaire." In a similar vein the road barriers at the junction of Henry Street and York Street (see Figure 3.12) restricted movement out of the Greater New Lodge area to the east, while the western section of the New Lodge had limited vehicle access points that fronted onto the Antrim Road and was located directly opposite the Girdwood Barracks where the British Army were stationed.[256] The cumulative impact of these defensive planning and design practices was to ensure that the Catholic community in the Greater New Lodge area was encircled or set into the type of "fixed moulds" identified by Hewins (see Figure 3.19).

FIGURE 3.17 One of the "environmental improvement" barriers blocking vehicular and pedestrian access from the northern side of the New Lodge onto Duncairn Gardens. The trees and the wall, visible in the background through the open gate, are part of the walled barrier surrounding the North City Business Park.

Photographs by Tim Cunningham, July 2015.

This strategy originated at the highest levels of government and was aimed primarily at containing violence. The benefits of this approach from a security point of view are obvious in that the easiest way to prevent the IRA and other paramilitary groups from moving weapons or explosives out of areas like the New Lodge was to close off roads and restrict movement, especially of vehicles. Certainly, members of the IRA were very aware of what was going on, not least given that they were directly impacted.

> The first socially engineered district was Twinbrook [in West Belfast] with one way in, and one way out, and that was done deliberately, and the open green areas … were there because they became a fire sight – in other words, if they trapped an IRA unit in that area, or rioters or whatever, they would have to cross a wide open space to get out … so they could contain them, in that area …[257]

The difficulty however was that this approach had the effect of fencing in or straight-jacketing a community that was growing in population and which was in dire need of housing. From a social perspective this was a disaster physically, socially, economically, and psychologically, disconnecting the New Lodge from the wider city ecosystem with increased overcrowding and concentrated poverty and inequality.

FIGURE 3.18 Fencing blocking pedestrian access from the New Lodge area to Duncairn Gardens with "anti-climb paint" augmenting the spiked metal barrier.

Photograph by Tim Cunningham, July 2015.

These security objectives often coincided with those of Unionist politicians who not only wished to protect Protestant territory from attack by the IRA, but who also wished to halt the advance of the Catholic community across the city to protect their electoral majorities. As outlined above, this had been a concern for Unionist politicians from the early decades of the nineteenth century.

The extent to which the security objectives of the British Army and the RUC overlapped with the interests of Unionist politicians is evidenced by the comments made by one Unionist Councillor on Belfast City Council to Brian Feeney who was at that time an SDLP representative for this area of the city:

> I can remember one Unionist Councillor coming to me, in all seriousness, I just couldn't believe it, in all honesty and openness and he said to me, look you know, we've finished now, this new gate that's going in at Hillman Street, opposite Tigers Bay, does that mean the whole New Lodge is sealed off now? And I said, well actually yeah, it does. But he was sort of asking me for clarification – you know, he didn't go to the next point which is, is there anywhere else we should be blocking off?[258]

The comments above illustrate an important point about the security-driven approach to planning the city, namely that it cut with the grain of the spatial politics of the city predicated upon protecting Protestant territory from Catholic encroachment which will be explored more fully in the next chapter.

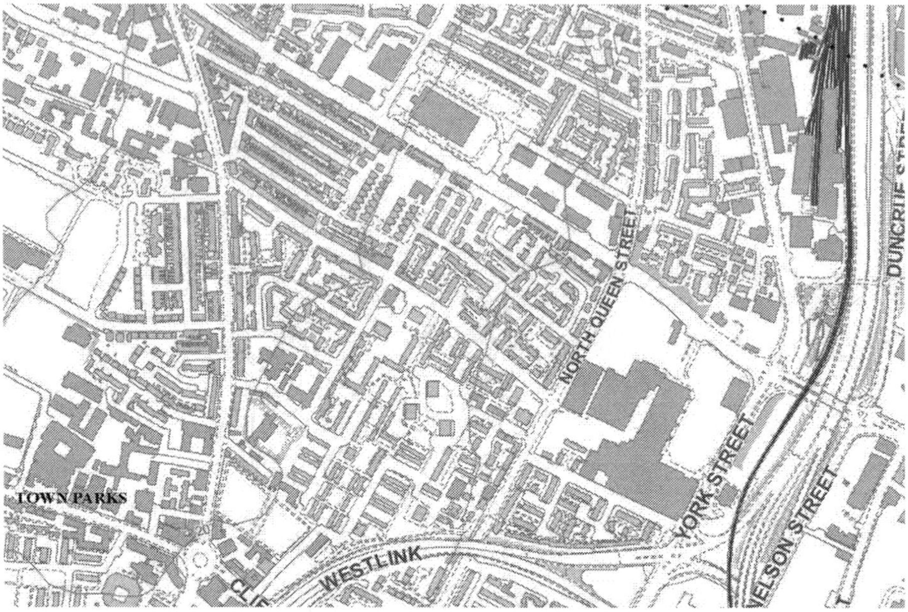

FIGURE 3.19 Map of the Greater New Lodge area including the cityside retail and leisure park. Although the old New Lodge area technically ends at North Queen Street, the Henry Street and Hill Street areas are considered part of the Greater New Lodge. For security and containment purposes, the Greater New Lodge was viewed as a single geographic and political entity. The map shows how the Westlink (A12) provided a barrier to the area to the south, York Street, and the M2 contained the population to the east, and street severance restricted access to the north along Duncairn Gardens. During the conflict, British Army were based in the Girdwood Barracks, which was located along the western side of the New Lodge on the Antrim Road. PRONI Historical Maps. Crown Copyright and Database Right 2025 - SpatialNI. Ordnance Survey of Northern Ireland.

At the same time, the planning framework established in the JWP Report was not exclusively targeted at the Catholic community. Another consequence of the security-driven approach to planning and housing provision relates to the emphasis in the original document of 1970 that "great importance" was attached to "a further reduction in the overall percentage of residents going back into the area after re-development" and the need "to provide more open space, particularly on either side of the new natural barrier." From a military perspective, this approach made sense – by reducing population density, and promoting more open spaces, military sightlines and fields of fire were significantly enhanced. This explains why the British Army covertly supported the demolition of many of the poorly designed system-built housing developments erected in the 1950s and early 1960s which provided rabbit-warrens and escape routes for insurrectionists and promoted a more enlightened housing design approach. Substandard units were typically replaced with less dense and higher quality semidetached (duplex) housing as part of a massive government-backed redevelopment program during the 1980s.

The promotion of open spaces and a reduction in the population returning to areas after development was difficult to achieve in the much more densely populated

Catholic areas, given the levels of overcrowding and limited options for that community to live elsewhere in the city. This did however become the default position for Protestant areas with devastating consequences (Figure 3.20).

The cumulative impact of the redesigning of the road network and housing developments is illustrated by a government report published in 2016 which examined the legacy of conflict planning in the Lower Oldpark area of north Belfast (see Chapter 4) and acknowledges the impact that changes to the built environment have had on isolating and segregating communities within the city.[259] These conclusions fit with the study by Appleyard and Lintell in San Francisco which showed that even a relatively limited deterioration of the quality of the outdoor environment can have a disproportionately severe negative effect on the extent of outdoor activities, including pedestrian use.[260]

The fortified structures captured in these images also highlight the Janus-like approach to urban development in the city. On one hand socially progressive initiatives like BAN and Targeting Social Need (TSN) were promoted as evidence of government commitment to addressing concentrated poverty and deprivation. On the other hand, the use of funds from these programs to fortify walls surrounding local communities and business parks sent subtle cues or what Goldhagen has identified as "primes" to local populations that they should remain within their own territory, and had the effect of promoting isolation, fear, and insularity. The message from the fortified walls of the North City Business Park for locals is very clear – stay away. This message is conveyed

FIGURE 3.20 The Protestant Lower Shankill area. Open spaces like this are indicative of the way in which many of these areas were hollowed out in the decades since the early 1970s. The image, on the gable wall of a house shows Martin Luther nailing his Ninety-five Theses to the church door in Wittenberg, widely regarded as a key event in precipitating the Protestant Reformation. Both Protestant and Catholic communities use wall art to enhance the visual culture of their areas. Photograph by Tim Cunningham, July 2015.

by the architecture to those who might wish to attack the premises but also to those looking for a job. Moreover, accessing employment and education opportunities outside the area was designed to be as difficult as possible. This coercive aspect of defensive architecture highlights one reason why area-based measures aiming to alleviate deprivation ultimately failed.

Jane Jacobs coined the term "shatter zones" to describe the consequences of the type of developments outlined above in which the social fabric of an area was shattered, reduced to an austere, barren wasteland devoid of life. Cullen adopted the term "desert planning" to describe urban areas characterized by wide streets and roads, with empty spaces where "life has literally been built out" of areas.[261] Several notable urbanists have identified how these terms apply to many areas of Belfast, notably the inner north of the city and the districts surrounding the Westlink.[262] The photographs of the Dunbar Link, the Westlink, and North Queen Street presented here illustrate how the security-led approach to redevelopment literally built life out of many sections of central Belfast. Gehl contends,

> To be able to move about easily and confidently, to be able to linger in cities and residential areas, to be able to take pleasure in spaces, buildings, and city life, and to be able to meet and get together with other people -informally or in more organized fashion – these are fundamental to good cities and good building projects.[263]

As the evidence outlined above clearly illustrates, the principles championed by Gehl and others, which are fundamental to designing a good city were the polar opposite of what the planning framework for Belfast sought to achieve from the early 1970s. Mainstream planning and architecture literature has long since identified the oppressive impact on communities of the kind of architecture and planning evident in Belfast during this period including reducing well-being and enhancing feelings of isolation as well as serving to concentrate social and economic inequality.[264] Collier has pointed out that recent research in neuroscience-enhanced social psychology has found that the modernist designs beloved of the planners reduced well-being by breaching common aesthetic values.[265] Physical exclusion and marginalization are mirrored by the high levels of social and economic deprivation that characterize many of these areas-such as the Lower Oldpark and the New Lodge. The phenomenon of severance, and the creation of "Shatter Zones," would in time be recognized as one of the most significant problems associated with urban road and other development schemes, and there are numerous examples of this surrounding the center of the contemporary city.[266]

The Forum for Alternative Belfast refers to the swathes of vacant and derelict land surrounding the core area of the city identified above as "the grey doughnut."[267] This "Grey Doughnut" is an important spatial legacy of the conflict and ought to be viewed as such. This streetscape, characterized by swathes of tarmac bounded by fences and other physical barriers continues to impede connectivity for pedestrians across the city and induces traffic demand especially at peak periods. The result is chronic traffic congestion, compounded by the fact that security imperatives ensured that Belfast is one of the few cities in western Europe where neither airport is directly connected to a rail link. Crucially however, in addition to destroying connectivity, efforts to straight-jacket communities were to have important consequences for distorting

FIGURE 3.21 Part of the grey doughnut that encircles the Belfast urban core. Fences and tracts of tarmac and other impermeable surfaces like this make for an inhospitable environment. The houses on the right are fenced off from the rest of the street with limited access. The Westlink runs perpendicular to these streets, invisible in the background due to its depressed design. The buildings on the left, including the offices of a local charity, are defended by railings. Photograph by Tim Cunningham, July 2015.

patterns of housing need, not least given the demographic imbalances that existed in the city in the early 1970s. This aspect of the legacy of the design practices of the conflict will be explored in the next section (Figure 3.21).

3.3.3 The Spatial Legacy of Conflict in a Contested City

A central contention of this book is that planning is fundamentally predicated upon the control and manipulation of space, and in a divided society where territory and identity are inextricably linked there is no such thing as neutral planning. This brute fact is all the more pertinent in a city like Belfast, where the late historian A.T.Q. Stewart observed that the "territorial imperative is extremely insistent."[268] In order to appreciate the full impact of the spatial legacy of the framework laid down in the original JWP Report, and how this approach exacerbated tensions around "turf," it is necessary to consider again Jacobs' notion of the city ecosystem and the processes that have shaped Belfast over time. One of the main consequences of the securitization of the urban environment from the 1970s was that it exacerbated problems arising from the mismanaged urban renewal that began in Belfast in the early 1960s highlighted in the previous chapter. In particular, the agenda put forward by the Joint Working Party served to further distort population imbalances between the two communities especially regarding housing needs.

For Belfast's Unionist community the outcome of the modernist planning agenda that began in the late 1950s was an exodus of the more educated, skilled, and upwardly mobile families from traditional "Protestant" heartlands.[269] Protestant exodus during this period would have adverse consequences both in terms of social capital and crucially, electoral support for political Unionism with Belfast now a more "Catholic city" than at any time in its history.[270] One particular difficulty for Protestant communities in Belfast is that the official government policy of "decanting" residents to the growth centers was followed, from the early 1970s, by another unofficial, but nonetheless important policy predicated upon the same principle, albeit with a security rather than economic imperative. This is evidenced by the fact that the original report of the Joint Working Party on areas of confrontation stated that it was "of great importance that there should be a further reduction in the overall percentage of residents going back into the area after re-development."[271] This objective is highly significant because over the years within the Protestant community there has been a strong suspicion that there was a deliberate policy of using redevelopment to promote an exodus from Belfast.[272] The report of the Joint Working Party not only confirms these suspicions but illustrates how the urban renewal programs of the 1970s coincided with the desire of the security forces to reduce local population densities to achieve security objectives. This phenomenon was especially problematic in Protestant areas, given that these were the very areas where people had begun to move away from in the early 1960s as part of the earlier Matthew Plan and were also the same areas that were subject to substantial redevelopment in the 1970s and 1980s.[273] Protestant population decline during this period was much more widespread than that experienced by the Catholic community.[274] The net effect of this was that the process of redevelopment during this period "simplified the social geography and made [the inner core of the city] more Catholic."[275] As with road construction, the security forces were adapting urban planning practices of the time that had already produced a destabilizing effect on working-class communities and accentuating the problem still further. Large-scale urban redevelopment allied with genuine concerns about safety therefore served to depopulate and destabilize Protestant communities from the 1970s onward. As one study concluded:

> The decline in the population of the inner city was greatly accelerated in the 1970s and 1980s by the flight of Protestants in particular to safer areas outside the boundary ... The simultaneous redevelopment of densely-crowded areas of working-class housing contributed to these changes; many of the scenes of urban desolation which formed the backdrop to television reports of the violence in the 1970s were caused by slum clearance rather than Semtex, while the availability of plenty of accommodation in new Housing Executive estates outside the boundary and well away from the violence encouraged some households to move voluntarily.[276]

While dispersal and depopulation became the de facto policy for inner-city and working-class Protestant communities in Belfast lasting several decades, the situation with respect to the city's Catholic community was quite different, not least because, as outlined in the previous chapter, there were fewer areas outside of north and west Belfast that Catholics could move to safely.[277] From the late 1950s onward many members of the Catholic community, like their Protestant counterparts, had sought to escape the chronic housing problems in Belfast by decanting to new developments such as Rathcoole that were constructed on the outskirts of the city as part of the dispersal policy that was central to the Matthew Plan.[278] Following the eruption of violence in the late 1960s many of those who had left quickly returned to safe havens in the traditional Catholic heartlands of the north and west putting additional pressure on already limited housing capacity in those areas.[279]

Further reduction in the overall percentage of residents going back into Catholic areas after redevelopment was therefore more difficult simply because Catholics had significantly fewer options in terms of where else they could go.[280] Another important factor was that the security forces believed that Catholic population growth across the city posed a security threat.

As the minutes of the SCH show very clearly, the default policy for containing the security threat in the 1970s and the 1980s was predicated upon containing the Catholic community within discrete boundaries, and this resulted in either housing developments being delayed, not built at all, or taking land that could be used for housing being developed for industrial or commercial purposes in order to provide a "buffer zone" between the two communities.

As noted above, this was a policy which Unionist politicians were more than willing to condone since it fitted with their historic aim of limiting the spread of Catholics throughout the city, partly, but not wholly, to protect their electoral interests:

> Apart from the 1940s and 1950s, no decade in the past 150 years has gone by without at least one summer of serious rioting ... In most cases the background was an attempt, individual or collective, by Catholics to assert their right of access to the city's housing stock and collective attempts by Protestants to contain them.[281]

The material examined during this research illustrates how many of the problematic practices regarding the sectarian nature of planning decisions that existed prior to the conflict remained in place. Both the Cameron Report and the Macrory Report had identified the sectarian nature of planning policies and the impact on the Catholic community as a source of concern for the civil rights movement in the 1960s.[282] Specifically, the Cameron Report had found that there were many cases in which councils in Northern Ireland during the 1960s had withheld planning permission, or caused needless delays, where they believed a housing project would be to their electoral disadvantage, that is, where the housing would be occupied by Catholics.[283] The report concluded that in certain Unionist-controlled areas, it was fairly frequent for housing policy to be operated not on the basis of actual housing need but maintenance of the current political preponderance in the local government area.[284] Similarly, the Macrory Report on the reform of local government in the region had the effect of removing responsibility for housing and planning from the hands of locally elected officials and instead centralizing these powers into the hands of government departments or new statutory agencies.[285] The most notable of these agencies was the Northern Ireland Housing Executive, established under the Housing Executive Act (Northern Ireland) 1971, which provided that public house building and allocation should be provided on the basis of an objective points system.[286] The establishment of a central housing agency and the adoption of a points system that sought to ensure housing allocation were based on an objectively defined waiting list became synonymous with the overall reform package introduced by the UK government to appease Catholic concerns in the early 1970s.

The difficultly however is that the secret planning agenda pursued by the security forces from the early 1970s onward, coupled with the sectarian geography of the city and the political composition of Belfast City Council, meant that the planning system reverted to the default historical option, namely, containment of the growth of the Catholic community in Belfast. According to one interviewee, an elected representative of the SDLP in North Belfast during this period,

> The fact and the truth of the matter is that the housing policy in north Belfast that the Housing Executive and the Housing Branch in the Northern Ireland Office pursued in the 1980s, was essentially the same policy as the UDA, i.e. they tried to curtail and circumscribe the expanding Catholic population in north Belfast … Essentially the authorities were following the policy of the UDA so that if you were to build Catholic houses then they would be burnt down, or they would be attacked, so don't try it. This is our territory, and we demand a wall to delineate it, and the Housing Executive and the NIO gave in to that and for the last 30 years they've been building those walls, they've been holding in the nationalist population that has been growing all the time.[287]

This claim is supported by evidence from the Standing Committee on Security Implications of Housing. One committee member, Jack Hermon, who subsequently went on to become Chief Constable of the RUC, cautioned against the emergence of a "green wedge" in north Belfast.[288] Hermon stated that if middle-class Catholics moved into certain areas this would in the long-term "draw in working class Catholics behind them" and that would open the door for the IRA.[289] Concerns were also expressed that the main "industrial and commercial areas [are] in potentially hostile territory," i.e. Catholic areas.[290] This highlights the extent to which security and sectarian imperatives aligned to limit Catholic residential mobility.

Evidence gathered during this research also shows that from the early 1970s, notwithstanding the measures that were introduced to address the sectarian planning and housing policies that existed under the Stormont government, various other techniques were deployed by Unionist politicians to limit the natural growth of the Catholic population in the city. For example, one interviewee describes how a desire to curtail Catholic housing in the south of the city impacted upon the development of the Gasworks site in the early 1990s. The site in question lay between Lower Ormeau, Donegall Pass, and the Markets area. If there had been residential development on the site housing demand in the area was such that the occupants of the housing would have been predominantly Catholic. According to the interviewee,

> That was going to be a big sticking point, and the decision was made not to have a residential development on the site, that was done because at that stage the site was owned by the Council and the Council was dominated by Unionists so in a sense they made the decision, we'll develop the Gasworks without any residential development ...[291]

This area also featured in the deliberations of the Standing Committee on Security Implications of Housing with the submission from the Security Forces expressing concern about the changing demographics of the Lower Ormeau Road (specifically the concern was that the Catholic population in the area had increased) and the security implications that might arise resulting from this. A Concerns were expressed that any additional housing in the area might lead to increased opposition among Catholic residents to traditional Protestant marches.[292] The letter concluded that

> while the problem has become less acute over the past year, the re-development of the area with perhaps an increase in the number of residents from the Markets area of Cromac Street, could escalate the problem ... [and that] ... discussion with the Security Forces will be essential before the advanced stage for this area is reached.[293]

This document confirms that the security objective at the time was predicated on Catholic containment to ensure that confrontations did not arise with respect to traditional Protestant marches, for example, which coincided with the desire of Unionist politicians to ensure that Catholic expansion outside of traditional boundaries was curtailed. Other documents highlighted above also illustrate how the default position of the security forces was to restrict the construction of much-needed housing for the Catholic community.

The difficulty from the early 1970s onward was that while the political reforms, which ensured that the housing executive was legally obliged through the introduction of a points system to ensure that the allocation of housing on an individual case-by-case basis was fair and equitable, there was no mechanism for ensuring that wider policies and practices within the planning system regarding where housing could be built were subject to the same oversight.[294] Crucially, important planning powers remained within the purview of local councillors and were open to manipulation according to sectarian biases., In a context in which Belfast City council and indeed all the local authority areas in the immediate vicinity of the city were controlled by Unionists, the default position was to block Catholic housing in the wider metropolitan area. Another example of this practice occurred in nearby Lisburn Borough Council where the Unionist leadership refused to grant planning permission for the Poleglass development which aimed to address chronic Catholic housing need in the west of the city. In this instance the high-profile nature of the development ensured that the Secretary of State, Roy Mason, intervened to allow the project to proceed. Lisburn Council subsequently refused to provide refuse collection to the residents and used other powers, such as withholding drinks licenses in the area, to frustrate the development. In the case of Poleglass, the security imperative required that the project succeed with public health laws invoked to ensure refuse collection. Military-inspired design ensured however that Poleglass had limited access points and was dominated by cul-de-sacs ensuring vehicles being pursued could not escape. Footpaths were reinforced and designed to be wide enough to carry the weight of army vehicles.[295]

In a context in which the two communities were isolated within distinct geographic areas, the issue of where housing developments were located and the securing of planning permission was crucial in determining wider patterns of housing inequalities. This was coupled with the widespread violence and intimidation that continued in Belfast through the 1980s and the 1990s with Catholic housing targeted in periods of serious inter-communal conflict.[296] For example, following the Anglo-Irish Agreement, a row of houses in Manor Street, adjacent to the Lower Oldpark that had been occupied by Catholic residents was burned to the ground.[297]

In the 1980s, the statutory body with responsibility for oversight of equality and human rights in Northern Ireland, the Standing Advisory Commission on Human Rights (SACHR) asked the Policy Studies Institute (PSI) to undertake a program of research in order to establish whether equality of opportunity existed between the two communities in the areas of employment, economic well-being and housing.[298] Part 4 of the PSI's report which dealt specifically with housing was published in June 1989 and highlighted significant inequalities between the two communities.[299] In terms of access to housing the report found that within the Belfast urban area "Catholic applicants had a considerably lower chance of being re-housed than Protestant applicants" with an almost 2:1 disparity between the two communities.[300] The issue of housing inequalities between the two communities was subsequently examined in a report published by the Standing Advisory Commission on Human Rights in 1990.[301] This report set out a series of recommendations, which aimed to identify the reasons behind the disparities in housing allocations between the two communities.[302]

During the 1990s, another substantial body of independent research, also carried out by the Standing Advisory Commission for Human Rights, along with another report by the Northern Ireland Affairs Committee had sought to examine continued social and economic inequalities between the two communities.[303] The research carried out during the 1990s focused on the extent to which disproportionate levels of social need and spatial disadvantage were concentrated within the Catholic community, especially in the north and west of the city.[304] At no point did this substantial body of research spanning over a decade consider the matters discussed in this chapter – namely the role of the security forces in shaping the built environment and the role of Unionist politicians in blocking housing developments. This highlights how public policy making and democratic oversight became subsumed within a security-led agenda that distorted patterns of spatial disadvantage within the city and allowed planning powers to be exercised without accountability. Moreover, the secret nature of the practices involved allowed the myth of neutrality to be maintained regarding the political impact of the reform measures of the early 1970s as evidenced by the SACHR Report of 1987 which stated explicitly,

> Since 1972, the Government of Northern Ireland has been under the direct authority of British Ministers and civil servants who have not been directly associated with either of the two sections of the community in Northern Ireland and who might be expected to have been free from any religious or political partiality of the kind alleged against previous Unionist administrations. A number of programmes have been instituted to ensure impartiality and equality in the allocation of governmental resources. In Belfast under the Belfast Areas of Need programme an objective assessment was made of levels of poverty and need in different parts of the city as the basis for the allocation of resources for housing renewal and other forms of urban redevelopment.[305]

It is clear from the sources examined in this chapter that the impartiality and neutrality alluded to in the SACHR Report were seriously compromised by the range of powers that councils retained regarding planning and the activities of the security forces. These factors contributed to the distortion of patterns of settlement in the city that led to increased pressure on waiting lists and patterns of housing inequality in Catholic areas. Incremental attempts by the Catholic community to expand beyond traditional territorial boundaries were frustrated by the vetoing of housing development by the security forces and Unionist politicians. The outcome of the security-led approach to the redevelopment of Belfast in the 1970s and 1980s vindicates the prediction of Hewins regarding the "straight-jacketing" of communities. For the Catholic community, which was growing in population with limited choices about where to live this "straight-jacketing" was especially problematic, increasing housing needs. This problem was further heightened by a political system in which decisions were made without input from the representatives of the Catholic community. The original JWP Report was compiled for the Unionist government at Stormont and presented to the then Prime Minister Brian Faulkner. As the documents from the 1980s show, political responsibility for the subsequent implementation of these findings was primarily in the hands of the security forces, accountable only to the U.K. government. In addition, some responsibility for planning remained in the hands of the Unionist-controlled Belfast City Council which allowed decisions to be made regarding where housing developments could be located. The common factor therefore was the lack of representation and input from representatives of the minority Catholic community. Ignatieff contends that segregation and hypersensitivity to turf is a phenomenon that afflicts many urban centers across the globe, and is not of itself necessarily problematic, so long as it is consensual and involves an element of "horizontal governance" that is based upon a degree of tolerance and mutual respect.[306]

As the material highlighted in this chapter shows clearly, the planning and design practices that characterized Belfast during the years of the conflict was not part of a "shared operating system." In fact, as the documentary evidence indicates, the planning and design of the city from the 1970s onwards was in fact characterized by a top-down, command-and-control style of governance driven by the security forces, especially the military. Neither was there "horizontal governance" of the type identified by Ignatieff but rather, one community, namely the Unionist community, was able to exert influence through its control of Belfast City Council. The frustrations felt by one interviewee as an elected representative of the Catholic community, and the example provided by anohter interviewee regarding the redevelopment of the Gasworks, as well as the documentary evidence, are indicative of the ways in which the historic power structures that had dominated Belfast retained influence. This was especially apparent regarding the desire on the part of the security forces to contain violence, which cut with the grain of thinking among elected leaders of Unionism who wished to contain the Catholic community within discrete boundaries. Moreover, the secret nature of the decision-making process meant that oversight by statutory agencies like the Standing Advisory Commission on Human Rights was nullified.

The fact that secret planning powers were exercised without accountability is even more significant, given that the 1980s was a period when the city of Belfast underwent an unprecedented level of housing redevelopment. As Bardon has pointed out the bulk of the Westminster subvention to Northern Ireland during this period was spent neither on direct aid to industry nor on the security forces – the overwhelming share was assigned to the public services and the most visible sign of this expenditure was the transformation of the region's public housing.[307] In 1981, the British government identified housing as its number one priority – with the result that in the following decade the Northern Ireland Housing Executive built 10,000 new homes in the city, "virtually all two-storey terraced or semidetached dwellings."[308] It is estimated that between 1982 and 1991 the Northern Ireland Housing Executive (NIHE) had spent £2.4 billion on new dwellings, maintenance, grants, and renovation.[309] The upsurge in public spending during this period was overseen by some of the more Keynesian-minded members of the U.K. government, typically referred to as "wets".[310] This security-led agenda involved redesigning urban space as part of a counter-insurgency strategy to render working-class enclaves more legible and illustrates how defeating the IRA rather than curbing public spending was the priority for the Thatcher government. Among the areas redeveloped during this period were some of the most notorious prefabricated high-rises built during the 1960s, including the "brutal concrete architecture of the Divis Flats, whose balconies and walls gave cover to the IRA gunmen."[311] At the same time however, this redevelopment took place in a way that entrenched and exacerbated isolation and segregation of the two communities with security objectives trumping all others.

As with other conflict legacy issues in Northern Ireland there is a contested view around the extent to which the city of Belfast was shaped by the conflict. Looking at the evidence considered above, the relationship between the planning and political systems in Belfast from the early 1970s was a complex one. This complexity is heightened by the fact that in a context in which serious levels of violence were perpetrated against the civilian population, communities themselves wanted to ensure that their built environment afforded them as much protection as possible. American and U.K. theories of urban design were already pointing in this direction by embracing the notion of defensible space in order to enhance community safety, which included not only a reduction in crime but also increased road safety through manipulation of the built environment.[312] The concept of defensible space, developed by Oscar Newman in the U.S. in the early 1970s, involves a range of mechanisms designed to bring an environment under the control of its residents.[313] On the micro-level, this resulted in windows and lights positioned on houses to enhance visibility. Newman argued that increased surveillance provided greater security and reduced anxieties and fears among residents. Another feature of defensible space is the minimization of through-flow which can be achieved by the use of culs-de-sac and other physical barriers to slow traffic movements. These design approaches are among the techniques that were being deployed in Northern Ireland during the 1970s and 1980s

> … the courtyard-type layouts, now almost universally employed by the [Northern Ireland Housing] Executive, owe something to the concept of 'defensible space' developed in the United States in recent years.[314]

The fact that the aims of the security forces and local communities on occasion overlapped is one reason why quite a number of commentators have argued that there is no evidence of security force involvement in the planning and design of the urban environment and that regeneration initiatives, including housing, were influenced solely by community household surveys and community consultation programs.[315] This counterargument is predicated on the notion that following reform of local government in Northern Ireland during the early 1970s, in response to allegations of discrimination under the previous Stormont regime, a technocratic ethos was established within the planning system that prioritized impartiality and neutrality.[316] From this perspective, the Westlink motorway "fortuitously made an effective barrier to possible sectarian confrontation."[317] Notwithstanding the publication of several leaked documents during the 1980s which appeared in the press the official line from government that was always to deny that the security forces had any specific role in the planning process in Northern Ireland and to stress the neutrality and objectivity of planning decisions in order to distance themselves from any hint that they were pursuing a security agenda.[318] The fact that much of the housing built during the 1980s onward was of higher quality than the system-built design of the 1950s and 1960s has also blinded observers to the negative aspects of security driven policy during this period. Typical of this approach is Brett who claims that housing from the 1980s onward, based around "neat culs-de-sacs and closes … are good to look at and good to live in."[319] As the next chapter will show, these developments may superficially have been good to look at, but they served to enhance isolation for the communities living in them.

In a context in which public consultation and security considerations were, on occasion, pointing in the same direction, a significant number of contributors to this debate concluded that the outcomes were ones that the residents themselves requested. In terms of the complexity of this process, one interviewee argues that

> it's not this huge conspiracy, it is a series of smaller tactics if you like … take the example of design, where a trend develops, and somebody who's a

> professional architect, knows that if they put in a particular type of plan, that it's going to be knocked back … so they start adapting their work to suit the prevailing trend, not because they know they're getting knocked back because of a British Army agenda, or an NIO agenda, but they know they're getting knocked back with a particular type of plan, so they start adapting to the criteria if you like that is set down, or the parameters that are quite subtly set down over a period of time, because remember this was a long conflict. So, there wasn't a big plan that they implemented in two years, or three years, or even five years, this was something that they did over a long period of time.[320]

Colin Scott summarizes this relationship as one in which subsystems relate to each other not through highly specified or materialized external rules but rather through working "with the grain of the understanding of ordering within other subsystems."[321] In the context of the built environment the desire of residents to have some protection against attack, the development of courtyard-type urban design models to limit traffic, and the need for substandard system-built tower blocks to be demolished cut with the grain of a security agenda designed to manage conflict. The fact that on occasion these interests coincided with solutions formulated by architects and planners allowed security officials and politicians plausible deniability regarding their role in shaping the built environment. Much of the detail behind defensive design and planning practices through the 1980s originated within the planning system, not among politicians or security officials, thus allowing the planning system to "own" the outcomes that reshaped the urban environment during this period.

As the material uncovered during this research shows, the Westlink did not, as Charles Brett argued, "fortuitously" separate communities. The fact that, on occasion, residents wished for the urban environment to afford them some degree of safety and protection does not mean that the significant role that the security agenda played in shaping the city over several decades ought to be minimized or discounted. Furthermore, as identified in the first section, the reforms of the early 1970s did not mean that sectarian biases were removed from the planning system but rather that they were reconfigured and played out in more subtle ways. The overall effect of this enhanced militarization of space- and security-driven approach to urban planning was to significantly increase material inequalities as well as heighten feelings of isolation, disorientation, and alienation in the absence of democratic accountability.

NOTES

1. DiMento, J.F.C. 'Stent (or Dagger?) in the heart of town: Urban freeways in Syracuse, 1944–1967 (2009) 8 *Journal of Planning History* 133–161.
2. Archer, D. "White men's roads through black men's homes": Advancing racial equality through highway reconstruction (2020) 73 (5) *Vanderbilt Law Review* 1259.
3. Bayor, R., *Roads to Racial Segregation: Atlanta in the Twentieth Century,* 15 J. URB. HIST. 3 (1988), p. 5. See also Bayor, R. *Race and the Shaping of Twentieth-Century Atlanta* (University of North Carolina Press (1997)) p. 58.
4. Bayor, R., *Roads to Racial Segregation: Atlanta in the Twentieth Century*, 15 J. URB. HIST. 3 (1988), p. 5. See also Bayor, R. *Race and the Shaping of Twentieth-Century Atlanta* (University of North Carolina Press (1997)) p. 58.

5 Mohl, R., Planned destruction: The interstates and central city housing. In J. F. Bauman, R. Biles, and Kristin Sylvian (eds.), *From Tenements to the Taylor Homes: In Search of an Urban Housing Policy in Twentieth-Century America* (University Park 2000) p. 230.
6 McNichol, D. *The Roads That Built America: The Incredible Story of the U.S. Interstate System* (Sterling 2006) p. 141.
7 Binkovitz L. 'How a Montgomery Highway Sought to Disrupt the Heart of the Civil Rights Movement'. Rice University. Kinder Institute for Urban Research. January 25, 2019.
8 Retzlaff R. and J. Zanzot, "The Interstates: Planned Violence and the Need for Truth and Reconciliation", The Metropole. https://themetropole.blog/2021/04/07/the-interstates-planned-violence-and-the-need-for-truth-and-reconciliation/#_ftn1.
9 Retzlaff, R. 'Interstate highways and the civil rights movement: The case of I-85 and the Oak Park neighborhood in Montgomery, Alabama' (2019) 41 (7) *Journal of Urban Affairs* 930–959.
10 Campanella, R. 'A half-century after Riverfront Highway Nixed, Shuddering at the New Orleans that Could Have Been.' August 2019. nola.com
11 Schneider, B. 'Why Is It So Hard to Kill This Freeway?' Bloomberg CityLab. April 12, 2023.
12 Rios, E. 'Could the US highways that split communities on racial lines finally fail?' (July 29, 2022) The Guardian.
13 Santana R. 'In New Orleans, documenting history of iconic Black street' (May 25, 2021). AP news. https://apnews.com/article/la-state-wire-new-orleans-lifestyle-trees-business-69e663d6bd4c98dbf518c244564979b2.
14 The Claiborne Avenue History Project. https://claiborneavenue.wordpress.com.
15 Alana, S. 'How to decimate a city' (November 20, 2015) *The Atlantic*. https://www.theatlantic.com/business/archive/2015/11/syracuse-slums/416892/.
16 ROCC Buffalo. 'The Vision'. https://roccbuffalo.org/?p=2873.
17 Ross-Brown, S. 'Transportation Secretary Foxx Moves to Heal Scars of Urban Renewal'. (September 30, 2016). The American Prospect. https://perma.cc/G8Z3-NSPP.
18 Foxx, A. 'Racism destroyed a thriving neighborhood in my hometown. Now the fix is in sight' (April 22, 2021). CNN. https://www.cnn.com/2021/04/22/opinions/american-infrastructure-spending-racism-foxx/index.html.
19 'Historic Charlotte Neighborhoods: History of the Brooklyn Community.' J. Murray Atkins Library. University of North Carolina Charlotte. https://guides.library.charlotte.edu/c.php?g=621704&p=4626874.
20 Rose, M. and Mohl, R. *Interstate: Highway Politics and Policy Since 1939*. (University of Tennessee Press 2012) p. 105.
21 Paull, E. *Stop the Road: Stories from the Trenches of Baltimore's Road Wars*. (Boyle and Dalton 2022).
22 Cassie, R. 'Road to Ruin' (February, 2023) Baltimore. https://www.baltimoremagazine.com/section/businessdevelopment/highway-to-nowhere-baltimore-expressway-demolished-black-neighborhoods/.
23 Maher, K. 'Black community cut off by Akron's 'Road to nowhere' seeks to undo damage' (September 20, 2023) *Wall Street Journal*. https://www.wsj.com/us-news/black-community-cut-off-by-akrons-road-to-nowhere-seeks-to-reconnect-a58cb054?mod=Searchresults_pos1&page=1.
24 Crane, R. 'Slouching towards Cincinnati' (August 24, 2022) Public Square. https://www.cnu.org/publicsquare/2022/08/24/slouching-towards-cincinnati.
25 Char, A. 'Groups across the country are fighting efforts to expand "harmful" highways' (October 15, 2022) *NBC News*.
26 Cohen, A. and Taylor, E. *American Pharaoh Mayor Richard J. Daley: His Battle for Chicago and the Nation*. (Little, Brown and Company 2000) p. 10.
27 Cohen, A. and Taylor, E. *American Pharaoh Mayor Richard J. Daley: His Battle for Chicago and the Nation*. (Little, Brown and Company 2000) 189 Cite Lewis, Divided Highways, p. 122.

28 Cohen, A. and Taylor, E. *American Pharaoh Mayor Richard J. Daley: His Battle for Chicago and the Nation.* (Little, Brown and Company 2000) 189 Cite Lewis, Divided Highways p. 189.
29 Briggs, J. 'Before the Highway: Saint Paul, Minnesota'. AARP Liveable Communities. aarp.org https://www.aarp.org/livable-communities/getting-around/info-2023/before-the-highway-rondo-saint-paul-minnesota-marvin-anderson.html.
30 *Restorative Rondo: Building Equity for All,* Past Prosperity Study. (Yorth Group 2020) https://reconnectrondo.com/wp-content/uploads/2021/01/Rondo-Past-Prosperity-Study.pdf.
31 Raboteau, E. 'How our Roads Hurt Us and Everything Around Us'. (September 10, 2023) The New York Times. https://www.nytimes.com/2023/09/10/books/review/crossings-ben-goldfarb.html?unlocked_article_code=1.4kw.3dmm.4rUPlhvKIKm8&smid=url-share.
32 Eldred, S. 'Marvin Roger Anderson was displaced from St. Paul's historically Black Rondo Neighborhood as a child when Interstate 94 was constructed. Now in his 80s, he sees promise in a plan to reconnect the area.' (March 31, 2022) sahan journal. https://sahanjournal.com/climate-environment/reconnect-rondo-rondo-neighborhood-st-paul/.
33 Estrada, G. 'If you build it, they will move: The Los Angeles freeway system and the displacement of Mexican East Los Angeles' (2005) 87 (3) *Southern California Quarterly, Fall* 287–315.
34 Estrada, G. Latino interchanges: Greater East Los Angeles in the freeway era. In R. Reft et al. (eds.), *Justice and the Interstates: The Racist Truth About Urban Highways* (Island Press 2023).
35 Meares, H. 'The thrill of Sugar Hill'. (February 22, 2018). Curbed. Los Angeles. https://la.curbed.com/2018/2/22/16979700/west-adams-history-segregation-housing-covenants.
36 Susaneck, Adam Paul. 'Segregation by Design' TU Delft Centre for the Just City. 2024. https://www.segregationbydesign.com/los-angeles/sugar-hill.
37 Reft, R. 'How Prop 14 Shaped California's Racial Covenants.' (September 20, 2017). https://www.pbssocal.org/shows/city-rising/how-prop-14-shaped-californias-racial-covenants.
38 Knickmeyer, E. 'The foundation of the wealth: Why black wall street boomed' (June 1, 2021) *AP News.* https://apnews.com/article/ok-state-wire-business-race-and-ethnicity-874fade343fc7515210117c4efe1335f.
39 Chatelain, M. 'A resilient black community 'built from the fire' of a massacre' (May 23, 2023) *New York Times.*
40 Luckerson, V. *Built from the Fire: The Epic Story of Tulsa's Greenwood District, America's Black Wall Street.* (Random House 2023).
41 Whitte, S and Nowell, S. *Youth-Driven Inquiry into the Tulsa Race Massacre.* (Oklahoma State University (2022)). English Journal 96–99.
42 Goble D. Final Report of the Oklahoma Commission to Study the Tulsa Race Riot of 1921. https://tulsareparations.z19.web.core.windows.net/FinalReport.htm.
43 Luckerson, V. Built from the Fire: The Epic Story of Tulsa's Greenwood District, America's Black Wall Street. (Random House 2023) p. 116.
44 Luckerson, V. Built from the Fire: The Epic Story of Tulsa's Greenwood District, America's Black Wall Street. (Random House 2023) p. 118.
45 Moreno, C. *The Victory of Greenwood.* (Jenkin Lloyd-Jones 2021).
46 Cortright, J. 'How highways finally crushed black Tulsa' (July 9, 2021) *Strong Towns.* https://www.strongtowns.org/journal/2021/7/9/how-highways-finally-crushed-black-tulsa.
47 Dembicki, G. 'A highway destroyed Tulsa's thriving Black Wall Street – now there's hope it could come back' (July 18, 2023) *The Guardian.* https://www.theguardian.com/us-news/2023/jul/18/tulsa-highway-black-wall-street?CMP=Share_iOSApp_Other.
48 Cortright, J. 'How highways finally crushed black Tulsa' (July 9, 2021) *Strong Towns.* https://www.strongtowns.org/journal/2021/7/9/how-highways-finally-crushed-black-tulsa.

49 Moreno, C. 'Black wall street's second destruction' (May 31, 2021) *Next City.*
50 Cortright, J. 'How highways finally crushed black Tulsa' (July 9, 2021). https://www.strongtowns.org/journal/2021/7/9/how-highways-finally-crushed-black-tulsa.
51 Gelfand, S. 'Greenwood residents, lawmakers call for removal of I-244' (September 2nd 2022) *News Channel 8 Tulsa.*
52 Mock, B. 'What really caused the destruction of Tulsa's "Black Wall Street"' (May 30, 2023) *Bloomberg.* https://www.bloomberg.com/news/articles/2023-05-30/in-built-from-the-fire-victor-luckerson-explores-tulsa-race-massacre-aftermath.
53 Bayor, R. *Race and the Shaping of Twentieth-Century Atlanta* (University of North Carolina Press 1997) p. 61.
54 Kruse, K. 'What does a traffic jam in Atlanta have to do with segregation? Quite a lot.' (August 14, 2019) *New York Times.* https://www.nytimes.com/interactive/2019/08/14/magazine/traffic-atlanta-segregation.html?searchResultPosition=1.
55 Sweet Auburn Works. https://www.sweetauburnworks.com/about-us/our-history/.
56 Wheatley, T. 'Atlanta hopes to heal neighborhoods torn apart by a highway' (January 5, 2022) *Axios.* Retrieved [August 1, 2024] https://www.axios.com/local/atlanta/2022/01/05/atlanta-neighborhood-torn-apart-highway].
57 Wiggins, D. 'Remembering sweet auburn before the expressway: What nostalgia reveals about the limits of postwar liberalism' *The Metropole,* https://themetropole.blog/2021/04/14/remembering-sweet-auburn-before-the-expressway-what-nostalgia-reveals-about-the-limits-of-postwar-liberalism/#_edn4.
58 Divided by Design, Smart Growth America. https://www.smartgrowthamerica.org/signature-reports/divided-by-design/.
59 Wiggins, D. 'Remembering sweet auburn before the expressway: What nostalgia reveals about the limits of postwar liberalism' *The Metropole,* https://themetropole.blog/2021/04/14/remembering-sweet-auburn-before-the-expressway-what-nostalgia-reveals-about-the-limits-of-postwar-liberalism/#_edn4.
60 Bayor, R. *Race and the Shaping of Twentieth-Century Atlanta* (University of North Carolina Press 1997) p. 54.
61 Henderson, J. 'Secessionist automobility: Racism, Anti-Urbanism, and the politics of automobility in Atlanta, Georgia' (2006) 293 30 *International Journal of Urban and Regional Research.*
62 Connolly, N.B.D. A World More Concrete: Real Estate and the Remaking of Jim Crow South Florida. (University of Chicago Press 2014).
63 Mohl, R. Race and space in the modern city: Interstate-95 and the black community in Miami. In A. Hirsch and R. Mohl (eds.), *Urban Policy in Twentieth-Century America* (Rutgers University Press, 1993), p. 134.
64 Going Overtown. https://goingovertown.org/listing/knight-beat-at-the-sir-john-hotel/.
65 Holden Luntz Gallery. https://www.holdenluntz.com/magazine/new-arrival/flip-schulke-muhammad-ali-boxing-underwater/.
66 Mohl, R. Planned destruction: The interstates and central city housing. In J. F. Bauman, R. Biles, and K. Szylvian (eds.) *From Tenements to the Taylor Homes: In Search of an Urban Housing Policy in Twentieth-Century America.* (University Park, Pa., 2000), p. 230.
67 Mohl R. Planned destruction: The interstates and central city housing. In J. F. Bauman, R. Biles, and K. Szylvian (eds.) *From Tenements to the Taylor Homes: In Search of an Urban Housing Policy in Twentieth-Century America.* (University Park, Pa., 2000) 103. Mohl, R. Making the second ghetto in metropolitan Miami, 1940–1960' (March 1995) 21 *Journal of Urban History* 398.
68 Mohl, R. 'Making the second Ghetto in metropolitan Miami, 1940–1960' (March 1995) 21 *Journal of Urban History* 398.
69 Rothstein, R. The Color of Law. (Liveright 2017) p. 127.
70 McFarlane, A. 'Race, Space, and Place: The Geography of Urban Development', 36 San Diego L. Rev. 295 (1999) 332.

71 Mohl, R. 'Race and space in the modern city: Interstate-95 and the black community in Miami. In A. Hirsch and R. Mohl (eds.) Urban Policy in Twentieth-Century America. (Rutgers University Press 1993).
72 Mohl, R. 'Race and space in the modern city: Interstate-95 and the black community in Miami. In A. Hirsch and R. Mohl (eds.) *Urban Policy in Twentieth-Century America.* (Rutgers University Press 1993).
73 How Interstate Highways Gutted Communities – and Reinforced Segregation, History.com.
74 'Older Adults Recall Life Before the Highway'. AARP Liveable Communities. https://www.aarp.org/livable-communities/getting-around/info-2023/before-the-highway.html.
75 Massay, D. and N. Denton, American Apartheid, (Harvard University Press, 1993) p. 161.
76 Massay, D and N. Denton, American Apartheid (Harvard University Press, 1993) p. 72.
77 Mohl, R. The Interstates and the Cities: Highways, Housing and Freeway Revolt. (Poverty and Race Research Action Council, Civil Rights Research 2002). p. 21.
78 Mohl, R. Planned destruction: The interstates and central city housing. In J. F. Bauman, R. Biles, and K. Szylvian (eds.) From Tenements to the Taylor Homes: In Search of an Urban Housing Policy in Twentieth-Century America. (University Park, Pa., 2000) p. 236.
79 There is evidence that corrupt building practices were also to blame for the increased costs of housing in Chicago. This book has largely overlooked the way in which corruption was linked to exclusionary practices both in the U.S. and the U.K. This topic is worthy of a separate study.
80 Hirsch, A. and Mohl, R. eds. *Urban Policy in the Twentieth Century.* (Rutgers University Press 1993) p.134.
81 Federal Housing Administration. Underwriting Manual: Underwriting and Valuation Procedure Under Title II of the National Housing Act. U.S. Government Printing Office. (Washington D.C. 1938). para. 937.
82 Federal Housing Administration. Underwriting Manual: Underwriting and Valuation Procedure Under Title II of the National Housing Act. U.S. Government Printing Office. (Washington D.C. 1938). para. 935.
83 Abrams C. Forbidden Neighbors: A Study of Prejudice in Housing. (Harper 1955) p. 236.
84 Al Jazeera has compiled an online resource "know their names" that lists the names of African Americans killed by police officers in recent years along with a description of the circumstances surrounding their deaths. https://interactive.aljazeera.com/aje/2020/know-their-names/index.html.
85 Cohen, P. 'Beyond Pandemic's Upheaval A Racial Wealth Gap Endures.' (April 9, 2021). The New York Times. https://www.nytimes.com/2021/04/09/business/economy/racial-wealth-gap.html.
86 'Over the past century, African-American life expectancy and education levels have soared: But racial gaps across these and other indicators remain wide.' (March 20, 2021) The Economist. https://www.economist.com/graphic-detail/2021/05/20/over-the-past-century-african-american-life-expectancy-and-education-levels-have-soared.
87 Coates, T-N. 'The Case for Reparations', June 2014. The Atlantic. https://www.theatlantic.com/magazine/archive/2014/06/the-case-for-reparations/361631/. Coates, T.-N. 'The case for reparations' (June 2014) The Atlantic 65.
88 Gerken, M. et al. 'Assessing the Legacies of Historical Redlining.' January 2023. Urban Institute. https://www.urban.org/sites/default/files/2023-01/Addressing%20the%20Legacies%20of%20Historical%20Redlining.pdf.
89 Jackson, K. T. *Crabgrass Frontier.* (Oxford University Press 1985) p. 203.
90 Norwood C., About Paterson:The Making and Unmaking of an American City. (Saturday Review Press 1974)
91 Jackson, K. T. Crabgrass Frontier. (Oxford University Press 1985) p. 216.
92 Jacobs, J, *The Death and Life of Great American Cities* (Modern Library 1993) p. 391.
93 Jacobs, J, *The Death and Life of Great American Cities* (Modern Library 1993) p. 385.

94 North End for Former West End: Where Would You Rather Live or Visit?' (June 24, 2016) http://www.janejacobsinthewoo.org/blog/2016/7/21/if-you-lived-here-youd-be-home-by-now-will-it-be-the-west-end-or-the-north-end.
95 'The Story of Back of the Yards.' wttw.com https://interactive.wttw.com/chicago-by-l/neighborhoods/back-of-the-yards.
96 'History of Back of the Yards.' Chicago Studies. The University of Chicago. https://chicagostudies.uchicago.edu/back-yards/back-yards-history-back-yards.
97 Chicago History Museum, The Newberry Library, and Northwestern University. The Electronic Encyclopedia of Chicago: Back of the Yards Neighborhood Council. (Chicago Historical Society 2005) http://www.encyclopedia.chicagohistory.org/pages/100.html.
98 Bayor, R., *Roads to Racial Segregation: Atlanta in the Twentieth Century*, 15 J. URB. HIST. 3 (1988), p.17.
99 https://dsl.richmond.edu/panorama/redlining/map/MN/StPaul/area_descriptions/D4#loc=13/44.9068/-93.1284
100 Fresh Air. 'Pullman Porters, Creating a Black Middle Class.' NPR (May 8, 2009) https://www.npr.org/transcripts/103933268.
101 Satter, B. *Family Properties: How the Struggle Over Race and Real Estate Transformed Chicago and Urban America.* (Metropolitan Books 2009).
102 Coates, T.-N. 'The case for reparations' (June 2014) *The Atlantic 65.*
103 Jacobs, J. *The Death and Life of Great American Cities* (Modern Library 1993) p. 391.
104 Scott, J. *Seeing Like a State.* (Yale University Press 2020) p. 61.
105 *Building the American City: Report of the National Commission on Urban Problems to the President of the United States.* (Washington, 1968) pp. 100–101.
106 Keynes, J. M. *The Economic Consequences of the Peace* (first published 1920, Penguin 1971) p. 14.
107 Jacobs, J. *The Death and Life of Great American Cities* (first published 1961, Modern Library 1993) p. xvii.
108 Collins, B. The Edwardian City. In J. Beckett et al. (eds) *Belfast: The Making of the City* (Appletree Press 1988).
109 Walker, B. and Dixon, H. *No Mean City: Belfast 1880–1914* (Friars Bush Press 1983) p. vii.
110 Buckland, P. *A History of Northern Ireland* (Gill and Macmillan 1981) p. 6.
111 Buckland, P. *A History of Northern Ireland* (Gill and Macmillan 1981) p. 6.
112 Hepburn, A. C. *A Past Apart: Studies in the History of Catholic Belfast 1850–1950* (Ulster Historical Foundation 1996) p. 121. Growth in the Catholic percentage of the population during this period was not linear. Hepburn observes that between 1861 and 1911 the Catholic share of the city fell from 34.1 to 24.1%, a decline of three tenths. He argues that it is 'highly likely' that the relative decline of Catholic migration to Belfast, and a differential out-migration of young adult Catholics, especially males, arose from a Catholic perception of reduced employment opportunities available to them in the city. A C Hepburn, *Contested Cities in the Modern West* (Palgrave 2004)169, Table 6.1.
113 Hepburn, A. C. *A Past Apart: Studies in the History of Catholic Belfast 1850–1950* (Ulster Historical Foundation 1996); *Boyd, A. Holy War in Belfast* (Anvil 1969).
114 Wright, F. Communal deterrence and the threat of violence in the North of Ireland in the 19th century. In J. Darby, N. Dodge, and A. C. Hepburn (eds) *Political Violence: Ireland in Comparative Perspective* (Ottawa 1990).
115 Maguire, W. *Belfast: A History* (Carnegie Publishing 2009) p. 170.
116 Bryan, D. In S. J. Connolly (ed.) *Belfast 400: Planning, Space, and Conflict* (Liverpool University Press 2012) p. 326.
117 Boal, F. *Shaping a City* (QUB 1995) p. 22.
118 Gaffikin, F., Morrissey, M and Sterrett, K. Remaking the city: The role of culture in Belfast. In W. Neill and H. Schwedler (eds) *Urban Planning and Cultural Inclusion: Lessons from Belfast and Berlin* (Palgrave Macmillan 2001) p. 148.

119 Gaffikin, F., Morrissey, M and Sterrett, K. Remaking the city: The role of culture in Belfast. In W. Neill and H. Schwedler (eds) *Urban Planning and Cultural Inclusion: Lessons from Belfast and Berli* (Palgrave Macmillan 2001).
120 Unlike many cities, Belfast Mayors are not elected directly but rather are elected by fellow councilors and serve for one year. This process has evolved into an informal arrangement by which Unionists and Nationalists rotate the position of Mayor and Deputy Mayor. In effect this represents an informal realpolitik that recognizes the demographic balance of the city. See https://www.belfastcity.gov.uk.
121 Shuttleworth, I. 'Are Northern Ireland's communities dividing? Evidence from geographically consistent census of population data, 1971–2001' (2009) 41 (1) *Environment and Planning A* 213.
122 Gaffikin, F., Morrissey, M., and Sterrett, K. Remaking the city: The role of culture in Belfast. In W. Neill and H. Schwedler (eds) *Urban Planning and Cultural Inclusion: Lessons from Belfast and Berlin.* (Palgrave Macmillan 2001) p. 148.
123 Murtagh, B. and Carmichael, P. *Sharing Place: A Study of mixed housing in Ballynafeigh, South Belfast* (Northern Ireland Housing Executive 2006).
124 Murtagh, B. and Carmichael, P. *Sharing Place: A Study of mixed housing in Ballynafeigh, South Belfast* (Northern Ireland Housing Executive 2006).
125 Gray, P., McAnulty, U., and Keenan, M. Moving towards integrated communities in Northern Ireland: New approaches to mixed housing' (2009) 9 (3) *European Journal of Housing Policy* 337–353.
126 Shirlow, P. Belfast: A segregated city. In C. Coulter and M. Murray (eds) *Northern Ireland After The Troubles: A society in transition.* (Manchester University Press 2008).
127 Gaffikin, F. and Morrissey, M. *Planning in Divided Cities: Collaborative Shaping of Contested Space* (Wiley-Blackwell 2011) pp. 197–229.
128 Northern Ireland Statistics and Research Agency. *Northern Ireland Multiple Deprivation Measures.* (Northern Ireland Statistics and Research Agency 2010) p. 3.
129 Gaffikin, F. and Morrissey, M. *Planning in Divided Cities: Collaborative Shaping of Contested Space.* (Wiley-Blackwell 2011) p. 218.
130 Among the measures include the Belfast Areas of Need (BAN) Programme, Making Belfast Work (MBW), Targeting Social Need (TSN), New Targeting Social Need (New TSN), and Renewing Communities. For a discussion of the impact and shortcomings of these programs see J Muir *Regeneration and Poverty in Northern Ireland: Evidence and Policy Review* (Sheffield Hallam University: Centre for Regional Economic and Social Research 2014).
131 Kennedy, L. *The Modern Industrialisation of Ireland 1940–1988* (Dundalgan 1989); Farrell, M. *Northern Ireland: The Orange State*, (Pluto Press 1976) pp. 227–256.
132 Buckland, P. *A History of Northern Ireland* (Gill and Macmillan 1981) p. 94.
133 Buckland, P. *A History of Northern Ireland* (Gill and Macmillan 1981) p. 94.
134 Department for Economic Development, *Report of the Joint Working Party on the Economy of Northern Ireland* (Cmnd 446, 1962) (*passim*).
135 Boal, F. 'Territoriality on the Shankill-Falls Divide in Belfast' [1969] 4 Irish Geography 41.
136 Farrell, M. *Northern Ireland: The Orange State.* (Pluto Press 1976) p. 230.
137 Matthew, R. H. *Belfast: Regional Survey and Plan.* (HMSO 1964).
138 Collier, P. *The Future of Capitalism: Facing The New Anxieties.* (Harper Collins 2018) p. 126.
139 Morrison, B. Planning the city; planning the region. In F. Boal and S. Royle (eds) *Enduring City: Belfast in the Twentieth Century.* (Blackstaff 2006) pp. 141–154.
140 Buckland, P. (1981) *A History of Northern Ireland.* (Gill and Macmillan 1981) p. 113.
141 Osborne, R. and Singleton, D. Political processes and behaviour. In F. W. Boal and N. J. Douglas (eds) *Integration and Division: Geographical Perspectives on the Northern Ireland Problem.* (Academic Press 1982) p. 177.
142 Matthew, R. *Belfast: Regional Survey and Plan.* (HMSO 1964) (*passim*).

143 The Housing (Miscellaneous Provisions) and Rent Restriction Law (Amendment) Act (Northern Ireland) (1956) gave the government of Northern Ireland the right to define certain properties as 'unfit' and mark them for redevelopment. Almost 100,000 houses were allocated under this Act, the majority of which were in the west of the city center (the Crumlin Road Shankill Road and Divis Street areas and to the south (Sandy Row, Donegall Road and The Markets). W Johnston, *The Belfast Urban Motorway: Engineering, Ambition and Social Conflict* (Colourprint Books 2014) 32–33.
144 Hepburn, A. C. *A Past Apart: Studies in the History of Catholic Belfast 1850–1950* (Ulster Historical Foundation 1996) p. 246.
145 Planning Advisory Board, *Housing in Northern Ireland: Interim Report of the Planning Advisory Board* (Cmnd 479, 1965).
146 Skeffington Report, People and Planning: Report of the Committee on Public Participation in Planning (HMSO, 1969) (*passim).*
147 Darby, J. and Morris, G. 'Intimidation in Housing', (1973) 111 (2) *Community Forum 7.*
148 Bardon, J. *A History of Ulster* (Blackstaff Press 2001) p. 295.
149 Boal, F. W. Contemporary Belfast. In J. C. Beckett and R. E. Glasscock (eds) *Belfast: The Origin and Growth of an Industrial City.* (BBC 1967).
150 Maguire, W. *Belfast: A History* (Carnegie Publishing 2009) p. 170.
151 Maguire, W. *Belfast: A History* (Carnegie Publishing 2009) p. 170.
152 Gold, J. *The Practice of Modernism: Modern Architects and Urban Transformation 1954-72*, (Routledge 2007) (*passim*).
153 Kynaston, D. *Modernity Britain: A Shake of the Dice.* (Bloomsbury 2014) p. 70.
154 Kynaston, D. *Modernity Britain: A Shake of the Dice.* (Bloomsbury 2014) p. 285.
155 Kynaston, D. *Modernity Britain: A Shake of the Dice.* (Bloomsbury 2014) p. 75.
156 Kynaston, D. *Modernity Britain: A Shake of the Dice.* (Bloomsbury 2014) p. 78.
157 There were however other important influences in Britain, notably the construction industry, see Dunleavy, P. *The Politics of Mass Housing in Britain 1945–1975.* Oxford University Press 1981) (*passim*).
158 Kynaston, D. *Modernity Britain: A Shake of the Dice.* (Bloomsbury 2014) p. 73.
159 Kynaston, D. *Modernity Britain: A Shake of the Dice.* (Bloomsbury 2014) p. 72.
160 Pluntz, R. *City Riffs: Urbanism, Ecology, Place* (Columbia University GSAPP 2017) pp. 29–32.
161 Disturbances in Northern Ireland: Report of the Cameron Commission Appointed by the Government of Northern Ireland (Cmd 532, 1969) 140.
162 Coyles, D. Hidden city. In D. Coyles and D. Wylie (eds) *Housing Plans for the Future.* (Steidl 2018) (book is unpaginated).
163 Disturbances in Northern Ireland: Report of the Cameron Commission Appointed by the Government of Northern Ireland (Cmd 532, 1969) 229.
164 Adams, (G. 1996) *Before the Dawn* (Heinemann 1996) 89. The word 'pram' is a U.K. term for stroller, while 'flat' refers to public housing tower blocks.
165 Other developments that also had to be demolished in the 1980s included the notorious 'Weetabix' flats on the Shankill Road – so called because they resembled cereal packets, as well as the Turf Lodge Flats in Belfast, both of which had to be demolished 'within a few years of their erection'. See C Brett, *Housing a Divided Community* (Queen's University Belfast 1986) Plate 30, 32.
166 Adams, G. *Before the Dawn* (Heinemann 1996) p. 93.
167 Ellison, G and Smyth, J. *The Crowned Harp: Policing Northern Ireland* (Pluto 2000) (*passim*).
168 McGarry, J. and O'Leary, B. *Policing Northern Ireland: Proposals for a New Start.* (Blackstaff 1989) p. 31.
169 Morison, J. and Livingstone, S. *Reshaping Public Power: Northern Ireland and the British Constitutional Crisis* (Sweet and Maxwell 1995) p. 126.

170 De Tocqueville, A. *The Old Regime and the French Revolution*, (tr A Gilbert, first published 1856, Anchor Books 1955). Tocqueville's thesis is distinguishable from the J curve: the idea that revolts and revolutions occur after long periods of rising expectations. Arguably the J curve was also relevant in this context with rising expectations from the Catholic community following the reformist promises of the O'Neill government in the early 1960s.
171 This is not to suggest a total absence of analysis of the wider structural factors that led to the outbreak of conflict in Northern Ireland in the late 1960s, see, for example, L O'Down, B Rolston, and M Tomlinson, Northern Ireland: Between Civil Rights and Civil War (CSE Books 1980).
172 Thayler, R. and Sunstein, C. *Nudge* (Yale University Press 2008) p. 3; See also O Yiftachel, 'Social control, urban planning and ethno-class relations: Mizrahi Jews in Israel's 'Development Towns' (2003) 24 (2) *International Journal of Urban and Regional Research 418.*
173 Birrell, W. *Labour Mobility in Northern Ireland* (New University of Ulster).
174 Craigavon: New City: Household Survey for Brownlow, Summer 1969.
175 Letter from St Anne's Unionist Association, 1965, PRONI, ref D2669/9.
176 McEldowney, O. Anderson, J. and Shuttleworth, I. Sectarian demography: Dubious discourses of ethno-national conflict. In K. Hayward and C. O'Donnell (eds) *Political Discourse and Conflict Resolution: Debating Peace in Northern Ireland.* (Routledge 2011) p. 160.
177 McCleery, M. 'The creation of the 'New City' of Craigavon: A case study of politics, planning and modernisation in Northern Ireland in the Early 1960s', 27 (1) *Irish Political Studies* 97.
178 Letter from St Anne's Unionist Association, 1965, PRONI, ref D2669/9.
179 Minutes of a meeting between the Prime Minister of Northern Ireland and Lord Glentoran with a deputation from the West Belfast Unionist Association, December 4th 1958, PRONI D2669/9, para. 2.
180 Cormack, R. and Osborne, R. *Discrimination and Public Policy in Northern Ireland* (Clarendon 1991) p. 10.
181 O'Dochartaigh, N. *From Civil Rights to Armalites: Derry and the Birth of the Irish Troubles* (Cork University Press 1997) p. xv.
182 Bardon, J. *A History of Ulster* (Blackstaff Press 2001).
183 Hepburn, A. C. *A Past Apart: Studies in the History of Catholic Belfast 1850–1950* (Ulster Historical Foundation 1996) p. 246.
184 Hennessey, T. *A History of Northern Ireland* (Gill and Macmillan 1997).
185 Whyte, J. How much discrimination was there under the unionist regime, 1921–1968?. In T. Gallagher and J. O'Connell (eds), (Manchester University Press 1983).
186 Draper, J. 'A University gerrymander? Coleraine and the Lockwood Report' (2014) 22 (5) *History Ireland.*
187 Whyte, J. How much discrimination was there under the unionist regime, 1921–1968?. In T. Gallagher and J. O'Connell (eds) *Contemporary Irish* Studies. (Manchester University Press 1983).
188 Hennessey, T. *A History of Northern Ireland* (Gill and Macmillan 1997).
189 Draper, J. 'A University gerrymander? Coleraine and the Lockwood Report' (2014) 22 (5) *History Ireland* 44–47.
190 Kynaston, D. *Modernity Britain: A Shake of the Dice* (Bloomsbury 2014) 28, 270, 276–277.
191 O'Dochartaigh, N. *From Civil Rights to Armalites: Derry and the Birth of the Irish Troubles* (Cork University Press 1997) (*passim*).
192 Farrell, M. *Northern Ireland: The Orange State*, (Pluto Press 1976) pp. 257–284.
193 Brett, C. *Housing a Divided Community* (Queen's University Belfast 1986) p. 33.
194 Rodgers argues that the decline of the "public realm" in modern cities is a prime culprit in the transition from "open-minded" to "single-minded" urban areas – from multifunctional

to monofunctional spaces and structures – which are seriously detrimental to urban community and identity. See R Rogers Cities for a Small Planet (Faber 1998) 9–10.

195 Collier, P. *The Future of Capitalism: Facing the New Anxieties* (Harper Collins 2018) p. 11.

196 Johnston argues that the visit to the U.S. should have sounded "alarm bells" both with respect to the impact of congestion and opposition from the local communities. See W Johnston, *The Belfast Urban Motorway: Engineering, Ambition and Social Conflict* (Colourprint Books 2014) 46–50.

197 Ignatieff, M. *The Ordinary Virtues: Moral Order in Divided World* (Harvard University Press 2017) p. 45.

198 For a discussion of the "revolutionary" impact of postwar urban redevelopment in the US see J Teaford, *The Metropolitan Revolution: The Rise of Post-Urban America* (Columbia University Press 2006); See also R Rothstein, The Colour of Law: A Forgotten History of How Our Government Segregated America (Liveright 2017) (17–38).

199 Both reports found for example that discrimination in housing and employment were important factors in contributing to the outbreak of violence. Disturbances in Northern Ireland: Report of the Cameron Commission Appointed by the Government of Northern Ireland (Cmd 532, 1969) 140; *Report of the National Advisory Commission on Civil Disorders* (Bantam Books 1968) 9–11.

200 *Report of the National Advisory Commission on Civil Disorders*. (Bantam Books 1968) pp. 12–13.

201 Massey, D. and Denton, N. *American Apartheid: segregation and the making of the underclass* (Harvard University Press 1993) (*passim*). Discriminatory attitudes and behaviors within white communities in the U.S. also played a significant role in perpetuating segregation during this period, see, for example, T Schelling *Micromotives and Macrobehaviour* (Norton 1978).

202 See above regarding the letters and representations to the Northern Ireland Government from the St Anne's Unionist Association. At the same time, there was strategizing within the Ulster Unionist Party regarding the impact of urban redevelopment on electoral districts. See, for example, Mulholland, M. Why did unionists discriminate?. In S Wichert (ed.) *From the United Irishmen to Twentieth-Century Unionism: Essays in honour of A.T.Q. Stewart*. (Four Courts Press 2004).

203 Ignatieff, M. *The Ordinary Virtues: Moral Order in Divided World* (Harvard University Press 2017) p. 36.

204 Stewart, A.T.Q. *The Narrow Ground: Aspects of Ulster 1609–1969*. (Blackstaff 1977) p. 182.

205 John Taylor was at the time an Ulster Unionist MP in the Northern Ireland Parliament representing the South Tyrone constituency and was Minister of Home Affairs. He went on to become an Ulster Unionist MP at Westminster, an Ulster Unionist representative in the European Parliament, and a member of the Northern Ireland Assembly, serving at all levels of government in Northern Ireland. He was also deputy leader of the Ulster Unionist Party 1995–2001 and became a life peer in the House of Lords in 2001, taking the title, Baron Kilclooney. He survived an assassination attempt by the Official IRA in February 1972 after being shot five times in the neck and head.

206 Government of Northern Ireland, *Future Policy on Areas of Contention, Second and Final Report of the Joint Working Party on Processions, (secret)* (Government of Northern Ireland April 1971 CAB/1634/1) para 39. Membership of this Joint Working Party included, among others, some of the most senior figures with responsibility for security matters in Northern Ireland at that time, including Major-General Acton of the British Army (Northern Ireland Headquarters), Assistant Chief Constable Baillie (representing the police force for Northern Ireland, the Royal Ulster Constabulary (RUC)) as well as senior officials, including Ken Bloomfield, from the NI Cabinet Secretariat. The other members of the group were J G Hill, Ministry of Home Affairs, W Slinger, Ministry of

Community Relations, D Perham, Ministry of Development, and A Hewins, Office of the United Kingdom Representative in Northern Ireland. The secretary to the group was R A H Miller, Ministry of Home Affairs.

207 Government of Northern Ireland, *Future Policy on Areas of Contention, Second and Final Report of the Joint Working Party on Processions, (secret)* (Government of Northern Ireland April 1971 CAB/1634/1) p. 39.

208 Government of Northern Ireland, *Future Policy on Areas of Contention, Second and Final Report of the Joint Working Party on Processions, (secret)* (Government of Northern Ireland April 1971 CAB/1634/1) p. 40.

209 Government of Northern Ireland, *Future Policy on Areas of Contention, Second and Final Report of the Joint Working Party on Processions, (secret)* (Government of Northern Ireland April 1971 CAB/1634/1) p. 41.

210 Government of Northern Ireland, *Future Policy on Areas of Contention, Second and Final Report of the Joint Working Party on Processions, (secret)* (Government of Northern Ireland April 1971 CAB/1634/1).

211 For discussion of the wider impact of the security forces on planning and architecture in Belfast during the conflict see Brown, S. Central belfast's security segment: An urban phenomenon (1985) 17 (1) *Area* 1–8 (*passim*); Cowan, R. Belfast's Hidden Planners (1982) 6 (56) *Town and Country Planning* 163–167 (*passim*); Pawley, M. *Terminal Architecture* (Reakteon Books 1998) pp. 93–111 (*passim*); Brand, R. 'Written and unwritten building conventions in a contested city: The case of Belfast (2009) 46 (12) Urban Studies 2669–2689 (*passim*). The full urban motorway was never completed, only the western section, which is named the "Westlink," officially designed as the "A12."

212 Gehl, J. *Life Between Buildings: Using Public Space*, (Island Press 2011) p. 72.

213 Johnston, W. *The Belfast Urban Motorway: Engineering, Ambition and Social Conflict.* (Colourprint Books 2014) p. 146.

214 Johnston, W. *The Belfast Urban Motorway: Engineering, Ambition and Social Conflict.* (Colourprint Books 2014) p. 146.

215 Forum for Alternative Belfast. https://forumfab.wordpress.com/wp-content/uploads/2009/06/emailfabb6.pdf

216 Faligot, R. *Britain's Military Strategy in Ireland: The Kitson Experiment* (Dingle 1983) p. 123.

217 Hillyard, P. Law and order. In J. Darby (ed.) *Northern Ireland: The Background to the Conflict* (Appletree Press 1983) p. 47.

218 Letter from Colonel Carter, Deputy Commander, 39 Airportable Brigade, (8 June 1971 LA/7/3E/14/46) 1.

219 Letter from Colonel Carter, Deputy Commander, 39 Airportable Brigade, (8 June 1971 LA/7/3E/14/46) 1 p. 4.

220 Department of the Environment for Northern Ireland, 'Belfast Urban Area Plan: Review of Transportation Strategy' (Department of the Environment for Northern Ireland 1978).

221 Department of the Environment for Northern Ireland, 'Belfast Urban Area Plan: Review of Transportation Strategy' (Department of the Environment for Northern Ireland 1978).

222 Johnston, W *The Belfast Urban Motorway: Engineering, Ambition and Social Conflict.* (Colourprint Books 2014) p. 110.

223 Forum for Alternative Belfast. https://forumfab.wordpress.com/wp-content/uploads/2009/06/emailfabb6.pdf.

224 Johnston, W. *The Belfast Urban Motorway: Engineering, Ambition and Social Conflict.* (Colourprint Books 2014) p. 138.

225 Dawson, G. 'Defensive planning in Belfast' (1984) (17) *Irish Geography* 27–41.

226 Bryan, D. In S. J. Connolly, (ed.) *Belfast 400: Planning, Space, and Conflict* (Liverpool University Press 2012) p. 326.

227 Herscher, A. *Spatial Violence* (Routledge 2019) p. 3.

228 Department of the Environment, *Belfast Metropolitan Area Plan 2015* (Department of the Environment 2004) Part 4, Vol 2, 23.

229 Cited in W Johnston, *The Belfast Urban Motorway: Engineering, Ambition and Social Conflict.* (Colourprint Books 2014) p. 138.

230 Jacobs, J. *The Death and Life of Great American Cities* (first published 1961, Modern Library 1993) p. 6.

231 Blackman, T., 'Planning inquiries: A socio-legal study' (1991) 25 (2) *Sociology* 311–327.

232 Blackman, T., 'Planning inquiries: A socio-legal study' (1991) 25 (2) *Sociology* 311–327.

233 Asmal, K. Dealing with differences: The South African experience. In F. Boal (ed.), *Ethnicity and Housing: Accomodating Differences.* (Ashgate 2000) pp. 27–33. See also Schindler, S. 'Architectural exclusion: Discrimination and segregation through physical design of the build environment (1934) 124 *Yale Law Journal.*

234 Rutherford, R. M. 'Belfast Urban area plan public inquiry' (September 1973), HMSO.

235 Government of Northern Ireland, *Future Policy on Areas of Contention, Second and Final Report of the Joint Working Party on Processions, (secret)* (Government of Northern Ireland April 1971 CAB/1634/1) p. 27.

236 Government of Northern Ireland, *Future Policy on Areas of Contention, Second and Final Report of the Joint Working Party on Processions, (secret)* (Government of Northern Ireland April 1971 CAB/1634/1).

237 Government of Northern Ireland, *Future Policy on Areas of Contention, Second and Final Report of the Joint Working Party on Processions, (secret)* (Government of Northern Ireland April 1971 CAB/1634/1) p. 28.

238 Government of Northern Ireland, *Future Policy on Areas of Contention, Second and Final Report of the Joint Working Party on Processions, (secret)* (Government of Northern Ireland April 1971 CAB/1634/1) p. 42.

239 Government of Northern Ireland, *Future Policy on Areas of Contention, Second and Final Report of the Joint Working Party on Processions, (secret)* (Government of Northern Ireland April 1971 CAB/1634/1).

240 Government of Northern Ireland, *Future Policy on Areas of Contention, Second and Final Report of the Joint Working Party on Processions, (secret)* (Government of Northern Ireland April 1971 CAB/1634/1) p. 49.

241 Government of Northern Ireland, *Future Policy on Areas of Contention, Second and Final Report of the Joint Working Party on Processions, (secret)* (Government of Northern Ireland April 1971 CAB/1634/1) p. 42.

242 Government of Northern Ireland, *Future Policy on Areas of Contention, Second and Final Report of the Joint Working Party on Processions, (secret)* (Government of Northern Ireland April 1971 CAB/1634/1) p. 51.

243 Letter from Eddie Simpson, Assistant Secretary, Housing and Urban Affairs, Department of the Environment (DOE) on 6 May 1981. It is not possible to identify all the members of the SCH from the evidence available, although the names and job titles of some of the more senior figures are listed in the minutes. As with the original JWP, the Security Force representatives on the SCH were among the most senior figures within the police and military establishment in Northern Ireland at that time – with the RUC represented by Deputy Chief Constable McAtamney and Assistant Chief Constable Chesney and the British Army represented by Brigadier Crowfoot (39th Infantry Brigade).

244 See, for example, Minutes of a meeting of the SCH Thursday 19 February 1981; Letter from ACC Chesney to Eddie Simpson, Assistant Secretary, Housing and Urban Affairs (DOE) on 26th February 1981. See also Minutes of a meeting of the Standing Committee on Security Implications of Housing on Thursday 1 April 1981; Minutes of a meeting of the Standing Committee on Security implications of Housing Problems in Belfast, 30 June 1980; Minutes of a meeting of the Standing Committee on Security implications of Housing Problems in Belfast, Thursday 19 February 1981.

245 Letter from ACC Chesney to Eddie Simpson, Assistant Secretary, Housing and Urban Affairs (DOE) on 26th February 1981.
246 Government of Northern Ireland, *Future Policy on Areas of Contention, Second and Final Report of the Joint Working Party on Processions, (secret)* (Government of Northern Ireland April 1971 CAB/1634/1) para 42.
247 Needham, R. *Battling for Peace* (Blackstaff Press 1998) p. 168.
248 Yiftachel, O. 'Social control, urban planning and ethno-class relations: Mizrahi Jews in Israel's 'Development Towns' (2003) 24 (2) *International Journal of Urban and Regional Research* 418.
249 Interview: Brian Feeney, historian and author. Former elected representative for the Social Democratic and Labour Party (SDLP) in North Belfast. (2011).
250 Dawson, G. 'Defensive planning in Belfast', (1984) *Irish Geography* (17) 36.
251 Dawson, G. 'Planning in the Shadow of Urban Civil Conflict: a case study from Belfast, Working Paper 24, May 1984 (University of Liverpool 1984) p. 36.
252 Interview: Brian Feeney, historian and author. Former elected representative for the Social Democratic and Labour Party (SDLP) in North Belfast.
253 Government of Northern Ireland, *Future Policy on Areas of Contention, Second and Final Report of the Joint Working Party on Processions, (secret)* (Government of Northern Ireland April 1971 CAB/1634/1) p. 5.
254 Government of Northern Ireland, *Future Policy on Areas of Contention, Second and Final Report of the Joint Working Party on Processions, (secret)* (Government of Northern Ireland April 1971 CAB/1634/1) p. 39.
255 Government of Northern Ireland, *Future Policy on Areas of Contention, Second and Final Report of the Joint Working Party on Processions, (secret)* (Government of Northern Ireland April 1971 CAB/1634/1) p. 39 p. 41.
256 See map of the area https://goo.gl/maps/bxk5untvDTJ7iqoZ7.
257 Interview: Gerry Kelly MLA, former member of Girdwood Advisory Panel and Sinn Féin elected representative for North Belfast since 1998. He is currently a member of Sinn Féin's Ard Chomhairle (National Executive Committee). (2011)
258 Interview: Brian Feeney, historian and author. Former elected representative for the Social Democratic and Labour Party (SDLP) in North Belfast. (2011).
259 Department for Social Development (2016) Building Successful Communities: Lower Oldpark and Hillview Action Plan, Belfast: HMSO. The legacy of conflict planning on the Lower Oldpark is discussed in some detail in chapter seven of this thesis.
260 Appleyard, D. and Lintell, M. 'The Environmental Quality of City Streets', (1972) 38 (2) *Journal of American Institute of Planners* 84.
261 Cullen, G. *Townscape*, (The Architectural Press 1961).
262 Gaffikin, F., McEldowney, M., and Sterrett, K. 'Creating shared public space in the contested city: The role of urban design', (2012) 15 (4) *Journal of Urban Design* 493.
263 Gehl, J. *Life Between Buildings: Using Public Space*, (Island Press 2011) p. 51.
264 Yiftachel, O. 'Theoretical Notes on 'Gray Cities': The Coming of Urban Apartheid?' (2009) 8 Planning Theory 88 (passim); P. Aelbrecht and Q. Stevens (eds), *Public Space Design and Social Cohesion* (Routledge 2019) (passim).
265 Collier, P. *The Future of Capitalism: Facing The New Anxieties* (Harper Collins 2018) p. 12.
266 Masters of Science in Urban Planning and Master of Architecture Columbia University GSAPP, 'Reconnecting Belfast' (Columbia University 2017) p. 68.
267 Masters of Science in Urban Planning and Master of Architecture Columbia University GSAPP, 'Reconnecting Belfast' (Columbia University 2017) p. 68.
268 Stewart, A.T.Q. *The Narrow Ground: Aspects of Ulster 1609–1969* (Blackstaff 1977) p. 182.

269 O'Dowd, L and Komorova, K., 'Regeneration in a Contested City: A Belfast Case Study', Divided Cities/Contested States Working Paper No. 10, (Queen's University Belfast 2009) p. 9.
270 Gaffikin, F., Morrissey, M., and Sterrett, K. Remaking the city: The role of culture in Belfast. In W. Neill and H. Schwedler (eds) *Urban Planning and Cultural Inclusion: Lessons from Belfast and Berlin.* (Palgrave Macmillan 2001) p. 148.
271 Government of Northern Ireland, *Future Policy on Areas of Contention, Second and Final Report of the Joint Working Party on Processions, (secret)* (Government of Northern Ireland April 1971 CAB/1634/1) para 14.
272 O McEldowney, K Sterrett, K. and F Gaffikin, (2001) 'Architectural Ambivalence: the Built Environment and Cultural Identity in Belfast', in W Neill and H Schwedler, (eds) *Urban Planning and Cultural Inclusion: Lessons from Belfast and Berlin* (Palgrave Macmillan 2001) 105.
273 The impact of redevelopment on Protestant depopulation is illustrated by the fact that between 1971 and 1978, the total population of the Shankill area declined from 46,363 to 33,297, a fall of 13,066 people or 28.1%. This compares to an overall Belfast average of a decline of 19%. These declines are explained mainly by the clearance of the old densely populated areas and their replacement with new housing at lower densities. In 1978, of those families who left the district, the most popular destinations were Newtownabbey, Bangor, and Antrim. Northern Ireland Housing Executive, *Draft Introduction to the Seven Districts Housing Statements* (Northern Ireland Housing Executive 1980) para 2.6 ENV/21/1/39/A.
274 Shuttleworth, I. and Anderson, J. 'Demographic Change Through Conflict: Belfast from the 1960s to 2001', Conference Paper, (2011), p. 3.
275 Doherty, P. and Poole, M. 'Ethnic Residential Segregation in Belfast, Northern Ireland, 1971–1991', 87 (4) *Geographical Review* 528.
276 Harrison, R. L. Population change and housing provision in Belfast. In P. A. Compton (ed.), *The Contemporary Population of Northern Ireland and Population-related Issues*, (QUB 1981); Singleton, D. A. Belfast; housing policy and trends. In Buchanan and Walker (eds.) *Province, City and People* (Greystone 1987).
277 Hayes, M. *Minority Verdict: Experiences of a Catholic Public Servant.* (Blackstaff Press 1995) p. 142.
278 Maguire, W. *Belfast: A History* (Carnegie Publishing 2009).
279 Elliott, M. *Hearthlands: A Memoir of the White City Housing Estate in Belfast.* (Blackstaff 2017) pp. 164–184.
280 Brett, C. *Housing a Divided Community.* (Queen's University Belfast 1986) p. 48.
281 Hepburn, A. C. Long division and ethnic conflict: The experience of Belfast. In S. Dunn (ed.) *Managing Divided Cities.* (Keele University Press 1994) p. 121.
282 Government of Northern Ireland, *Disturbances in Northern Ireland: Report of the Cameron Commission* (Cmd 532, 1969) p. 37, p. 229.
283 Government of Northern Ireland, *Disturbances in Northern Ireland: Report of the Cameron Commission* (Cmd 532, 1969) p. 139.
284 Government of Northern Ireland, *Disturbances in Northern Ireland: Report of the Cameron Commission* (Cmd 532, 1969) p. 140.
285 Macrory, P. A. Review Body on Local Government in Northern Ireland 1970: Report', June 1970.
286 Housing Executive Act, 1971 as amended by Article 22 of the Housing (NI) Order 1981.
287 Interview: Brian Feeney, historian and author. Former elected representative for the Social Democratic and Labour Party (SDLP) in North Belfast. (2011).
288 Cobain I (2020) *Anatomy of a Killing: Life and Death on a Divided Island.* Granta. P. 103.
289 TNA CJ 4/1982, The Guardian, 13 March 1982.
290 Cobain I (2020) *Anatomy of a Killing: Life and Death on a Divided Island.* Granta. P. 104.

291 Interview: Kyle Alexander, Director Maze/Long Kesh Redevelopment Project at the time of interview. Former member of the Crumlin Road Gaol/Girdwood Park Advisory Panel. (2011).

292 Letter from ACC Chesney to Eddie Simpson, Assistant Secretary, Housing and Urban Affairs (DOE) on 26th February 1981.

293 Letter from ACC Chesney to Eddie Simpson, Assistant Secretary, Housing and Urban Affairs (DOE) on 26th February 1981.

294 Documents also show that sometimes the aims of the security forces and Catholic residents were aligned. For example, on one occasion Unionists members of Lisburn Borough Council sought to 'oppose and obstruct the Poleglass development, both by spreading rumours that tenants owed large rent arrears and by stalling over the granting of pub licences. DCC McAtamney stressed that the police attached high importance to the provision of pubs in the area; the demand was otherwise likely to be met by clubs with paramilitary associations.' Note of the fifth meeting of the Standing Committee on Security implications of Housing Problems in Belfast, Thursday 19 February 1981.

295 Cobain, I. *Anatomy of a Killing: Life and Death on a Divided Island.* (Granta 2020).

296 Shirlow, P. Belfast: A segregated city. In C. Coulter and M. Murray (eds), *Northern Ireland After the Troubles: A society in transition* (Manchester University Press 2008).

297 Bardon, J. *A History of Ulster* (Blackstaff 2001).

298 Standing Advisory Commission on Human Rights, *Religious and Political Discrimination and Equality of Opportunity in Northern Ireland: Report on Fair Employment*, Cm 237 (SACHR 1987), p. 1.8.

299 Smith, D. and Chambers, G. *Equality and Inequality in Northern Ireland Part 4: Public Housing Policy Studies Institute* Occasional Paper 47 (Policy Studies Institute 1989) p. 5.

300 Smith, D. and Chambers, G. *Equality and Inequality in Northern Ireland Part 4: Public Housing Policy Studies Institute* Occasional Paper 47 (Policy Studies Institute 1989) p. 5.

301 Standing Advisory Commission on Human Rights, *Religious and Political Discrimination and Equality of Opportunity in Northern Ireland* (Cm 110, 1990) (*passim*).

302 Standing Advisory Commission on Human Rights, *Religious and Political Discrimination and Equality of Opportunity in Northern Ireland* (Cm 110, 1990) (*passim*) p. 9.13.

303 Smith, D. and Chambers, G. *Equality and Inequality in Northern Ireland Part 4: Public Housing Policy Studies Institute London* Occasional Paper 47 (PSI 1989) (*passim*); Standing Advisory Commission on Human Rights, *Religious and Political Discrimination and Equality of Opportunity in Northern Ireland* (Cm 110, 1990) (*passim*); Standing Advisory Commission on Human Rights, *Employment Equality: Building for the Future* (Cm 3684, 1997) (*passim*); Northern Ireland Affairs Committee, Session 1998-99 House of Commons Fourth Report, Vol.1 95–1; E. McLaughlin and P. Quirk (eds.) *Policy Aspects of Employment Equality in Northern Ireland.* (SACHR 1996) (*passim*).

304 Shuttleworth, I. Shirlow, P., and McKinstry, D. Vacancies, access to employment and the unemployed: Two case studies of Belfast and Londonderry. In E. McLaughlin and P. Quirk (eds) *Policy Aspects of Employment Equality in Northern Ireland.* (Standing Advisory Commission on Human Rights 1996) pp. 27–49; Sheehan, M. and Tomlinson, M. Long-term unemployment in West Belfast. In E. McLaughlin and P. Quirk (eds) *Policy Aspects of Employment Equality in Northern Ireland.* (Standing Advisory Commission on Human Rights 1996) pp. 51–88.

305 Standing Advisory Commission on Human Rights, *Religious and Political Discrimination and Equality of Opportunity in Northern Ireland: Report on Fair Employment*, Cm 237 (SACHR 1987), p. 3.3.

306 Ignatieff, M. *The Ordinary Virtues: Moral Order in Divided World.* (Harvard University Press 2017) p. 45.

307 Bardon, J. *A History of Ulster* (Blackstaff 2001) p. 303.

308 Morrison, B. Planning the city: Planning the region. In F. Boal and S. Royle (eds) *Enduring City, Belfast in the Twentieth Century.* (Blackstaff Press 2006) p. 151.

309 Shannon, C. *Building a Better Belfast*, (1991) pp. 58–59. J'TOWN Pamphlets 363. 50941671 SHA 65.

310 Chris, Patten, who served as an environment minister during this period, has stated that Northern Ireland 'seemed to have become a place to which those whose Thatcherite sympathies could not be wholly trusted were exiled… It was suggested that since I was keen on public spending, I should be given lots of money to distribute in Northern Ireland before being weaned off it with some future return to a domestic, mainland (sic) department'. See Patten, C. *First Confession: A Sort of Memoir.* (Allen Lane 2017) p. 158. Other notable 'wets,' who served in Ministerial office in Northern Ireland in the 1980s included Jim Prior and Richard Needham.

311 Patten, C. First Confession: A Sort of Memoir. (Allen Lane 2017) p. 170. Documents from the period also show that Brigadier Crowfoot was of the view that the replacement of the Turf Lodge flats with two-storey dwellings would 'broadly assist the army'. Note of the fifth meeting of the Standing Committee on Security implications of Housing Problems in Belfast, Thursday 19 February 1981.

312 Gaffikin, F, McEldowney, M., and Sterrett, K. 'Creating shared public space in the contested city: The role of urban design' (2012) 15 (4) *Journal of Urban Design* 505.

313 Newman, O. *Defensible Space: Crime Prevention Through Urban Design*, (Macmillan 1972) p. 3.

314 Brett, C. *Housing a Divided Community* (Queen's University Belfast 1986) p. 102. Sir Charles (Charlie) Brett was a significant figure in Northern Irish public, professional, and cultural affairs for several decades. He was for a time Chairman of the Northern Ireland Labour Party and served on numerous public bodies including the Northern Ireland Housing Executive, of which he was Chairman. He was probably best remembered for his dedication to the conservation, restoration, and re-use of the built heritage; and as an architectural historian, an expert on that heritage. He was one of the founders of the Ulster Architectural Heritage Society.

315 Morrison, B. The culturally inclusive city: The Belfast potential. In W. Neill and H. Schwedler (eds.), *Urban Planning and Cultural Inclusion: Lessons from Belfast and Berlin* (Palgrave Macmillan 2001) p. 202; Anson, B. 'Response to architecture in no-man's land' (1984) 37 (180) *Architects Journal* 47; Murtagh, B. *The Politics of Territory: Policy and Segregation in Northern Ireland*, (Palgrave 2002) (passim).

316 Murtagh, B. Planning for anywhere: Housing policy in Northern Ireland, housing studies (1998) 13 (6) 833; Bollens, S. *Urban Peace-Building in Divided Societies: Belfast and Johannesburg* (Westview Press 1999); Ellis, G. The city of the black stuff: Belfast and the Autism of planning. In M. McEldowney, M. Murrary, B. Murtagh, and K. Sterrett (eds.), *Planning in Ireland and Beyond: Multidisciplinary Essays in Honour of John V. Greer*, (QUB 2005).

317 Brett, C. *Housing a Divided Community* (Queen's University Belfast 1986) p. 77.

318 Letters from the DOE concerning alleged security force involvement in planning were leaked to 'The Guardian' (13.3.82), and to the Irish News (15.3.82). Subsequent articles in the press concerning this matter are found in 'The Belfast Telegraph' (15.3.82); 'Built Environment' (17.3.82), 'The Belfast Newsletter' (15.3.82); 'The Irish News (23.3.82); and 'The Guardian' (6.4.82, 13.4.82); Scope (1982) Who designs the City? Belfast: Scope.

319 Brett, D. "Geologies of site and settlement" in N. Allen and A. Kelly (eds.), *The Cities of Belfast*, (Four Courts Press 2003) p. 23.

320 Interview: Gerry Kelly former member of Girdwood Advisory Panel and Sinn Féin elected representative for North Belfast since 1998. He is currently a member of Sinn Féin's Ard Chomhairle (National Executive Committee). This description is similar to what Lindblom describes as 'incrementalist decision-making', C Lindblom, The Market System: What it is, How it Works, What to Make of it (Yale University Press 2002).

321 Scott, C. Regulation in the age of governance: The rise of the post-regulatory state., In J. Jordana and D. Levi-Faur (eds.) *The Politics of Regulation: Institutions and Regulatory Reforms for the Age of Governance*. (Edward Elgar 2004) p. 153.

4 The Legacy of Coercive Architecture and Planning in Belfast

Case Study – The Redevelopment of the Crumlin Road Gaol and Girdwood Barracks

4.1 DIVISIVE ARCHITECTURE AND THE POLITICS OF TERRITORY IN NORTH BELFAST

4.1.1 The Girdwood Redevelopment Project: Background

Following serious disturbances in North Belfast in 2001 that centered around the high-profile "Holy Cross" protest,[1] the U.K. Government commissioned an independent review of community relations in the area chaired by the former Moderator of the Presbyterian Church, Rev Dr. John Dunlop.[2] The Review team conducted a six-month research and consultation project in North Belfast, publishing a final report (the "Dunlop Report") in May 2002.[3] A key theme of this study was that North Belfast was divided into a large number of distinct communities, most of which – particularly those in the more deprived areas – are either exclusively Protestant/Unionist or Catholic/Nationalist territory where

> the boundaries or interfaces between these oppositional communities are fractured zones where hostility and antipathy are maintained and renewed through violence and disorder.[4]

This represents the legacy of military urbanist design processes that reconfigured the planning and architecture of the city over several decades, using the road network as well as commercial and residential development to provide barriers between the two communities through the creation of defensible space. As indicated in the previous chapter the net effect of this strategy was to contain the growth of the Catholic community within specified territorial boundaries while at the same time promoting the depopulation of Protestant areas, which in turn increased the demographic imbalance between the two communities in the city.

DOI: 10.1201/9781003208129-4

> The sectarian geography of the place is like a patchwork quilt of communities living side by side with interfaces and walls and fences, and various other boundaries separating them ... and these two communities ... exhibit very different needs in respect of housing. On the Catholic side, it's generally speaking a young growing population, with lots of kids around, very big waiting list for housing ... people living with relatives, this type of thing, in overcrowded conditions. On the Protestant side, the opposite would be the case in respect of that type of need ... Older people, living in older houses.[5]

Demographic imbalances fueled fears within the Protestant community that the shift in territorial identity in North Belfast was one-directional i.e. that Protestant areas were being "taken over" by the Catholic community, which sought to drive the Protestant working class out of the city.[6] On the other hand, within the Catholic community there was a feeling that housing inequalities were ignored, leaving Catholics trapped behind security barriers euphemistically referred to as "peace lines," which closed off vacant brownfield sites that could otherwise be developed to provide much-needed accommodation.

> North Belfast is notorious for having I think about 18 miles of peace walls that delineate and designate territory, and so every community in this constituency claims its land. You see it for example dramatically at the Cupar Street interface, where there are acres and acres and acres of available land on the Shankill Road [Protestant] side, and crowded housing on the other [Catholic] side, but of course the wall will not, and cannot be moved, even to meet housing need, because to move it or to take it down would inevitably mean the expansion of green territory, and the lessening of orange territory.[7]

The capacity of territorial boundaries to impact adversely upon housing inequalities was also highlighted in the North Belfast Housing Strategy (NBHS), published by the Northern Ireland Housing Executive (NIHE) some months prior to the Dunlop Report.[8] The NBHS identified significant differentials between the two communities regarding patterns of housing need in North Belfast. For example, although Catholic households made up about half of the population of North Belfast overall, they comprised approximately three-quarters of the waiting list for houses; as of March 2000, the composition of the waiting list was such that 60% of Catholic applicants (727) were classified as "urgent need" compared with 36% of Protestant applicants (153). The NBHS document also found that although the housing waiting lists had fallen for both communities, the reduction had been much greater for Protestants (–39%) than for Catholics (–12%).[9]

Although there is some dispute, particularly within the Unionist community, about the details of the findings within the NIHE strategy document[10] there is general acceptance that disproportionate housing need exists for the Catholic community.

> The demand for social housing is very much a green demand in North Belfast at the minute.[11]

Significant disagreements arise however regarding the appropriate means by which public agencies ought to address Catholic housing inequalities, not least with respect to land availability for additional residential provision. The NIHE had previously completed a review of its land holdings as well as a comprehensive assessment of possible "opportunity sites" that could be used for new housing. The review concluded that given the pressing housing needs in North Belfast, at least part of the Crumlin Road Gaol/Girdwood site, which had just become available following the closure of the British Army military base, should be used to provide additional housing, both social and private.[12]

> The sectarian geography creates a very inefficient housing market ... if that land was available to everyone and there weren't the fears and the worries about security, and people are willing to live together, and all of that, then it would be a much more efficient housing market, and there would be more land readily available that could be developed quickly, and it would be less impact on the public purse, going out acquiring new land, but we have always taken the line, and we still insist upon this, this isn't within our gift ... This is not something that the housing executive or indeed any individual agency can change, this requires political and community, agreement ...[13]

The retention of territorial boundaries, and their consequent impact upon spatial inequalities and community relations, was an issue that the Dunlop Report picked up on concluding that there could not be integration of the two communities unless relationships between them improved.[14] The Report also concluded that to demand change of the territorial boundaries between the two communities without working to improve relationships and trust would be experienced as a threat and contrary to the objective of consolidating the nascent peace process.[15]

Taken together, the Dunlop Report and the NBHS highlight the spatial legacy of the Northern Ireland conflict in this area of the city in which an intersection of socio-economic and identity-based inequality exists within the context of a growing Catholic population in need of housing and a declining Protestant population fearful that they were being expelled from their traditional areas. Moreover, these reports illustrate the extent to which territorial boundaries serve to keep these two communities apart, forming a "patchwork quilt" of differing political identities that distort the allocation of limited public resources in a geographical area experiencing some of the highest levels of deprivation in Northern Ireland. To address these challenges the Dunlop Report sought to encourage the government to develop a major site to serve as a symbol of hope and economic regeneration for North Belfast that would generate investment, increase economic opportunity, improve the environment, and lift the spirits of the wider community. In this context the Report noted that the Crumlin Road Gaol, which had closed in 1996 along with the adjacent Girdwood Barracks which had been used as a base for the British Army since the early 1970s, had the potential to meet a wide range of needs for the communities of North Belfast. It was in this context that the Draft Masterplan for Crumlin Road Gaol and the former Girdwood army base was drawn up.

Following the publication of the Dunlop report, ownership of the former Crumlin Road Gaol in North Belfast was transferred to the new power-sharing government of

FIGURE 4.1 Aerial photograph of Crumlin Road Gaol and the former Girdwood army barracks. The Gaol, with its radial panopticon is at the bottom of the image. Beyond the car park, in front of the Gaol, the new Girdwood Community Hub is visible along with playing fields. This area was a brownfield site when this study commenced. The Protestant Lower Oldpark community is located to the left of the Girdwood site and playing fields. The Catholic New Lodge area is to the right of the Girdwood site. Google Images.

Northern Ireland that was established following the 1998 peace agreement. The Office of the First Minister and Deputy First Minister (OFMDFM) was given responsibility for the site in August 2003 under the "Reinvestment and Reform Initiative" which had been established by the UK Treasury to help boost the Northern Ireland economy following the ending of the conflict.[16] In February 2005 the UK Government announced its decision to close the adjacent Girdwood Army Barracks, and the following September the then Minister for Social Development David Hanson[17] announced that the development of both sites would be taken forward through a Master planning process. In early 2006, ownership of both sites (comprising some 27 acres) was acquired by the Northern Ireland Department for Social Development (DSD) from the UK Ministry of Defence (Figure 4.1).

To take forward the development of the combined site an overall Masterplan was required in line with the site's zoning under the draft Belfast Metropolitan Area Plan (BMAP). In March 2006 David Hanson, the U.K. government Minister with responsibility for the site, established an advisory panel comprising political, community, and statutory representatives which was tasked with overseeing the production of a Masterplan and making recommendations.

Building Design Partnership (BDP) (London) Ltd was selected as the preferred planner for the site in February 2007 and following consultation with statutory bodies, voluntary agencies, and community groups, proposals were presented to the advisory panel. By this time, a new devolved Northern Ireland Government at Stormont had replaced the Direct Rule administration from Westminster and the new Minister for Social Development, Margaret Ritchie of the Nationalist Social

Democratic and Labour Party (SDLP), launched the Draft Masterplan for a period of public consultation in October 2007 on behalf of the DSD and the OFMDFM.[18] The fact that control of the site was the responsibility of DSD and OFMDFM was highly significant ensuring that any progress on the development could only take place with the agreement of the political leadership of both main communities. This was to prove decisive given the subsequent opposition to the proposals from the Unionist community, especially among elected representatives of the Democratic Unionist Party (DUP).

Public consultation on the draft Masterplan commenced on October 16, 2007.[19] At their first meeting, the advisory panel unanimously agreed on the following mission statement:

> To create a regeneration project of international significance which brings maximum economic, social and environmental benefits to the local and wider community and in so doing creates a vibrant, inclusive and diverse environment which attracts present and future generations of people to live, work and visit.[20]

To deliver this vision, the advisory panel agreed on several development principles, these included a desire for a "transformational shared future scheme" that would address pressing needs in the locality and the wider area. Bringing people into the area, both as residents and visitors, was considered fundamental to the success of the Plan. To achieve this the Plan sought to maximize connectivity and access to facilitate and encourage movement throughout the site and between the site and the rest of the city. It was agreed from the outset that opportunities for the development of the site should embrace principles of equality, based on objective assessment of need and a "Shared Future" i.e. that the site needed to be developed in such a way that there would be access and use by both communities.

The Masterplan proposed that the Crumlin Road Gaol, designed by the Victorian architect Charles Lanyon, should serve as the anchor for the site, acting as a landmark attraction that would be visited by people from across the city as well as those from further afield. This was to be located within a safe and attractive environment which was easily accessible by car, public transport, or on foot. The Gaol would be combined with mixed use development (including retail, housing, workshops, and training facilities) for the largest part of the site creating a new "city quarter" based upon the concept of "sustainable living" with a new access road, the ARC Route, at the heart of the site linking the Crumlin Road with the Antrim Road.[21]

4.1.2 The Planning System and Telescoping Division

Subsequent disagreements illustrate how the architecture and planning that underpinned the Masterplan served to generate political disagreement among local communities. Within the Unionist community, one of the chief objections to the Masterplan was that the social and economic benefits of the proposals were focused almost exclusively on the geographic area *within* the boundaries of the redevelopment site, thereby ignoring the needs of deprived Protestant communities immediately adjacent.

> The regeneration of the site was seen as if it was a site that could be developed conceptually with a wall around it. The consultants … were brought in and asked 'how would you regenerate this site?' when actually the political question was, 'how can this site be regenerated so that everybody in a much, much wider area, can benefit from it?'.[22]

The original Masterplan had in fact claimed that the proposals would bring "maximum economic, social and environmental benefits to the local and wider community" and that the redevelopment of the site would create a vibrant, inclusive, and diverse environment "which attracts present and future generations of people to live, work and visit."[23] The Plan also claimed that the provision of a range of mixed tenure housing on the site would ensure that "existing residential areas adjacent to the site would eventually be strengthened and diversified."[24]

> The Girdwood/Crumlin Gaol site was seen as a potential development site which could help to lift that whole area of North Belfast, economically and socially.[25]

The difficulty was that within Protestant communities immediately adjacent to the site there was a high degree of skepticism that these objectives of wider regeneration would be achieved.

> All the buzz words were used … but the flaw in that thinking and it is a profound one, is that the context of regenerating a much wider area was not part of the brief. The assumption was made that if we can get these 27 acres agreed, everybody will be happy.[26]

The question of where planners ought to "draw the line" around a development site is one that arises in any regeneration context, as do tensions between a need to stay focused on a particular site and the relationship between that site and its wider hinterland.

> You don't just look within the line of the site, whether it's Laganside, the Maze, or any major regeneration scheme, your prime responsibility is to secure the regeneration of that site but to do that you've got to understand the hinterland, and you've got to be able to have ways that you can improve those areas as well, and that also reflects your mind set because that means you're respecting and acknowledging the wider problems of the area … at the same time though you don't want it to go too big, so that you lose your focus.[27]

These problems were heightened in North Belfast, where territory is rigidly demarcated along sectarian lines. The boundary line of the regeneration site, and the question of how far benefits extend directly beyond that line assumed much greater political salience, particularly within a context in which the Protestant community already felt under threat.

> If you telescope these issues down to a relatively small geography, you are going to create these pressure points … it's going to be very difficult for people to see the wood for the trees, and it's going to translate into a zero sum game … if we get this, you lose out, we get our housing, you're going to lose some land that you maybe thought was yours … telescoping of spaces … can be counterproductive in that way.[28]

This view was endorsed by the senior government official with responsibility for regeneration of the site who felt that focusing on a narrow geography served to increase tensions.

> North Belfast itself in terms of population and the issues is a small area, but by looking at North Belfast, you're looking at the overall community needs. What was happening, given the very nature of North Belfast, we started to bring it down to very small, geographical areas, and given the territoriality in North Belfast, you were never going to shift some of the people to think outside those territorial areas.[29]

In this case, the "telescoping" effect was further magnified by the design and visual graphics of the Plan with a front cover showing an isolated and discrete "island" of redevelopment and prosperity, illustrated in color without any natural hinterland (see Figure 4.2). The presentation of the Plan clearly served to concentrate and highlight spatial divisions in an area characterized by hypersensitivity to territory and defensible space.

This cover image of the redevelopment plan was not lost on the representatives of the surrounding Protestant communities who could point to the fact that they were literally, as well as figuratively, airbrushed from the future vision for the area and that the economic and social benefits of the development were weighted toward "Catholic" needs through additional housing on the site.

> Unless they're seeing regeneration in their own communities, then … Girdwood will be … a flagship development that sails past them. That's the big issue there.[30]

In Northern Ireland, where paintings, symbols, and imagery carry particular significance, this was a problem that at least one member of the advisory plan was aware of in advance.

> The original Laganside concept plan, which was 1987 … it had exactly that again, the Waterfront area was in color, with lovely designs of buildings, and the rest of the city was grey, but it was the start, because people at that stage were only thinking about the economic and physical development of the Waterfront …

FIGURE 4.2 BDP, "Crumlin Road Gaol and Girdwood Barracks Draft Masterplan" (BDP 2007) Front cover of document.

> our later plans … show the surrounding area and linkages … The same interestingly is occurring with the Maze here … it's the architects … if they're designing a building, they only ever just show the building, I always want to see the surrounding street.[31]

A common feature of contemporary urban development is concern about the tendency for economic regeneration to precipitate "gentrification," with local communities adversely impacted by rising property prices that in turn lead to displacement.[32] Within the context of North Belfast, which is characterized by micro-territorial claims, "where the Ulsterman carries the map of … religious geography in his mind almost from birth,"[33] the issue assumes much greater salience.

There is another important temporal aspect to the proposals that significantly impacted how they were viewed by the two communities. Although the Masterplan was drawn up to provide a future vision for the city, the proposals were viewed from the perspective of the processes that had shaped the area over the previous decades. Within the Catholic community, the Masterplan represented an opportunity to address long-standing housing needs utilizing land that had previously been taken over by the

British Army in the 1970s. From the perspective of the Protestant community, the proposals represented further Catholic encroachment on a community that had experienced several decades of population decline and already felt under threat. Failure to appreciate the extent of opposition to the Plan from within the Protestant "hinterland" of the Lower Oldpark was due largely to the way in which the proposals were considered to be weighted in favor of Catholic community needs. This in turn fed into a narrative of territorial loss and community decline.

> With us, it's been very strongly felt that we're not looking at Girdwood until you look at Lower Oldpark and Lower Shankill, because they have been just left to lie for decades ... any trickle down isn't going to be big enough and significant enough, and it's also a bit daft ... how can we go into a community and say there's no resources to do what you've needed for decades but we've found a hundred million for ten yards across the road.[34]

Although there were separate plans by public agencies to address deprivation within these areas, the Girdwood Masterplan was viewed as such a large development that it would swamp or crowd out these other more modest proposals. Where socio-economic inequality and identity politics intersect around ownership and access to territory, differences were further "telescoped" by the fact that the Girdwood site was located directly between the Catholic New Lodge and the Protestant Lower Oldpark areas.

> If you've spoken to anyone in Lower Oldpark you'll know that there are strong feelings there about Girdwood ... it sits on a fault-line almost.[35]

One of the ways in which the proposals had sought to extend benefits to the surrounding hinterland was by using the physical development on the site to leverage jobs and training to the local communities through "contract clauses," which would require supply chain companies to engage with the new "Learning Hub."[36] There was also a strong emphasis within the original Masterplan that the new leisure facility and playing fields would be accessible to local communities as well as those from other parts of the city. The difficulty was that the Learning Hub, Community Hub, and Children's Centre were originally located at the southern i.e. Catholic end of the site and given the sectarian geography of the area, less likely to be used by members of the Protestant community in the Lower Oldpark which lay beyond the northwestern section of the site (see Figure 4.1). Moreover, the differing demographic profiles of the two communities meant that employment opportunities targeted at training young people in the construction industry and other skills would tend to favor the Catholic community rather than the smaller and older Protestant community. Where distrust and suspicions run deep, even the topography of the land was viewed as being tilted in favor of one community.

> It's on the other side of Clifton Park Avenue, it's backed onto by the Mater, by St Malachy's and goes down to the lower Antrim Road, so its natural streetscape, took it down the road rather than up the road.[37]

From the perspective of the Unionist community, the key to breaking the Girdwood logjam was to bring forward proposals that would *specifically* address the problems "up the road" i.e. the depopulation and dereliction of the Lower Oldpark area immediately beyond the line of the regeneration site. The Girdwood Masterplan illustrates how conventional planning orthodoxy, which is reliant upon the presentation of a draft plan for a development site, complete with color illustrations and boundaries increased tensions as the proposals came to be viewed through the lens of zero-sum sectarian politics. As the next section will show, the belief within the Protestant community was that the Masterplan provided a Trojan horse for continued Catholic expansion in this area of the city.

4.1.3 Urban Design, Conflict and the Politics of Territory

Jane Jacobs viewed planning as fundamental to the "ecology of cities" and identified a city ecosystem as being composed of physical-economic-ethical processes active at a given time within a city and its close dependencies.[38] An important element of the physical processes active within the city ecosystem at any given point is the architectural philosophy that shapes the urban realm, which "will either create an order that affronts or one that enhances the quality of life."[39] Urban design serves an important function, either facilitating or impeding the movement and the flow of people and resources, and can either bring communities together or keep them apart. This illustrates what Yitachel has characterized as the emancipatory or socially coercive aspects of architecture and urban planning.[40]

Fundamental to the success of the Masterplan was the need to bring people into the area both as residents and visitors, and to achieve this the proposals aimed to maximize connectivity and access to facilitate and encourage movement throughout the site and between the site and the rest of the city. Although the Masterplan made no explicit reference to the defensive design practices of the past it clearly aimed to reverse this approach, transforming the "architectural scabs and scars" that characterized conflict urbanism into more "democratically productive spaces."[41] To achieve this the Masterplan placed a high degree of importance on the use of quality public realm in ensuring a successful urban environment by encouraging movement through effective signposting, lighting, planting, and paving.[42] Trees of differing heights and maturity, responding to the scale of adjoining development would line all public routes to reduce the impact of the vehicular environment, while variety in materials would help to reduce the speed of vehicles passing through the site and therefore enhance safety for pedestrians and cyclists. Reference to the use of color and texture to enrich the environment in the Masterplan contrasts with the verdict of Bollens that much of the planning in the city during the conflict delivered "greyness where colour matters."[43] The urban design philosophy of the Masterplan, which sought to maximize connectivity between this area and the rest of the city and provide a quality living environment, also contrasts with a government report on the nearby Protestant Lower Oldpark area (see Reflexive Planning and Conflict Transformation section) that found "a low quality living environment for residents" that was inward looking and defensive.[44]

Crucially, however, this reconfiguration within the planning system, and the use of architecture to facilitate movement and bring people together was not driven by "transformational" objectives per se, nor by political agreement between the two communities for a new vision for this part of the city, but rather was a function of the economic basis of the Masterplan that was contingent upon making the area attractive to private investors. This was to prove decisive in terms of hardening opposition to the proposals from within the Protestant community and will be explored in the next section.

A major difficulty for the Girdwood Masterplan was that the primary driver behind the more emancipatory and transformative design aspects of the proposals was the economic imperative to finance the development through the sale of apartments. From the perspective of those living within or representing the Unionist community within the Lower Oldpark area, which lay outside the Girdwood site boundary but adjacent to the proposed mixed-use development (see Figure 4.2) the new design philosophy that lay at the heart of the Girdwood Masterplan was viewed with horror.

> There were principles [in the Masterplan] that you would apply in every other city in the world, but whenever you come to North Belfast, it spooks the hell out of people. Freedom of access is a wonderful design principle, and it's all about getting inclusion and connection to a site. In North Belfast, freedom of access means somebody can come and attack my house …[45]

Concerns that the design of the plan was too reliant on international best practice and failed to take account of the territorial nuances of the local area were not restricted to members of the Unionist community, as the views of the North Belfast Housing Manager make clear.

> Where housing would go on the site … if you look at Clifton Park Avenue, with Lower Oldpark, the Crumlin Road and the Lower Shankill … the initial plan saw new housing going right on the interface … which wouldn't necessarily, you would think, be a tremendously good idea.[46]

This view was also endorsed by the senior government official with responsibility for taking forward the regeneration of the site who felt that the urban design at the heart of the proposals served to increase rather than allay fears within the Protestant community by locating the new housing in direct proximity to existing interfaces and by the presence of a new road that had the effect of dividing the area into two segments.[47]

> BDP designed what would have seen as a model estate, on how a model estate should be developed, and that was the one where you had the road basically dissecting the site … whenever I looked at this, I came in just after that was finished, and I said that's not going to happen … because it is dividing the site into Protestant and Catholic [territory] … and it's just not going to wash.[48]

Fear that the redevelopment of the site would lead to encroachment on an already vulnerable and beleaguered Protestant community in the nearby Lower Oldpark was a key factor in the rejection of the proposals by Unionist politicians and representatives. These feelings were allied to what was viewed within the Unionist community as an unhealthy and crude attachment to innovation and international precedent, which ignored the spatial realities of the local area and failed to recognize legitimate concerns within that community.

> The core mistake they made was international precedent, their answer to criticisms always was, well, all the other cities in the world do things like this. Yes, wonderful, fine, we're not every other city … it was always used as a fait accompli … well Barcelona does that so you're being localized and parochial and a bit of an idiot, and you can't see the bigger vision and the grander vision and all the rest of it so just sit down ….[49]

From the Unionist perspective, the problem was not innovation or international comparators per se but rather the way in which these models were applied in a "one size fits all" approach to urban development.

> We tend to define innovation in Northern Ireland terms as copying somewhere else … innovation would have been … how can we actually get a Masterplan for the site that deals with the localised issue of fear.[50]

The disagreements around the design aspects of the Girdwood Masterplan illustrate two competing and mutually exclusive visions both of which were shaped by decades of conflict. On the one hand, within the Catholic community and among those engaged in economic development for the site the aim was to open the north of the city to new people and new investment. Within the Protestant community, where access means someone "can attack my house," the desire was to preserve the territorial integrity of communal boundaries. Again, this highlights the capacity of architecture and planning orthodoxy to generate "irritants" within the political system in a context where there is heightened tension around ownership of land and territory in a post-conflict context.[51]

4.1.4 International Precedent and Politics of Elsewhere

Several members of the Advisory Group felt frustration at what they perceived to be a failure on the part of those closest to the ground to embrace a new vision for North Belfast and move forward. From this perspective, examples of best practice from beyond Northern Ireland that had been used to illustrate the potential of the site fell victim to localized and sectarian mind-sets that were closed to viewing the economic potential of the site.

> It didn't matter that we went and looked at Oxford, that we brought in examples of what happened in Brighton, that we talked about these matters and it didn't matter what we said about the Gasworks … It was a lack of maturity, and that's why who marches where, and what flag is flying where, was the

> dominant paradigm that we were in … if we were sitting in Manchester, that would not have been the discourse. The discourse would have been, here is 26 acres, here's a fantastic potential site to light the fuse for North Belfast, to become the absolute big anchor, a new site for the new future, and none of them had the vision to grasp that, none of them …[52]

Some of the views expressed here would tend to lend weight to the objections of Unionist politicians and community representatives who felt there was an unwillingness to embrace genuine fears and concerns. As outlined in the previous chapter, a common criticism of official planning documents over the past 40 years in Northern Ireland has been a failure to recognize the link between identity and territory and adopt a "planning for anywhere" approach.[53]

> If you looked at planning documents here, over the years, like the Belfast Urban Plan, you could be forgiven for thinking that they're talking about planning in Nottingham or somewhere …[54]

There were aspects of this approach within the Masterplan which was not helped by the fact that the design consultants were London-based and viewed as less familiar with the complexities of local issues than they ought to have been. At the same time, there was a recognition by a number of stakeholders that the subtle nuances of North Belfast spatial politics present a particular challenge.

> I found that BDP were ok about doing the drawings and doing the numbers but not very good at how you would develop a site that would maximise the benefits and gain the confidence of the two communities around there … they had urban planning skills but not community planning skills … although maybe they had experience of community planning … [but not with] … North Belfast community planning.[55]

These comments further illustrate the difficulties that arise when applying technical planning and design skills to an area in which territory and identity are inextricably linked, and where hypersensitivity to turf has the capacity to reignite political conflict. Conventional approaches to urban design might well suggest the location of a new "city quarter" within the largest area of the site and an "arc route" running through the development. The difficulty is that the spatial legacy of the conflict guaranteed that such an approach was viewed as a direct attack on the Protestant community of Lower Oldpark.

> North Belfast is an area of micro-territorial claim and what people don't realize is once you threaten the occupation of their territory, you actually strip them of their confidence and their own identity.[56]

Further evidence that deference to technocratic planning orthodoxy blinded some government officials to legitimate concerns from local communities is provided by a senior public official with responsibility for managing the project and bringing forward the Masterplan.

> That road that just bisected the site ... I said why was this done, and the answer was 'well ... they're the planners and they know'. And I kept asking and challenging ... and then it got to the stage I suppose, it was a bit of frustration from me, and I said ... tell me what you think my job is, is my job, to regenerate the Crumlin Road Gaol/Girdwood site, or is my job, to try and sell this plan to people. And they said, no, your job is to sell this plan. I said, well, I don't see my job that way.[57]

Frustration with the urban design elements of the Masterplan mirrors some of the difficulties that arose in the 1960s when leading figures in the planning profession such as Sir Robert Matthew arrived in Belfast and applied techniques developed in Britain and the U.S. with disastrous consequences. Further examples of the way in which the recursive processes that shape contemporary planning orthodoxy have the capacity to provoke rather than alleviate conflict, especially in places like North Belfast, are outlined in the next section.

4.2 THE ECONOMICS OF REDEVELOPMENT VERSUS THE POLITICS OF PEACEMAKING

This section focuses on the political dimensions of planning and architecture in the contested territory of North Belfast, highlighting how decisions around housing provision and attempts to address socio-economic inequalities are shaped by the imperative to prevent a return to conflict in the region. By considering the Crumlin Road Gaol/Girdwood Army Base in the context of other redevelopments that have taken place across the city in recent decades this section shows how the gridlock that developed around the Masterplan was a function of wider political developments, notably the consociational arrangements at the heart of the Good Friday/Belfast Peace Agreement.

4.2.1 The Economics of Regeneration and a Divided City

Since the 1980s, urban development in the UK, as elsewere, has been predicated upon attracting private sector investment and a reduced role for state funding. This approach was central in shaping the Masterplan for the site. Key to understanding the proposals for economic regeneration of the Girdwood site was a financial appraisal paper that concluded the gross development costs for the project would be in the order of Stg231 Million (including land costs), with a net realization of Stg266 Million leaving a Stg34 Million surplus with a profit of 15.10%. The financing model assumed that private developers would pay for the infrastructure works (road, basement car park, and sports pitch) as part of the first phase and that land would not transfer to the developer until works were completed. The financial appraisal also assumed that there would be housing in the first phase, which would pay for the infrastructure development so that construction costs would not be carried over the lifetime of the project. Without the inclusion of private housing in the first phase, the infrastructure cost would be carried for the full implementation period potentially adding up to Stg68 Million to the cost of the plan.

There were several key objectives of the Masterplan. The first, and most important from the point of view of the Plan itself, was to attract private investment from the sale of high-end residential accommodation that would appeal to young families and professionals. In addition to profits generated from the sale of private residential units, the Plan also sought to attract investment through the inclusion of shops, restaurants, and cafés, with the redeveloped Gaol complex serving as the economic anchor that would draw people and money into the area. The Plan also included several measures designed to help address social and economic inequalities, including social housing provision and several schemes designed to promote employment in the area as well as a community hub. The entire costing of the project however, including the infrastructure works (i.e., a new road, basement car park, and sports pitch), was contingent upon profits generated from the private residential development on the site. This approach followed conventional economic regeneration practice with an anchor tenant serving as a flagship for the project, private residential accommodation providing the financial leverage for the scheme, and some additional proposals that sought to address social and economic deprivation. One member of the Advisory Panel summarized the processes that lay at the heart of the Masterplan as follows.

> How to resolve the problem at Crumlin Road/Girdwood was as clear as the nose on your face for anybody who had come from an urban regeneration framework. There is a basic economics to doing regeneration, the public sector puts a certain amount of money in for the infrastructure, then you need private sector anchors, and then, within that context, you can put more social economy or more projects in that will help the local community.[58]

Central to the economics of the Plan was the need for people to be brought into the area who would be living on the site who would purchase or rent the new residential units and spend money in the cafes, bars, and restaurants provided, as well as avail of the new facilities in the refurbished Gaol complex.

> The overall regeneration of the site from a straightforward theoretical regeneration basis is that you need people living on a site like this to give it life … and we would have been totally on board with this … as being a mixed tenure site.[59]

The Plan stated explicitly that "it is hard to conceive the successful regeneration of a large inner-city urban site without the vitality of people living there,"[60] a view shared by public officials and several members of the Advisory Panel.

> I was very strongly of the view, and the master planners were also of the view of that if you don't put residents on this it's a dead site, so you have to have people living on the site, absolutely.[61]

For some members of the Advisory Panel, there was also a sense that this approach represented a new vision for this area of the city which had in effect become a "no-go" zone not only for visitors from outside Northern Ireland, but also for residents of other parts of Belfast.

> If you walk a mile from the front of the City Hall south, you end up in Queen's University and look at what the resources and assets that are there. If you walk a mile north you end up in Crumlin Road Jail, now look at the assets that are there, and now the contrasts are incredibly stark.[62]

The economic aspects of the Plan also cut with the grain of a buoyant Belfast property market at that time which had led to high demand for private residential accommodation both from owner-occupiers and the private rental market.

> The idea of very high-end apartments and housing, close to the Gaol, within a ten-minute walking distance to the city centre, in any normal city that would have been snapped up at that time.[63]

The mixed tenure housing model proposed for the site falls within the framework for social housing provision favored by the NIHE that has moved away from large single tenure developments, which have come to be regarded as creating barriers to social mobility and concentrating poverty.

> The old days where you build big housing estates of 300–400 houses, all social housing, are simply gone, and that's not perceived as the way to go ...[64]

Although the economics of the proposals were predicated upon the sale of private residential units there was an attempt to strike some balance between attracting households with higher income levels and addressing some of the socio-economic inequalities experienced in the wider area. The desire to ensure that the redevelopment of the site went some way toward addressing housing needs in local communities was allied to a strong sense that simply extending the boundaries of the working-class communities in the Protestant Lower Shankill or the Catholic New Lodge was not the answer for this part of the city.

> Simply reproducing the New Lodge is not actually going to benefit the residents of the New Lodge either, we actually have to do something far better than that and the Lower Shankill ... There's a real imperative to do something which actually offers some kind of long-term possibility that there will be employment, that there will be resources and assets, and Crumlin Road/Girdwood needs to be also producing some of that and for that to happen.[65]

Although the Masterplan reflected how economic development in Northern Ireland had moved away from an approach to regeneration that was predicated upon large-scale public investment to a model that reflected a much more limited role for the state, the Plan ultimately required approval from local politicians. It was at this point that local imperatives around ownership and access to territory conflicted with conventional economic development practices.

For those seeking to reach a compromise on a way forward for the Masterplan, the difficulty with the Girdwood proposals was that there were in fact two very competing visions about how the area ought to be developed, one of which, as explicitly set out in proposals, was predicated upon bringing people onto the site. This approach however was in direct contrast to the dominant view within the Unionist community in the area, which was predicated upon restricting access to the site and essentially keeping people out.

> Access is, and for the last 30 years has been, a dirty word, not a good word, and then you get this big plan … When you walk in with that plan to the Lower Oldpark, they were looking at the map going, they can get in there, they can get in there, they can get in there … now they can get in here, here and here, and they're building houses here, here, and here. We're for the hills.[66]

A major impediment to achieving agreement on the Plan was the way in which the demographic imbalance and different patterns of housing need between the two communities skewed the allocation of social housing on the site in favor of the Catholic community. This in turn cemented Unionist opposition to the proposals.

> It is as base as this, if there's more housing then it becomes Catholic because the waiting list is 95% Catholic, there isn't anywhere near the same proportionate need for housing in Protestant Unionist areas …[67]

The Plan acknowledged that the issue of residential development on the Gaol/Girdwood site was "extremely contentious" and noted that the housing issue had been debated by the advisory panel at various community and cross-community meetings as part of a wide and inclusive consultation process.[68] A key concern of the Unionist community was that the site would cease to be "neutral" if housing were to be included in the proposals, while a major concern for Nationalists was that the site had to be used to help address the significant housing inequalities experienced by the Catholic community.

> There is very little land available in Catholic areas to build new housing, although we are forever looking for land … that's one of the reasons why Girdwood is seen, on the Catholic side, as an opportunity to tackle the high levels of housing need within that community, a windfall site if you want to call it that.[69]

Efforts to address housing inequalities between the two communities were viewed as a zero-sum game with the Catholic community pushing for housing on the site to address patterns of inequality in that community while concerns within the Unionist community about Catholic expansion into traditional Protestant territory meant that the issue of housing became in fact a deal-breaker for securing their agreement to redevelop the site. To try and break the logjam around the housing issue, the Masterplan had recommended further discussions about how housing might be delivered in a way that addressed both Unionist and Nationalist concerns.

To achieve this, the Masterplan outlined details of an innovative public housing development located outside Enniskillen in County Fermanagh which had been developed and completed under the principles of "the Shared Future" community relations strategy. Within the Carran Crescent development properties had been allocated to applicants on the Common Waiting List in accordance with the rules of the NIHE i.e. based on housing need with a pre-condition that community balance should not exceed 70% from either of the two main community groupings in Northern Ireland. A second pre-condition of the scheme was that tenants should agree to the development of a "Neighbourhood Charter." This was a way of ensuring that neither the Catholic nor Protestant community would dominate the site through the display of flags, emblems, or other symbols.

The Carran Crescent development was able to ensure that the allocation of housing was balanced between the two main communities and that the 70% threshold for either community was not exceeded because the housing waiting list in the area was balanced between Catholics and Protestants. While acknowledging that the case study site in Co. Fermanagh was very different from the one in North Belfast and that the numbers of houses involved there were relatively small, the Masterplan proposed that this new "Shared Future" approach to housing development demonstrated that it was possible to manage the allocation of housing to create a mixed (i.e., cross-community) social housing development and that this could be of relevance for determining how housing might be delivered in the future in North Belfast. The Carran Crescent example was an illustration of the way in which thinking around the need to address housing inequalities was linked to efforts to maintain good relations between the two communities, first, by ensuring balance in terms of allocation and, second using the Neighborhood Charter.

The Masterplan for the Crumlin Road/Girdwood site in North Belfast posed the question as to whether similar principles might be usefully deployed in the Girdwood development.[70] In several respects this was somewhat disingenuous given the very different demographics in North Belfast. As noted above, in the Enniskillen development the waiting list for allocation of social housing was fairly evenly balanced between the two communities thereby enabling a balance of Protestant and Catholic tenants on the new site. Within North Belfast however the demographic imbalance between the two communities and the disproportionate levels of housing inequality among the Catholic community ensured that such an approach would be untenable. By law, the Housing Executive is required to allocate housing from the Common Waiting List on the basis of an objective points system and therefore any efforts to manipulate the housing waiting list in order to choose residents on the basis of their community background, regardless of motive, would be deemed direct discrimination and unlawful.[71] This principle prohibited any efforts to introduce quotas or limits that would restrict the proportion of Catholics who would be permitted to live on the site after redevelopment.

The biggest problem in North Belfast however was not the legal impediment to introducing quotas – laws can after all be amended but rather the demographic imbalance that meant housing need was skewed almost exclusively in favor of the Catholic community. This is an important point because claims about the need for a "Shared

Future" model of housing, similar to the Carran Crescent development, have a long pedigree in Northern Ireland, going back several decades.[72] The difficulty with most contributions to this debate is that the challenge is often posited from the perspective of legal impediments in the form of equality laws and the need for effective policing to address the issue of intimidation, while failing to explain the demographic challenge.[73] Public officials, politicians, and those closest to the ground were however acutely aware of the fact that it was demographic imbalances that presented the main challenge to shared future-type social housing developments.

> We're interested in promoting the concept of Shared Future, but the question is how do you do it in a place like Girdwood? If you stick a couple of hundred houses into the Girdwood site, and a lot of them are social housing, the need isn't there on the Unionist side, especially if you're also trying to regenerate these areas, the people that would get housing in here, the need, as measured through the waiting list would be predominantly Catholic, so it's very difficult in a one-off scheme, to have that sharing.[74]

Certainly, there was little doubt among elected representatives and other stakeholders about the difficulties of securing agreement on residential development for the site given the history around this issue and the extent to which debates around housing inequality have become polarized.

> Some of these things nearly become controversial, almost by stealth, but, housing has always been a huge difficulty in North Belfast.[75]
>
> The big thing of course, which is the usual hand grenade in North Belfast, is houses.[76]
>
> They can't agree about the residential use … that's right to the stark reality of North Belfast.[77]

Key to understanding the strength of feelings around this issue is the demographic imbalance and differential levels of housing inequality alluded to in the Dunlop Report, the NBHS, and elsewhere. As outlined in chapters three and four, these imbalances have their origins in the planning practices of the 1960s, the population movements that occurred in Belfast during 1969–73, and the defensive planning practices of the 1970s and 1980s. The past five decades have been characterized by a decline in the Protestant population in Belfast and a corresponding growth in the Catholic population. This is one reason why efforts to address housing inequalities within the Catholic community have come to be viewed by Unionists as something of a Trojan Horse to drive Protestants from the city which in turn impacted on the way the proposals were viewed.

> If you do not have a balanced demography, then by definition, the regeneration will favor the majority constituency in that area … So, inevitably, when the land becomes available, and you have a nationalist community in dire need of housing, and relatively few Unionists or Loyalists in dire need of housing, inevitably, the availability of a significant piece of land becomes, and I think

> it was Alban Magennis [of the Catholic/Nationalist SDLP] used the phrase, a 'windfall site' for housing. That was a most unfortunate term. Even if it was true it was still a very unfortunate term because what that meant was that the leaders of the nationalist community were claiming territory for themselves.[78]

It is also clear that the politicization of territory in areas like North Belfast meant that redevelopment proposals invariably became embroiled in other contested issues such as that of parading. Given that the Girdwood site sits on the route of the annual Protestant "Tour of the North" Parade, as well as parades commemorating the anniversaries of the Battle of the Boyne on July 12th and the Battle of the Somme on July 1st, the question of whether new housing on the site would be allocated to Catholics assumed even greater significance from the perspective of those involved.

> In the late 1990s, prior to Masterplan, we were brought in to do a development brief for Crumlin Road Gaol/Girdwood, and they'd been talking about it before that … and basically everybody knew at that stage that if you got new land like Girdwood for housing it was essentially going to be Catholic housing. From the Protestant point of view, that is from the residents of lower Clifton Park Avenue and Lower Crumlin Road, this was Catholic housing coming more and more in their direction and creating a new flashpoint … and they weren't happy with the prospect even that housing would come onto the front of the Crumlin Road, opposite the courthouse, because they felt that would mean you'd suddenly have new residents who'd be objecting to their tradition when they were marching down the Crumlin Road, the usual thing.[79]

Disagreements over residential development on the Girdwood site were given added significance by growing housing need within the Catholic community and several decades of housing provision based upon the retention of historical boundaries. As outlined in the previous chapters, since the early 1960s Unionist politicians had repeatedly acted upon concerns about the need to limit Catholic housing provision to protect the identity of traditional Protestant communities. In this context, disputes about the need for housing on the Girdwood site were part of a recurrent theme regarding redevelopment in Belfast.

> It's perfectly reasonable for citizens looking for some kind of accessible housing … to be asking public authorities what that's going to look like and where it's going to be … But in not doing so, what you do is turn every single inch of land in North Belfast into the same row, every single one.[80]

As ever in Belfast, in addition to the issue of parading, micro-territorial claims impinge on land use planning in another important respect centered around the relationship between houses and votes.

> Aside from all the stuff about not letting Catholics get houses and not letting them move in, it's not just as simple as bigotry, or we don't like Catholics, or we're not doing that … because if you were to … allow nature to take its

> course, demolish all the walls, build houses ... then you would reach the amazing position where there would be no Unionist MPs in Belfast.[81]

Sinn Féin had sought for several years to replace the then sitting Member of Parliament, Nigel Dodds of the DUP from the North Belfast parliamentary constituency, which in turn exacerbated tensions around new housing developments.[82]

> You've got Gerry Kelly there from Sinn Fein who obviously is very keen to take the seat and you've got Nigel Dodds holding on, and quite frankly housing estates here there or anywhere is the key thing that will change those demographics ...[83]

In addition to representation of the North Belfast parliamentary constituency, political control of Belfast City Council was also a factor in shaping the discussions.

> Belfast Council is very evenly split between nationalists and unionists, with the balance of power held by the Alliance Party at the minute ... But that part of the North Belfast constituency within the Belfast City Council area would be very finely balanced, so obviously ... if the population of that particular part of Belfast were to grow, it could in theory swing not only that seat, but it could swing control of the City Council as well.[84]

Sinn Fein had sought to allay these fears by arguing that since regeneration tends to decrease rather than increase populations, the regeneration of the New Lodge would benefit the Unionist community – in effect making Gerry Kelly's job of winning the seat more difficult.[85]

> The unionists had this big thing about votes, and we tried to argue many, many times, by saying, look, the New Lodge ... has to be regenerated, and you lose at least a third, sometimes up to two thirds of the population, so you're not talking about more votes coming in, you're not talking about more people coming in, you're talking about being able to expand in modern housing terms, so that people can have a bit of a garden and all of that, but, it never worked. ...[86]

One senior civil servant within the OFMDFM adopted a slightly more nuanced opinion, but one which would be inclined to lend more support to the Sinn Fein view by stating that given the pressures on housing within the Catholic community, any additional social housing would go to existing tenants in the area, many of whom were already living in overcrowded conditions with relatives.[87] Among Nationalists and Republicans, much weight was given to the proposed boundary changes for the North Belfast parliamentary constituency that would potentially make the position of the Unionist MP more secure by moving the overwhelmingly Protestant Shankill area into the North Belfast parliamentary constituency and reducing the number of parliamentary constituencies in Belfast from four to three.[88] Crucially however, one DUP representative, while acknowledging the role of electoral outcomes in terms of heightening tensions, was dismissive when asked about the likely impact of boundary changes making an agreement on housing more tenable: "no, but I can see why somebody would think that."[89]

Opposition to Catholic housing by Unionist politicians fearful that they might lose their electoral majority has been a perennial issue in Belfast, where political conflict is shaped by demography and where populations "are voters as well as occupiers of land."[90] As the Protestant population of the city continues to decline and the Catholic population continues to expand, thereby increasing patterns of housing inequality, these issues continue to retain significance.

At the same time, it would be a serious error to conclude that concerns about electoral majorities are the only factor in explaining the nature of territorial claims. Although voting patterns and electoral boundaries serve as a catalyst for heightening tensions around an already toxic issue, other more fundamental "micro-territorial claims" ensure that even when electoral demographics were not relevant, the territorial imperative remains strong. This is evidenced by the fact that the longest waiting lists for social housing within Belfast are in the west of the city, a Sinn Fein electoral stronghold. The fact that Unionist politicians have long since given up hope of winning an election in west Belfast has not in any way diminished the strength of feeling within the Protestant community about the need for adherence to traditional territorial boundaries in that area of the city.[91] This would lend credence to Stewart's claim that conflict in Northern Ireland is very much concerned with the relationship of people to land, giving rise to "a territorial imperative that is extremely insistent."[92]

It is also important to note that intra-community political tensions also played a role in heightening tensions around the proposed development of the site, a factor that was especially evident within the Catholic community as the SDLP and Sinn Fein sought for political supremacy.

> The SDLP were fighting every street with Sinn Fein in North Belfast, so the decisions were being taken on voting they weren't being taken on 'What is good for this site?' Margaret Ritchie was never going to push that agenda because she had all her SDLP activists in North Belfast saying 'we need the SDLP not Sinn Fein delivering this, and if you start saying we're not going to have social housing on the site we'll get this on the doorsteps'.[93]

These developments served to increase tensions with the Unionist community when the SDLP made several announcements about the future of the site. Speaking at the SDLP annual conference in February 2010, Minister for Social Development Margaret Ritchie announced that 200 new social housing units were to be built on the site.

> I was bringing some people around the Gaol on Saturday morning and they told me that they'd just heard on the news that Margaret was down at the party conference and had announced that there was going to be social housing on the Girdwood site … That then, I think, moved people away from 'is there something here we can negotiate with' to 'why even bother. You're going to do this, we're not going to do it' …[94]

A year after the publication of the original Masterplan in March 2011, by which time Margaret Ritchie had been replaced as Minister for Social Development by her party colleague Alex Atwood, another similar announcement was made, this time indicating that 220 new houses would be provided on the site. In neither case did any house construction actually take place – both announcements required cross-party support by virtue of the fact that responsibility for redeveloping the site was shared between DSD and OFMDFM.

Before the matter could proceed any further Alex Attwood of the SDLP was replaced as Minister for Social Development by Nelson McCausland of the DUP and one of his first actions as Minister was to cancel all plans for new social housing on the site and announce a review of all housing provision in the area. The announcements from the two SDLP Ministers proved to be highly controversial, not least given the obvious lack of political agreement that was essential in order for the housing to go ahead in the first place. Controversy over the announcements within the Unionist community was matched by skepticism on the part of the senior officials about the wisdom of the announcements.

> Margaret Ritchie … she made the decision, oh, I'm going to put houses on this … and then instructed the housing people in DSD to start working on proposals about how you could get housing on it. But even though they were instructed to do that, it's like putting together a plan for getting gold out of seawater.[95]

Certainly, outside of the SDLP, there was a strong sense that these political announcements had served merely to increase polarization around an already toxic issue.

> The SDLP, played an extremely foolish game with houses on that site, they promised twice that there was going to be 200 houses built, and they were ghost houses, they didn't exist, they were never ever going to happen in the way that they had proposed or described. And that has not helped, because one, it's basically got the Unionist community's backs up, and made a promise to communities in housing need, in nationalist areas, that wasn't going to be kept.[96]

There is little doubt that within the Unionist community announcements from SDLP Ministers that social housing was to proceed caused significant difficulties, not least given the timing of the announcements.

> The 200 houses were to be built in this financial year and they didn't even have planning permission … the Minister didn't have the legal authority to say, I'm going to build houses on these sites, but two Ministers in a row did it, and did it two and a half months before an election, both times … So those are the sort of things that have been going on all around it and messing it all about.[97]

Concerns and doubts about the wisdom of the announcements also extended to Sinn Fein.

> What the SDLP did … they just made announcements, and that's what made the situation worse, because their announcements meant nothing … because the Unionists of course came straight out and said, well, you can make all the

> announcements that you want, but it has to go to the Executive and it isn't getting through … so it really was just a propaganda act, and a propaganda act which of course wound the situation up worse[98]

The difficulty was that in a context in which housing was already considered "a hand grenade" any limited hope that existed of political compromise or agreement on the provision of residential accommodation on the site quickly evaporated as the prospect of short-term electoral gain trumped incentives for compromise.

4.2.2 Consociations, Consensus, and Conflict Resolution

The economic model that underpinned the Girdwood Masterplan reflected a recursive approach to urban regeneration that lay at the heart of other similar developments across the city over the past several decades, notably, the Castle Court and Laganside Developments in the city center in the 1980s, and the Gasworks redevelopment in south Belfast in the 1990s. Several of the key figures who had been involved in these earlier projects were directly or indirectly involved with taking forward the Girdwood project. Roy Adams, who played a central role in the Castlecourt development was the main figure behind drawing up the Girdwood proposals, while Kyle Alexander, the former Chief Executive of Laganside Development Corporation was a member of the Advisory Panel.

An important difference between the Girdwood Masterplan and the earlier developments at Castlecourt, Laganside, and the Gasworks was that there had been a significant reconfiguration within the political system in the intervening period. As Northern Ireland transitioned from conflict the political system moved from one of direct rule from Westminster to power-sharing by local political parties. At the heart of the political arrangements is the consociational power-sharing mechanism enshrined in the Northern Ireland Act 1998.[99] One of the core elements that defines consociations as distinct constitutional arrangements is parity through cross-community power-sharing – i.e. arrangements that encourage or oblige communities to make public policy and law *jointly*, through executive, legislative, judicial, policing, and bureaucratic institutions that ensure that the relevant groups are represented adequately in such institutions.[100]

Although responsibility for the site oscillated over two decades between UK and devolved ministers during periods when the NI institutions were suspended, the common theme was the need to secure the agreement of the two communities on the ground. In realpolitik terms, this has meant Sinn Fein and the DUP. Although the SDLP held the DSD ministerial portfolio for two periods and both Margaret Ritchie and Alex Attwood announced large numbers of social housing for the site that lacked cross-community support, these proposals lacked the legal authority to proceed since they failed to secure approval of Unionist political leadership. There is little doubt that both SDLP Ministers understood their proposals would be vetoed by the representatives of the DUP in the OFMDFM and their announcements regarding social housing on the site were clearly more about political posturing and point scoring than any objective reality that the site could be developed in a way that did not bring the Unionist community on board.[101]

The Girdwood redevelopment illustrates the difference between de jure and de facto consociational governance. Even when devolved government in Northern Ireland was suspended the political imperative has been to secure cross-community agreement on a way forward. Moreover, even when a government minister unilaterally announced redevelopment proposals, these were circumscribed by procedures requiring cross-community approval. The cross-community consociational arrangements at the heart of the Belfast Agreement, given legal effect by the Northern Ireland Act have been subject to a significant degree of criticism, not least because these arrangements, it has been argued, entrench sectarianism and group divisions.[102] The evidence from this case study suggests that the opposite is in fact the case, namely, that the consociational arrangements evident in the Northern Ireland Act merely give de jure and legal recognition to the de facto political reality of a divided community. The single theme emanating from the empirical work carried out on this case study is that political reality dictates that redevelopment within the contested territory of North Belfast can only take place with cross-community agreement. This is further evidenced by the way in which Belfast City Council, which is not subject to formal consociational governance arrangements, and which now has responsibility for planning in this area of the city, also operates on the basis that redevelopment of North Belfast takes place subject to cross-community consensus. In other words, in so far as the urban reality of Belfast is concerned, the political system is characterized by formal sharing of power, whether the legal system formally requires it or not. This approach supports Ignatieff's argument regarding the need for horizontal governance and a shared operating system through political coalition building for cities that are ethnically diverse.[103]

The need to prevent a return to violence and secure cross-community agreement highlighted how the new political arrangements at the heart of the Belfast Agreement ensured that the economic model, which had shaped earlier redevelopments in the city in the 1980s and 1990s, came unstuck. Some members of the Advisory Panel who had played a role in previous regeneration efforts in the city expressed frustration at the way in which local politicians failed to recognize the economic opportunities that were presented by the proposals. Several interviewees contrasted the post-conflict politics of redevelopment, evidenced by the disagreements around the Girdwood site with the political arrangements that prevailed during the 1980s and 1990s when Richard Needham, Environment Minister in the Conservative Government under Margaret Thatcher, played a significant role in the redevelopment of the city. Needham had given political approval for these earlier developments at Castle Court and Laganside in the face of much opposition. Several of those interviewed contrasted the Needham approach with the way in which local politicians failed to agree on a way forward.

> If there had not been the equivalent of Richard Needham there, somebody with the power and the balls to take the decision, I'm not sure what would have happened.[104]

Failure to secure political agreement on the Draft Masterplan 1 certainly led some members of the Advisory Panel to lament the absence of the earlier approach that "got things done."

> Richard Needham made Castle Court happen, and we didn't have a Richard Needham … if you'd had a Minister who was prepared to say 'I don't care, we're going forward, and you can fight to the cows come home, we're making this happen', but you didn't have that.[105]

Others adopted a more nuanced approach, recognizing the limitations of the Direct Rule model of governance while also conceding the advantages of a single entity with a clear remit and someone with political clout to take difficult decisions.

> You had a focus behind Castle Court, you had a certain set of people who were responsible who went ahead and did the thing, then they moved on up the river and set up Laganside, they set up a corporation for it, and the corporation had a remit that said within that area you're given a fair leeway to kind of do something and co-ordinate it …. Now, there are democratic deficits to all of that, there is an issue about how accountable these people are and how can you get wider involvements and so on but nevertheless the virtue is that it has a clear remit, that it has a clear budget line to do something.[106]

At the same time it is important to recognize the limitations inherent with this "democratic deficit" especially in the context of the politics of Belfast. As outlined in the previous section, similar disagreements arose in the early 1990s regarding the redevelopment of the Gasworks site in south Belfast. In that case concerns about Catholic encroachment onto Protestant territory led Unionist councilors to quietly remove social housing development from the site.

The power-sharing arrangements established by the Belfast/Good Friday Peace Agreement changed this dynamic.

> With Girdwood, politically things had changed, if Girdwood/Crumlin Road had been in the 1990s, with a Unionist dominated Council, they would have simply said, no, there won't be any residential here, let's get ahead and develop it …[107]

Concerns about the tendency of consociational politics to lead to gridlock ought to be balanced therefore against that which had gone previously, namely a command-and-control form of Direct Rule from Westminster and Unionist controlled Councils that in some cases amended development plans to further their own political ends with adverse consequences for the minority Catholic community. The Gasworks development, unlike Girdwood, went ahead because proposals for "Catholic housing" on the site were quietly removed from the original plans due to the dominance of Unionists in Belfast City Council. This example needs to be considered alongside attempts by Unionist politicians to undermine the Poleglass housing development in southwest of the city in the 1970s, which further illustrates how politicians manipulated planning policies for political objectives i.e. to limit and contain the growth of the Catholic community.

The gridlock around the issue of housing on the Girdwood site reflected the changed political reality that saw elected representatives of the Catholic community involved in political decision-making at a high level for the first time. In a context in which political identity, population imbalances, and patterns of inequality are built into the fabric of the local landscape, and where parading and election outcomes are all contingent on where residential development may, or may not be located, deliberations on the Girdwood Masterplan can appear to be a case of the irresistible force encountering the immovable object.

> The public sector as such lacked leadership because once you allow a lot of vested interests from the political and community side to sit round the table and to be part of the process then you're never going to get to agreement, they became the Palestinian/Israeli peace talks, over an urban regeneration site ....[108]

It is also clear however that in a post-conflict context, local politicians increasingly have control over shaping the urban landscape as political negotiation replaces urban conflict.

> I suppose the Girdwood project is an example where it's the local politics has been key in terms of what has been done, or not done on the site, but in some ways that's not unique, because I think more and more in NI politics comes into play and whether or not there is going to be residential uses on the site has been the key issue.[109]

Although members of the local Catholic community, and their representatives, had concerns about the Masterplan, among the political leadership of that community there was a desire to reach agreement and embrace what was viewed as an unprecedented opportunity to regenerate this area of the city. Regarding the presence of housing on the site, Gerry Kelly of Sinn Fein indicated a willingness to compromise both with regard to the number and location of housing on the site.

> ... the Catholic community have needs ... in terms of housing ... [but] 'I'm not talking about housing all over the site ... it doesn't profit us to have another community fearful ... it doesn't profit us to have another interface where Catholic homes could be attacked'.[110]

Within the Protestant community, and especially among the political leadership of Unionism, the inclusion of housing within the Plan *per se* was viewed as a deal-breaker. In a context in which the post-conflict political arrangements ensured that the Masterplan would only proceed with the support of both Unionist and Nationalist politicians, this meant that by the time the proposals were issued for consultation they were already in effect, dead in the water. For one member of Advisory Panel, who played a key role in drawing up the Masterplan, there is little doubt as to why the proposals ultimately failed.

> Where things went wrong, in my opinion, was the intransigence of the DUP .... I was seeking to get agreement on the seven principles, let's not get into the

> detail, let's just agree on the principles … but I could not get them to agree the wording on housing. And I went to see Nelson McCausland and Diane Dodds separately, from the others, on several occasions, trying to get the wording of the clause, or the principle changed to a point where they would agree to it, but they never would. And at the end, they said, we can't agree to any of this, and I said, if you agree to this vision statement which talks about people living in the place, how can you live in a place where there's no housing.[111]

The same interviewee argued that DUP rejection of the proposals arose from a mixture of concern about protecting votes and how the Masterplan was perceived by their constituents on the ground.

> I am certain that there was a votes issue in North Belfast, which was concerning them. Whereas most of the housing, whatever I said, would end up going to Catholic people, and secondly, I don't think they could sell it to the nearby Protestant wards which they were representing.[112]

The view that it was a combination of factors concerning the Unionist community rather than one single issue that led to rejection of the Plan was also shared by the senior public servant with responsibility for taking forward the regeneration of the site.

> To be honest with you, I thought it was all those other factors which prevented us from getting to what was the real issue, which is how do we develop this site, that meets [the] needs of both communities.[113]

In a city in which territory and identity are inextricably linked, and where the reflex action of Unionist leaders for the past two hundred years has been to halt the growth of the Catholic community across the city, there is a strong suspicion within some elements of the Nationalist community that the primary motive behind opposition to the original Girdwood Masterplan was an atavistic sectarian desire to keep Catholics, literally and figuratively, in their place.

> Everything else that has been presented as a reason for not building housing on Girdwood is completely spurious, the DUP, in the shape of Nigel Dodds and Nelson McCausland, all of them, all their excuses and explanations are completely contrived to prevent those acres being filled in, because the demand for housing is enormous.[114]

To attribute failure on the part of the DUP to agree to the Girdwood Masterplan because of a sectarian reflex to keep the Catholic community at bay would be somewhat simplistic. First, as one member of the Advisory Panel, Roy Adams, has pointed out above, he doubted that the DUP could sell the plan to their voters in the surrounding areas like the Lower Oldpark. This leads to the second point, namely, that in so far as the Unionist community were concerned, within the contested streets of North Belfast, where territory, votes, and identity are inextricably linked, the Girdwood

Masterplan did not offer much by way of a solution to the very real patterns of deprivation, dereliction, and depopulation facing communities in areas like the Lower Oldpark.

The Girdwood Masterplan illustrates the complexities involved in moving from an urban regeneration model based on the command-and-control politics of Direct Rule toward a model predicated upon consociational power sharing and consensus. In many respects, this case study illustrates Ignatieff's wider point that

> If human rights activism wills the end of a successful transition, it must also will the means, that is, the less than perfect political processes that make a democratic transition possible .... It is also a fact that outsiders do not have the power, if we ever did, to determine how fast or slow any local transition proceeds anywhere. We have no ultimate standing on the question of questions: Whose place is this? Who rules?[115]

This is especially apposite in a context in which the design of a street has the capacity to precipitate conflict and where

> Slow demographic change continues to extrude into local political geography, shaping school openings and closures, residential housing markets and disputes over access to public housing, and controversies over electoral districting.[116]

4.2.3 Conclusions

The distinction between community planning and North Belfast community planning, and the failure of the design aspects of the Masterplan to recognize the importance of local tensions illustrates the inherently political nature of urban planning in Belfast and the degree to which urban design has the capacity to provoke fears and precipitate conflict. In particular, the case study shows how key principles that lie at the heart of effective urban planning, specifically the desire to promote connectivity and encourage movement failed to gain traction in a context in which access is "a dirty word and connectivity translates as 'someone can come and attack my house." In many respects, the competing approaches toward urban design favored by the two communities mirror the competing approaches to housing provision.

For a growing and expanding Catholic population, facing high levels of housing inequality the imperative is to facilitate the growth of that community through additional residential units and an urban design philosophy predicated on opening the area up and redeveloping tracts of vacant land. For the Protestant population of North Belfast, which is older, and does not have the same demand for housing, the imperative is to protect territory. This in turn favors an urban design philosophy predicted upon the retention of walls, barriers, and clearly defined territorial boundaries. The next section will show how the subsequent development of the site, which involved a political compromise, sought to balance these two competing and in many respects mutually exclusive planning visions for the city.

4.3 REFLEXIVE PLANNING AND CONFLICT TRANSFORMATION

The section examines how the legacy of defensive architecture and planning, combined with efforts to prevent a recurrence of political conflict, led to the emergence of proposals that were quite different from those outlined in the Girdwood Masterplan. These alternative proposals, negotiated by local political elites from both communities, run contrary to much planning and architectural orthodoxy and highlight the contradictions and compromises inherent in reaching agreement on redevelopment plans in a divided city. This section also highlights the dangers of failing to take heed of local nuances around ownership of territory where the political imperative is to prevent a return to conflict.

4.3.1 Beyond the Girdwood Masterplan

Although political disagreement halted progress on implementation of the original Girdwood Masterplan, it was economic developments that dealt the ultimate blow to the proposals. The global economic downturn of 2009, and the collapse in the market for residential accommodation needed to finance the Girdwood redevelopment ensured the vision for a new city quarter on the site was finished. For some, this was a missed opportunity, symptomatic of the kind of sectarian zero-sum approach to urban development and politics that still characterizes certain areas of the city.

> At the heart of this is 'I don't want to share my territory with you' … the folks in the Lower Oldpark have said to me explicitly, they do not want any redevelopment of Girdwood while their area is untouched. That is the road to complete stalemate … Unless I'm getting something, you're getting nothing, ok, got it … this is barren community dynamics and politics. This is a desert area, so we leave the whole area desert, as long as you don't get one over on me, we'll keep it a desert … that's where we're at, and that's where political representation has brought us to.[117]

On the other hand, for those who were most opposed to the original Masterplan, notably the leadership of the DUP, the changed economic circumstances were viewed as an opportunity to develop proposals more aligned with political reality.

> It's now probably going to be a bit more piecemeal in its development … just out of practicality … it's probably going to take longer and be done in smaller pieces which isn't necessarily that bad, because, getting it done right, and showing that the sky hasn't fallen in, might be a device for getting people to relax around the site and get it developed properly.[118]

With a lack of political agreement about where to go next regarding the development of the site it was the public sector, in the form of Belfast City Council that stepped in to fill the void, submitting a funding application to the Special EU Programs Body for a new community facility in the area.

> We put the [community facility option] in front of them, as much as a catalyst as anything … what we were saying is … if we could get some sort of public services hub … we didn't quite know what it was at the time … we felt that it could work as a catalyst to actually get people together and talk about it which then might feed into a broader conversation.[119]

In 2012, the EU announced that the funding application had been successful and that £16 Million would be made available under their peace and reconciliation program for a new "community hub." In May of that year, the Minister for Social Development, Nelson McCausland of the DUP, who had by this time replaced Alex Attwood of the SDLP, announced that a revisedGirdwood Masterplan Conceptual Framework (GMCP) had been agreed.[120] This revised "Conceptual Framework" differed significantly both in content and approach from the original proposals – this time there was no document issued for public consultation or Equality Impact Assessment, nor indeed a final published detailed plan. The sensitive negotiations around the future of the site required a deliberately vague announcement and presentation to the media with a notable absence of official documentation around the proposals.[121]

Unlike the previous Ministerial announcements from Margaret Ritchie and Alex Attwood regarding social housing provision, the "Conceptual Framework" secured cross-community support in the Northern Ireland Executive, and therefore had the necessary political agreement in order to proceed.[122] Ultimately, a Community Hub, which was, under the original proposals, to have been situated within the refurbished Gaol complex was relocated on land originally set aside for mixed-use development, i.e. the area to the northwest of the site adjacent to the Protestant Lower Oldpark. No mixed tenure residential development was envisaged but some new social housing was provided at the southern end of the site adjacent to the Catholic New Lodge, with additional new build social housing situated just beyond the line of the original site boundary within the Protestant Lower Oldpark area. The politics of territory in North Belfast ensures that the new social housing developments will benefit each of the two communities with the Community Hub acting as a buffer in the middle of the site.

The Conceptual Framework approach exemplifies how contemporary urban planning in North Belfast is driven by a desire to prevent a return to conflict in an area of the city where hypersensitivity to turf continues to determine what kind of redevelopment is possible. Lack of detail regarding the detail of the final plans for the site was crucial for affording both sides the political space to negotiate around how best to maximize the benefits of new development in the area for their own communities. The compromises at the heart of the Conceptual Framework highlight the difficulties associated with addressing systemic and structural conflict legacy issues in Belfast and raise some important questions about the means and ends required to prevent the politics of territory precipitating a return to violence.

One difficulty with the Conceptual Framework approach is that rather than "getting it done right … and developed properly," there is a strongly held view, especially outside the DUP, that the redevelopment that took place in the area perpetuates some of the most egregious defensive and sectarian planning practices of the past. At the same time, there is an equally strong view within the local Unionist community, and indeed among some other stakeholders, that the final proposals represent the kind of compromise necessary to deliver political stability and prevent a return to violence.

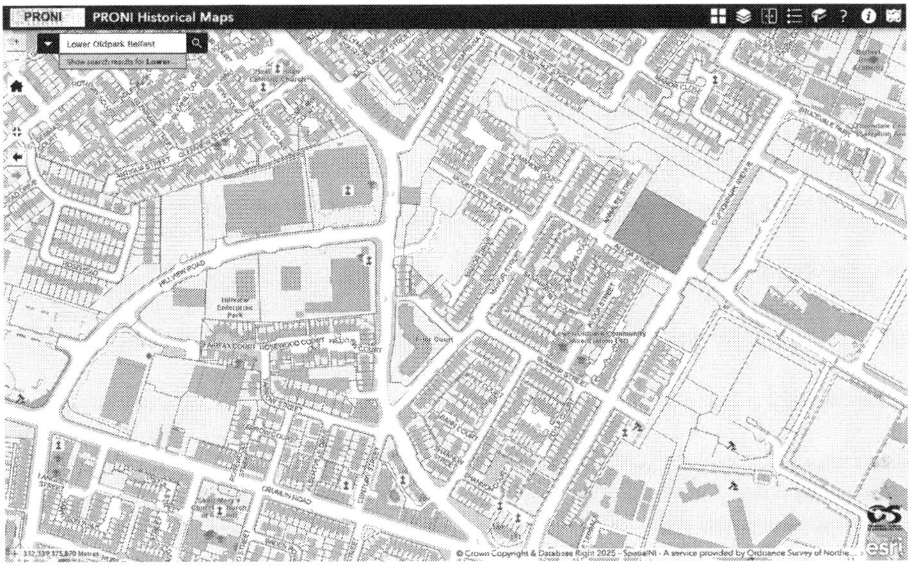

FIGURE 4.3 The Protestant Lower Oldpark area. A security barrier aka "Peace Wall" runs along the length of Rosevale Street (upper center right of image), closing the area off to the north and providing a buffer between Lower Oldpark and the Catholic Cliftonville area. The Crumlin Road Gaol and Girdwood site is visible as empty space on the right-hand side of the image and provides a buffer for the Lower Oldpark to the east. The Hillview Enterprise Park is visible as open land on the left side and provides a buffer between the Protestant Lower and Catholic Oldpark and Ardoyne to the west. The area is characterized by large tracts of empty space and a disconnected street network that replaced the original Victorian street patterns in the area. The intention was to significantly restrict movement within the area and create a disorienting sensation for outsiders, thereby enhancing security. Crown Copyright & Database Right 2025. SpatialNI. Ordnance Survey of Northern Ireland.

To understand why these alternative proposals came about, it is necessary to appreciate how conflict planning and architecture shaped areas like Lower Oldpark (see Figure 4.3), a small and isolated Protestant community protected against Catholic encroachment by a range of defensive architectural and planning designs.[123]

The geography and demography of Lower Oldpark help explain why the proposals in the original Masterplan to provide mixed tenure housing on the Girdwood site proved so controversial. The fear was that the area would be "overwhelmed" by an expansion of the Catholic population into territory that had hitherto provided them with a buffer zone.

> I always thought probably the solution to Girdwood was to show that you had a comprehensive plan that was also going to deliver something in the surrounding community areas, and I think that was especially the case on the PUL side … to demonstrate you were working in a wider area.[124]

> My negotiation theory training was always saying, look, there is the cake, whenever cutting the cake doesn't work, try to make the cake bigger … you're not getting your deal on this, well, let's look around, can we see something bigger happening in some way.[125]

One of the biggest challenges facing attempts to regenerate areas like the Lower Oldpark is that like many other inner-city Protestant communities in Belfast the spatial legacy of the conflict has left the area in geographical isolation with high levels of socio-economic deprivation augmented by a demographic "time bomb." By the early 2000s the population in the Lower Oldpark was estimated at 711 persons, of whom a disproportionate number were older people. High levels of benefit dependency and anti-social behavior, low levels of educational attainment, and issues of sectarian unrest and paramilitarism mean that this is one of the most deprived communities in Northern Ireland.[126] In a context in which ethno-national conflict is shaped by demography[127] the political imperative within areas like the Lower Oldpark is to address high levels of socio-economic deprivation and population decline while retaining territorial identity.

Social deprivation and population decline has in fact been a concern of elected representatives from the Unionist community across the city for several decades. Although the most egregious urban renewal practices of the 1960s and 1970s have ceased, suspicions remain within the Protestant community that economic development is something of a Trojan Horse aimed at precipitating their removal from the city.[128] Part of the problem is that urban redevelopment necessitates some form of population reduction, something that the statutory body with responsibility for social housing provision concedes. Citing the example of the redevelopment of the Unionist "Village" area of South Belfast, a senior official with the NIHE concedes,

> If you have a big area, for example the Village, which is one which we're in the middle of at the minute, there's about 700 houses to be taken out, and they are to be replaced with about 300, and it's done in phases … at each phase, people are required to move out of their homes, so some will move to other areas and not return. The process of redevelopment in itself, which can take many, many years from the beginning to the end, has almost built into it the process of people moving out of the area ….[129]

Suspicions that population decline within the Protestant community has been socially engineered by redevelopment have been fueled by the experience in the "Torrens" area, located less than a mile from the Lower Oldpark. Like the Lower Oldpark, Torrens was a traditional working-class Protestant enclave that underwent substantial redevelopment during the 1980s.

> As the troubles progressed, people moved out and didn't want to live in the area, but they had more choice … There were a lot of empty properties in Torrens but there would have been maybe 70–100 families living in it … about 2005 it had gone down to nearly a dozen, ten, and when it reached that point,

> the families came to see us as a group and said look we want out, we can't stick it anymore, so they were rehoused into areas that they wanted to go ... they eventually left, and the land subsequently was developed and it was Catholics who moved into it.[130]

The experience of Torrens is crucial to understanding the continued hypersensitivity to turf within the Protestant community particularly in the north of the city. As with many aspects of life in Northern Ireland the factors behind the decline of the Protestant community of Torrens are politically contested. Within areas like the Lower Oldpark there is a strongly held view that Torrens is evidence of a conspiracy, involving the NIHE and the wider Catholic community, including the Irish Republican Army (IRA), to drive working-class Protestants from the city. In fact, "there won't be another Torrens"[131] has become something of a rallying cry for the Unionist community in the area and accentuated the political imperative for Unionist representatives to protect Protestant territory from Catholic expansion. Further examples of this hypersensitivity to turf, or "territoriality,"[132] is provided by the comments from one representative of the DUP, interviewed for this study, who alluded to a conversation that had been overheard by a member of his party in which a senior official in the NIHE expressed a view that the organization was happy to "let the Lower Oldpark die."[133]

As outlined in Chapter 4, the Joint Working Party on security in the early 1970s had considered it "to be of great importance" that there should be a reduction in the overall percentage of residents going back into areas in the city after redevelopment.[134] This would appear to vindicate Unionist concerns that there was indeed a high-level strategy in place to depopulate Protestant areas beginning in the early 1970s but that the NIHE, who were absent from the discussions, was not actively involved in the deliberations around implementation of this policy. A further twist to this is provided by the fact that the original Joint Working Party committee was prepared under the auspices of John Taylor, the then Ulster Unionist Minister for Home Affairs, and presented to Brian Faulker, then Prime Minister. This would indicate awareness and approval of this policy at the highest levels of leadership within the Ulster Unionist Party but not necessarily among local Unionist communities. Although local Protestant communities may have lacked the details of the wider security policy they were able to observe how urban development played out in practice – in the same way that Catholic communities were able to see the impact of street closures and road realignments. The legacy of Torrens in shaping the Unionist response to the Girdwood Masterplan was especially powerful given that the Lower Oldpark area is in such close geographical proximity and resembles Torrens both in its geographical isolation and demography. As outlined in the previous chapters the proposals contained within the Girdwood Masterplan, predicated upon conventional theory and design practice, singularly failed to take account of the legacy of Protestant depopulation and deprivation and fit a narrative within the Unionist community which held that significant redevelopment was taking place on the "Catholic side" while the Protestant community in the Lower Oldpark would be "left to die."

Outside of the Unionist community, there is an alternative explanation for the processes that have shaped areas like Torrens. Among government officials and representatives of the Catholic community the view is that Torrens became socially and economically unsustainable and that residents of the area exercised choice and moved to more desirable housing elsewhere. From this perspective the declining population in areas like Torrens and the Lower Oldpark can be traced to the urban renewal programs that began in the early 1960s which ensured that many traditional Protestant communities have passed the "tipping point" beyond which they are sustainable. Consequently, social mobility and the availability of better accommodation outside the city in towns such as Newtownards and Carrickfergus encouraged a Protestant exodus from the city beginning in the 1960s (see Chapter 4). This phenomenon is common to many cities in Britain and elsewhere and is typically referred to in U.S. contexts as "white flight." From this perspective, the exodus of the Protestant community from the north of the city is explained partly as the result of political violence, but also by secular factors that characterized urban policy across many post-industrial cities. Outside of the Unionist community there is little sympathy or support for the view that the declining Protestant population in the north of the city can be attributed to any form of conspiracy or hidden agenda on the part of either the Catholic community or public officials. These differing perspectives illustrate how the contested nature of the urban legacy of the conflict continues to impact upon post-conflict redevelopment. There is little doubt however that a security-led policy from the early 1970s onward that was predicated upon reducing the population of the city and increasing the number of open spaces had a significant impact on destroying social capital and the sustainability of local Protestant communities. The culprit in this instance was not the NIHE but rather security-led policies that operated at higher levels of government and included military officials.

To fully appreciate the spatial legacy of the conflict from the perspective of Protestant communities in areas like Torrens and Lower Oldpark, which are characterized by high levels of social housing, it is necessary to consider also the wider population changes that have taken place within the private housing sector housing market. As members of the Protestant community have moved away from owner-occupied housing in areas like the Cliftonville Road and the Antrim Road these properties have typically been purchased by upwardly mobile Catholics moving onto the property ladder.

> I remember canvassing the Cliftonville Road 25 years ago and there was this sense then that the great unwashed were coming up from the New Lodge and moving into flats and all the rest of it... occasionally you would knock the door of this very nice detached house on the Cliftonville Road and an elderly woman would come out who was clearly a Protestant who had been overwhelmed and left behind in the great tidal wave that came up from the New Lodge … mahogany door, brass, all the rest of it, and you just felt that time had passed her by, and at that time there were half a dozen of them on the Cliftonville Road, and in some respects there are a number of other places like that in North Belfast.[135]

The difficulty for those living in traditional Protestant social housing developments therefore is that they have in effect become "surrounded" by "the other side" as members of their own community have left traditional heartlands. Upward and outward mobility and individual choices within the Protestant community played a major role in undermining attempts to sustain traditional communities in the city something that political representatives of Unionism are acutely aware of.

> Right now, if you regenerate, there's a high risk, if we got in with people, got them jobs, got them working, then bye bye. Getting on means getting out in too many of these communities ...[136]

In a context in which communal identity is a proxy for political affiliation, urban redevelopment policy in Belfast is inextricably linked with electoral representation.

> I would make a general point in terms of regeneration, if people are still wanting to leave where you are representing, then you've a problem that you need to solve in your area ...[137]

This shows how the spatial legacy of conflict has been impacted by secular economic and social factors, notably the tendency for younger Catholic families to purchase property in the private housing market from older Protestants and the fact that for many people in the Protestant community a surplus of good and affordable housing in "Protestant" suburbs has further encouraged population movement. Efforts to address this issue on hte part of Unionist politicians is explored in the next section.

4.3.2 Identity, Mobility, and the Politics of Territory

One way in which the spatial legacy of the conflict continues to impact upon plans for economic regeneration is illustrated by the desire on the part of Unionist politicians in areas like the Lower Oldpark to address patterns of deprivation and population decline while at the same time seeking to ensure that social mobility does not facilitate further Protestant exodus from this area of the city. Achieving this objective requires the reversal of a demgraphic trend that has been established over several decades where the city ecosystem has been characterized by Protestant population decline.

> We need people to be staying ... to give those communities a bit of life ... young people ... who can use the shops and facilities because if the trend continues ... then those are going to be absolutely dying communities, and you're not reaching a critical level for core services like schools and shops.[138]

A significant barrier to achieving this objective is that in many Protestant communities new social housing provision has not been a priority for public agencies given the lack of demand for housing as populations continue to decline and waiting lists diminish as residents move elsewhere. Public housing provision in Belfast, as elsewhere, is driven largely by demand and therefore a "chicken and egg" situation has developed in which Unionist representatives contend that areas are in decline because of lack of new housing and public agencies argue that lack of demand negates new house build. For representatives of the Protestant community, the lack of new build social housing, tracts of vacant land and derelict housing stock sends a clear message

> There's an important sub-text here that gets missed … you're asking me to offer a dismal vision to the community that I primarily represent … you're in decline, you will continue to be in decline … and this is where you're back into the broader definition of regeneration … Why could regeneration not be about those communities growing again, as opposed to, well you're Prods, you can go and live in Newtownabbey and Carrick?[139]

A key political objective of the DUP, and in particular political representatives like the former Social Development Minister Nelson McCausland, is to offer an alternative to this "dismal vision" in order to reverse the exodus of members of the Protestant community from traditional heartlands in the north of the city and deliver economic regeneration that will in turn help to secure Unionist electoral majorities and begin to rebuild Protestant communities. This was a key objective of the housing-led approach to regeneration as framed within the "Building Sustainable Communities" (BSC) strategy, launched by Nelson McCausland on 28 October 2013[140]

> You've got a lower demand for housing [in Protestant areas] because people are willing to go and move to East Belfast or wherever it is, so the idea is to try to build better housing, nicer housing, and using that to either attract those people back into the community who have left or to encourage people to stay within the community, for people like Nelson McCausland that would be very much the agenda …[141]

The housing and regeneration debate illustrates how socio-economic inequalities, notably housing need and dereliction, are manifest in different ways across the two communities with Protestant areas characterized by low-medium housing demand, tracts of vacant land and dereliction while Catholic communities are characterized by overcrowding, a shortage of available land, and high levels of housing demand. Efforts to address spatial inequalities are therefore highly politicized, given that the two communities are characterized by very different patterns of *inequality.* Efforts to address socio-economic deprivation and housing inequality, as measured by overcrowding and time on the housing waiting list, will, de facto, disproportionately benefit members of the Catholic community. Efforts to address socio-economic deprivation and dereliction will, by definition, disproportionately benefit members of the Protestant community. Similarly, initiatives that target inequality, which are based on increasing education and employability are viewed within the Protestant community as potentially destabilizing given that there is a much older population demographic in these areas.

Given that local politicians are acutely aware of these differences it is unsurprising that very different approaches and priorities for addressing inequality and the spatial legacy of the conflict exist across the political parties. The debate outlined above also leads on to another important point, namely, the link between urban design, inequality, and conflict. In a context in which the fear within the Protestant community is of a growing and expanding Catholic population, the use of urban design to protect traditional territory becomes even more crucial.

Allied to high levels of deprivation, depopulation, and isolation, the Lower Oldpark exemplifies another important spatial legacy of the conflict, namely the defensive design practices that shaped the planning system of Belfast during the

1970s and 1980s. Barricaded by the Hillview Enterprise Park to the west, the peace wall to the north, and the Girdwood site to the East, the only "friendly" territory for the community of the Lower Oldpark is to the south, which fronts onto the overwhelmingly Protestant Crumlin Road (see Figure 4.3 lower left side of image).

The Lower Oldpark Action Plan summarized some of the design issues facing the area and provided a damning indictment of the way in which several decades of military urbanism and defensive architecture impacted upon the local community. The report found that the current road network and street design had lost much of the pre-1960s layout

> with the streetscape mirroring the divisions in wider society - inward facing avenues, defensible homes and dead-ends …. The original street pattern of housing fronting onto roads and back garden areas serving as private spaces has been replaced with closely knit residential pockets i.e. cul-de-sac developments or courtyards which provide no through access to main roads.

This is clearly visible in Figure 4.4, which shows how the disconnected street network limits movement both within the Lower Oldpark and between the Lower Oldpark and its wider geographic hinterland through "inter" and "intra" communal barriers.[142] This can be contrasted with Figure 4.5, which shows how the historic street grid prioritized connectivity.

The report went on to highlight how the changes that were brought about to the street patterns during the 1970s and 1980s ensured that "direct routes which residents

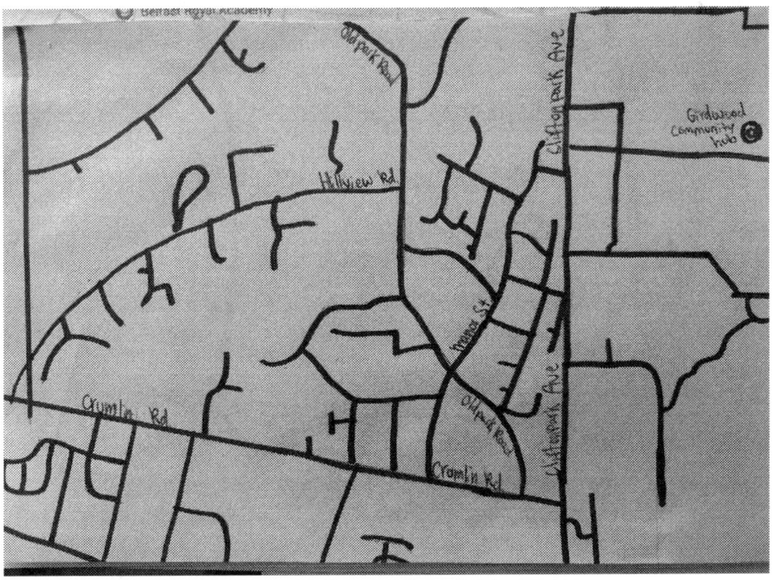

FIGURE 4.4 The Protestant Lower Oldpark area. The disconnected street network visible in this image above impedes movement both within the Lower Oldpark and between the Lower Oldpark and its wider geographic hinterland, creating both inter- and intra-community barriers. Drawing by Cara Cunningham.

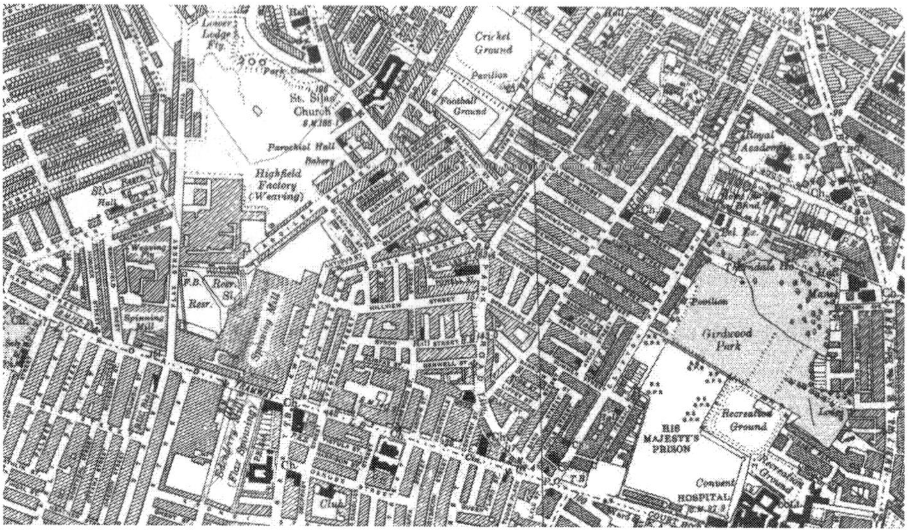

FIGURE 4.5 The historic street grid of the Lower Oldpark and its surrounding hinterland. Security imperatives ensured that connectivity and density were significantly reduced, with several streets removed altogether and others closed. Crown Copyright & Database Right 2025. SpatialNI. Ordnance Survey of Northern Ireland.

could have previously walked or driven on no longer exist, making it difficult to navigate the area and access services and facilities elsewhere …"[143] The net effect of the cleared sites, the peace wall, and the street network is

> a low quality living environment for residents and for visitors to the area the impression is created that this is territory that is closed to outsiders …. This sensation is augmented by a streetscape that is hard in appearance with little street planting and a dominance of hard paved surfaces with few landscape features to aid in softening the visual appearance of the environment and very little differentiation in materials to help denote public and private spaces and to enforce any sense of route hierarchy.

The BSC Area Analysis report describes Hillview Road as "unsightly and unwelcoming … dominated by security style fencing and gates, along with a high peace wall"[144] while Century Street is described as "dark and dismal … could benefit from new housing and perhaps the re-opening of the street giving better access to the Crumlin Road."[145]

The Lower Oldpark BSC Action Plan proposed restructuring the existing layout of the area through rebuilding or remodeling existing properties to remove design weaknesses and improving the streetscape by ensuring that new development addresses the street. The Plan also recommended the creation of new streets that would make it easier to move around and navigate the area.[146] These proposals illustrate again the important role that urban design can play in addressing, or indeed promoting patterns of social and economic disadvantage with the Action Plan setting out a framework that aims to enhance connectivity and remediate some of the coercive architecture created by the defensive design tactics of the 1970s and 1980s.

The DSD Design Team produced concept proposals to upgrade and improve the local public realm of the area including decorative entrance features to Hillview Road at Crumlin Road and Oldpark; resurfacing of footpaths using natural stone paving bands, grass verges and asphalt; decorative street lighting, railings, gates and street furniture; and semi-mature tree planting. This plan also envisioned that new housing development, beginning at Manor Street and Cliftonpark Avenue on cleared and gap sites would assist in addressing design weaknesses with respect to "lack of supervision of back alley spaces and car parking courts.. as well as offering an important opportunity to maintain a feeling of change and growth within the community."[147] In addition to presenting a physical symbol and foundation of change the initiative also aims "to lift the community's confidence in the process of regeneration providing them with a sense of empowerment to embrace further change."[148] The BSC program therefore aims to begin to address some of the key demands made by the Protestant Unionist Loyalist (PUL) community during the course of the consultations on the Girdwood Masterplan.

> The demand [was made] both publicly and certainly to me … that there should be no development whatsoever on the Girdwood site until the contiguous area in Manor Street was properly developed and the Protestant community restored and regenerated so that there would be a balanced community in the wider area.[149]

While the original Girdwood Masterplan was predicated upon bringing people into the area by increasing connectivity between the site and the rest of the city, the "Conceptual Framework" approach encapsulated within the BSC Action Plan is firmly predicated upon respecting the integrity of existing inter-communal territorial boundaries. This is explicitly spelt out in the BSC Plans[150] which state that the Hillview Enterprise Park provides "a necessary interface barrier" and that the existing 'peace line' at Manor Street "could be developed to include commercial/retail property."[151] Similarly, the location for the first new housing units on Cliftonpark Avenue[152] is referred to as a "significantly strategic location for demonstrating ongoing investment in the area," with "early delivery of new housing on these visually prominent sites" presenting "a strong indication of the commitment to delivering new housing in the area."[153] Within the context of the Belfast territorial imperative the significance of these visually prominent sites for new housing and urban design framework is that there will be no encroachment from the Catholic community along Cliftonpark Avenue but rather "shoring up" new housing on the "Protestant side," as illustrated in Figures 4.6 and 4.7.

The BSC Strategy therefore has two discrete and mutually reinforcing aims. The first aim is to provide a framework for opening access and increasing mobility and connectivity *within* the PUL community of the Lower Oldpark to address the problematic streetscape and road network that was put in place in the 1980s. The second aim is to consolidate and enhance existing boundaries *between* the Lower Oldpark and the Catholic communities in surrounding areas to "hold the line" and provide reassurance of a future Protestant presence in the area.

Figures 4.6 and 4.7 highlight the most controversial aspect of the approach adopted by Unionist politicians regarding redevelopment of traditional Protestant communities, namely, that housing-led regeneration sits alongside the preservation of existing territorial boundaries. Rather than increasing connectivity to "bring new people into the area," which was the aim of the original Girdwood Masterplan, the retention of territorial boundaries illustrates how defensive urban design around the Lower Oldpark provides a de facto exclusion mechanism that prevents encroachment from the Catholic community and the natural population growth up the Antrim Road. The retention of communal boundaries illustrates how the defensive urban design model favored during the conflict continues to be the default position for the Unionist community with the recursive design practices of the 1970s and 1980s deployed in order to protect territory.

FIGURE 4.6 Additional social housing, located at the "Protestant end" of Cliftonpark Avenue, situated on the right of the photograph, behind the trees and fence. The location of housing here is significant in that it confirms that the territorial boundaries of the Lower Oldpark are being respected and that the area is not "being left to die." The fending and trees confirm the barrier effect that this Protestant territory. Photograph by Tim Cunningham 2018.

FIGURE 4.7 Cliftonpark Avenue looking north. The fencing on the left of the image illustrates how urban design has been deputized to help solidify and shore up the Protestant territorial boundary lines of the Lower Oldpark. This fencing joins up with the older and more solid security barrier (aka "Peace Wall") located in the background on the left side of the image and running perpendicular to the newer fencing. The fencing on the right surrounds the new Belfast Community Hub, funded largely by EU Peace and Reconciliation monies. Photograph by Tim Cunningham 2018.

4.3.3 Systemic Inequality, Deprivation, and Urban Design

Outside of the DUP there is a high degree of skepticism about the value of seeking to regenerate areas while retaining territorial boundaries and through providing additional social housing in areas where there is a lack of demand and where existing residents are "voting with their feet."

> Some in the PUL community say the housing executive needs to take the lead in regenerating these areas, and begin to fix these houses up, and people will come … we have never been comfortable buying into that because … if you suddenly build new houses, for which there is no waiting list, who's going to move into them … it's going to be people from other houses, and they will leave those other houses empty, or people maybe will come back in from Newtownabbey, or whatever, but leave vacant properties there …[154]

In addition to the potential displacement effect of social housing tenants and waste of public resources, more fundamental concerns have been expressed at the likely profile of applicants for housing in areas that interface with the Catholic community and are known to have high levels of deprivation, low levels of educational attainment and associations with loyalist paramilitarism.[155] Several studies have highlighted the extent to which interface areas in Protestant communities in North Belfast have become "dumping grounds" for undesirable, anti-social "problem tenants" including people with criminal convictions, mental health problems, and other behavioral issues. Sinn Fein representatives express concern about this issue.

> One of the difficulties is that Nelson McCausland has a view that he can repopulate Protestant areas ... Where I think he is flawed ... and I don't have a principled difficulty with what he would like to do because why would you not try and reinvigorate some of these old Protestant areas ... but he's got to deal with the paramilitaries .... and it's a post-conflict thing, the loyalist paramilitaries are turning into drug-dealing criminal gangs, and that's the biggest difficulty, and if you don't deal with that, how are you going to attract people back into these areas, to populate them?[156]

DUP politicians concede that there are significant difficulties with convincing members of their own community to return to areas like Lower Oldpark where success and social mobility are equated with living in more affluent suburbs on the outskirts of the city.

> To some degree there's almost become a culture of it, you're successful now when you move out, it's a sign of your success .... People can complain about the other community, or this public agency, or that public agency, whatever they like, my comment is, in North Belfast, the Unionist community has to have a serious discussion with itself. Why is it when somebody does good the first thing they do is head to Carrick? They're now going to have to commute at extra cost, the house price differential disappears, so they're not getting a cheaper house, so why do these people want to leave?[157]

For some members of the Girdwood Advisory Panel, the retention of territorial boundaries is a major factor in ensuring continued Protestant exodus from traditional areas, and the cause of, not the solution to, the problem.

> There's no evidence that interfaces have protected the demographics of Protestant communities, absolutely not, they've driven everybody who had a choice, out. Catholic communities had far less choices and therefore stayed ... their best bet is to attract a few more people to move there and frankly, as long as you have massive great walls you have absolutely zero possibility of that.[158]

Notwithstanding these concerns, the NIHE has conceded a willingness to

> explore taking some calculated risks around beginning to fix up properties that are derelict, to try to generate a waiting list.[159]

For a public agency with limited resources, "generating a waiting list" when there are many areas in the city with existing lengthy housing waiting lists is a highly controversial approach to social housing provision. Moreover, there is also the legal difficulty of providing housing for Protestant areas when Protestants are not as highly represented on housing waiting lists. The legal problems were surmounted by the adoption of the "Building Sustainable Communities" (BSC) strategy, launched by Nelson McCausland on 28 October 2013.[160] The BSC strategy allowed for new social housing to be located in the Lower Oldpark area under the rubric of a pilot scheme predicated on housing-led regeneration.[161] The NIHE concedes that if there had been "no building successful communities there would have been no new build in Lower Oldpark," given the lack of demand in the area.[162] The BSC strategy is specifically aimed at regenerating communities experiencing blight, dereliction, and decline through a housing-led approach to regeneration.[163]

> Under BSC ... new homes are built that would not have been built using the traditional approach, with BSC you do things you don't normally do ....[164]

Six BSC "pilot areas" were selected for this housing-led approach to regeneration including the Lower Oldpark,[165] all of which experienced problems such as stigma, blight, void stock, anti-social behavior, low or medium projected housing need and vacant government land that could be developed to bring new housing back into the area.[166] The BSC strategy in North Belfast is targeted predominantly at Unionist communities – unsurprising given that this program is very much a creation of the DUP, and in particular, the then Minister for Social Development, Nelson McCausland.

> BSC came directly from the Minister ... Only Nelson was the Minister it would never have happened[167]

The "Conceptual Framework" that replaced the original Girdwood Masterplan was therefore not only a political compromise but also provided legal cover for the proposals given that the established framework for site location of new housing was not followed. The "Conceptual Framework" did subsequently generate a legal challenge with a Catholic public housing tenant, Ellen Doyle, who had been on a housing waiting list for 13 years taking a judicial review of the proposals outlined by Nelson McCausland in 2013.[168] Lawyers for the plaintiff argued that the motive behind the revised Girdwood proposals was sectarian – designed to halt the advance of the Catholic community in the New Lodge.[169] An out-of-court settlement was agreed which included an assurance that there would be additional social housing at the southeastern end of the Girdwood site that allowed the revised and unpublished "Conceptual Framework" for the area to proceed.[170] In effect the legal challenge provided leverage in the negotiations about the amount of additional new build housing that would be located on the eastern end of the site that would be accessible to Catholics without fundamentally altering the overall "Framework" of the regeneration.

As outlined in the previous chapter, the introduction of a "points system" and a common waiting list based on the notion of objectively determined housing need was one of the key legal reform measures introduced in the early 1970s when the

NIHE was established, the purpose of which was to address the sectarian housing allocation that had taken place under Unionist controlled councils between 1921 and 1971. The *Doyle* case reflected both the controversial nature of the Conceptual Framework proposals and the limits of law in this area, where complex issues of urban regeneration interface with identity and where a reluctance on the part of courts to stray into areas of public policy that are generally regarded the purview of elected representatives.

In terms of assessing the "Conceptual Framework" approach to regeneration of the Girdwood/Lower Oldpark area several preliminary conclusions can be drawn. Perhaps unsurprisingly, the DUP, the main architects of this approach, are quite positive about the program and are of the view that following new house build on Cliftonpark Avenue interface violence has reduced, which is a

> … direct result of the intervention … Building houses was a statement to the Unionist community that they had a future in the area …[171]

The NIHE also concedes that a combination of new social housing and the work of a number of community workers from the Lower Oldpark Interface Programme has ensured that "the Lower Oldpark has got really settled."[172] Certainly, the new housing delivered as a result of the "Conceptual Framework" for the area has helped allay Unionist fears "that Lower Oldpark was going to be another Torrens,"[173] which in turn has had a positive short-term impact on reducing inter-communal tensions and violence.

Initial anecdotal evidence also suggests that some of concerns about levels of demand and occupancy have not been realized. According to the DUP 30 new social housing units were built in the area as a result of the BSC program all of which were filled right away with people moving from the Mid-Shankill confirming in their view that "there was a waiting list but it was people who were not directly in the area … when you build them they will come …."[174] The Housing Executive have also conceded that in spite of "initial skepticism about BSC … it has worked to an extent" with some unexpected positive outcomes from the program not least of which is that there has been an increase in diversity with some new housing units occupied by ethnic minorities "because that's where the available housing is …"[175].

One potential consequence of this type of approach to regeneration that was cited is that members of ethnic minority communities, facing longer waiting lists in overcrowded Catholic areas, may have an opportunity to avail of social housing in traditionally Protestant areas that "traditional Protestants" no longer find desirable. Furthermore, according to the Housing Executive, there is no evidence to date of displacement of tenants to the Lower Oldpark from other areas or tenants moving into the area who had previous issues with anti-social behavior.[176] They do concede however that this has been partly a result of an overall increase in housing need generally in the area. In other words, more Protestants in need of housing overall has led to more "needy" tenants moving to the Lower Oldpark.

There is a need for caution regarding these early indications of success however not least given this is a pilot program. That the BSC is a creation of the DUP and implemented by the DSD would cast some doubt about partiality with respect to initial assessments of success. A two-stage process has been outlined for monitoring the impact of the BSC initiative that requires collecting output/outcome measures relating to specific projects and interventions on an annual basis and periodic neighborhood monitoring which will measure impacts of the regeneration process on neighborhood baseline conditions over three yearly intervals. As yet, limited data has been released to allow for a more robust examination of the extent to which BSC has succeeded in delivering the outcomes that have been set. Until such time as this data becomes available, there must remain some doubt as to whether the pilot scheme for the area can be deemed a success. Ultimately, any decision about whether or not the new house build program continues in the Lower Oldpark area in its present form will be a political one, highlighting again the way in which urban planning and efforts to address housing inequalities continue to be politicized amid the contested space of North Belfast. Within the Unionist community, suspicions remain about the sustainability of the initiative regardless of its success.

> 'the Housing Executive don't like the concept ... they're hoping that a nationalist [Minister] will take over [DSD] and scrap the idea ....'[177]

Notwithstanding initial positive feedback that the "Conceptual Framework" for the area has succeeded in delivering positive outcomes with respect to helping the Lower Oldpark/Cliftonpark Avenue interface "get settled"; providing reassurance to the PUL community that they will not face "another Torrens"; and addressing some of the more problematic streetscape and design features of the 1970s and 1980s; much more fundamental questions remain about the long-term economic sustainability of the program and this approach to regeneration generally.

Questions arise about the extent to which the retention of territorial boundaries will serve to undermine the need to attract inward investment, private sector development, and crucially, new people. There must also be some doubt as to what economic leverage the addition of this number of social housing units can provide in an area experiencing the level of deprivation and dereliction that exists in the Lower Oldpark with an existing residential population of 711, a disproportionate number of whom are older people. This was an issue that Roy Adams addressed during discussions around the original Girdwood Masterplan.

> There were exchanges that took place in meetings in the Lower Oldpark ... where I said, your community is dying, putting in a few more housing executive houses, or even 100 more housing executive houses, isn't going to change the fact that its dying. Its dying, because you need a new dynamic, new blood, new investment, new everything, you need a transfusion of something that is more than just more subsidized housing ...[178]

Several other members of the original Girdwood Advisory panel were equally skeptical about the extent to which housing-led regeneration predicated upon maintaining the integrity of sectarian boundaries could provide a long-term basis for sustainable economic and social development.

> Social housing does not bring regeneration, if it did, half of Belfast would be regenerated. We had the best track record in social housing in Western Europe ... Did it regenerate anything? No, it just gave poor people better houses to live

> in, it didn't get them jobs, it didn't bring money into those communities, it just brought better environments to the houses that they lived in ... the economics of regeneration are that you have to ... create wealth in those communities[179]

An examination of the data on multiple deprivation across Belfast illustrates that "interface areas" are among the most deprived areas of the city and that the most deprived PUL communities in Belfast are those immediately adjacent to poorer Catholic communities. What emerges from the data is a pattern in which the poorest sections of both communities, in areas which have experienced the highest levels of violence, are competing for limited resources in a zero-sum game. One interviewee describes this phenomenon in the context of the New Lodge and the Protestant community immediately adjacent in the Tigers Bay areas.

> What we're actually seeing is the decay of places like Tigers Bay into essentially derelict zones and everybody who gets options is moving to somewhere where that decay is not the everyday reality they have to face. So, what you face is you've the New Lodge sitting there full, looking for additional resources and you have Tigers Bay empty and kind of fighting a rear-guard action which is ultimately like a bath with the plug out as water is going in and the whole scenario is that everybody is poor.[180]

Outside of the Unionist community, there is a view that the key to regenerating areas like inner-city North Belfast lies not with preserving territorial boundaries but by removing barriers and using mixed tenure residential developments and other economic anchors to serve as a generator for economic development.

> There comes a point when there's too many old people, there's too many drug dealers and you need to get all those people out of that area, you need to flatten it, turn it into a car park, and re-house those people in mixed communities that are on the upward trajectory.[181]

Political imperatives not only continue to shape urban planning policy but also highlight two very different frameworks for promoting economic regeneration. For the Protestant community, the retention of territorial boundaries is a sine qua non for any new redevelopment program in order to ensure that the territorial identity of these areas remains the same, while outside that community territorial boundaries are viewed as the problem not the solution.

4.3.4 Inequality, Defensive Design, and the Politics of Territory

An important theme of the original Girdwood Masterplan was economic "sustainability," and fundamental to this was the provision of new housing with the Plan claiming that mixed tenure quality residential development would respond to housing need in the local and wider community and provide a sustainable framework within which communities could become established and grow. The original Masterplan conceded that the role of design was crucial in creating a sense of security and in that context ensuring natural surveillance and activity throughout the day and evening by providing a mixture of uses and tenures would make streets and spaces feel safer. Quality design was also considered essential for ensuring the sustainability of the development and in this context the Plan stated that new public spaces and routes should be attractive, safe, uncluttered, and work effectively. For this reason, a coherent and clearly defined urban structure was proposed characterized by a network of interconnected routes defining blocks of housing, open spaces, and other uses. Legible, easy-to-understand places providing recognizable routes, intersections, and landmarks were an essential element of the design strategy.

Notwithstanding the very welcome and necessary improvements made to the streetscape and new housing on Cliftonpark Avenue and Manor Street the actual redevelopment that took place on the Girdwood site and the immediate surrounding area is the antithesis of that envisaged in the original Masterplan which aimed to create a vibrant, inclusive, and diverse environment "which attracts present and future generations of people to live, work and visit."[182] The actual redevelopment ensures that Catholic social housing at the southeastern end of the site and social housing for the Protestant community in Lower Oldpark beyond the northwest of the site is "buffered" by the Girdwood Community Hub in the center of the new development, as illustrated in Figure 4.8.

FIGURE 4.8 New social housing is visible on the left-hand side of the image at the southern end of the site, with the new Girdwood Community Hub visible on the right. The open land in between serves to act as a buffer zone. The social housing in this image will be largely occupied by members of the Catholic community given its location on the site. Photograph by Tim Cunningham (2018).

The original Girdwood Masterplan had situated the Community Hub between two wings of the redeveloped Gaol with mixed-use residential development on the largest area of the site. Under the revised "Conceptual Framework," the Hub was moved from the Gaol exercise yard to the original site for mixed tenure housing on Cliftonpark Avenue. The significance of the decision to change the location of the Hub, taken by DUP Minister Nelson McCausland, was not lost on one senior official who pointed out how the different location limited the number of houses that could be located on the site.

> Nelson came in and said ok, well I support this, but I want it in the middle of the site. So, you put it in the middle, and you put the access roads in, that limits the space for housing … certainly the numbers [of houses on the site] could be reduced significantly.[183]

Walking around the redeveloped Girdwood site one is struck by the sterile and fragmented feel of the design, the antithesis of the vibrant "city quarter" envisioned by the original Masterplan and unlikely to bring "maximum economic, social and environmental benefits to the local and wider community."[184]

> What I didn't want to see happening was a segmentation of the territory, what I was looking to do was to achieve something that was a truly holistic thing that was more than the sum of its parts, and it could have been a shared space, right now I don't think it will ever be a shared space, except insofar as, people from both sides of the community go to the Mater Hospital, and people on both sides might go to visit the Crumlin Road Gaol.[185]

From a defensive planning perspective the community hub and new streetscape halt any further Catholic encroachment from the New Lodge toward the Lower Oldpark and the new house build delivered under the auspices of BSC provides assurance that the integrity of Protestant territory will be respected. The design of the new facility and surrounding streets mirrors the worst aspects of desert planning that characterized urban design during the conflict, with life "built out" of the area (see Chapter 3). The original Masterplan had proposed that the Crumlin Road Gaol serve as the "anchor" for the site combined with "mixed use development" based upon the concept of "sustainable living" with "garden city blocks" providing the basis for the urban design structure to create an "authentic urban feel." Instead, a publicly financed leisure facility serves to act as a barrier between the two communities, very ably assisted by high fences, open spaces, and a road network that restricts movement from one community to the other with "no access" the order of the day. These two approaches highlight again the important role that urban design can play in either increasing connectivity and boosting the economic and social life of an area or, conversely, amputating an area of a city using defensive architecture and planning practices. The adoption of the latter approach, evident in Figures 4.9 and 4.10, illustrates how the planning system in the city continues to be steered by the imperative to keep communities apart rather than bring them together.

FIGURE 4.9 The Girdwood Community Hub viewed from the 'Protestant' end of the site at Cliftonpark Avenue. The fencing and green spaces illustrate the defensive design and how the Hub serves as a buffer between Catholic and Protestant territory. Photograph by Tim Cunningham 2018.

A striking feature of urban policy in Belfast in recent decades has been efforts on the part of Unionist politicians to repopulate traditional heartlands with Protestants from other areas of Belfast and surrouding suburbs. The former Director of Urban Regeneration at Belfast City Council recalls how demand for increased social housing has been something of a default position for Unionist politicians across the city for several decades

> The answer from Unionist politicians was always for the housing executive to build sheltered housing for elderly people … Sorry, that doesn't work economically, no shop is going to locate in an area that is full of old people, or maybe one shop will. You are not going to regenerate the area if you do that …. [186]

Research shows that pressure from Unionist politicians has been applied to pursue an approach like that contained within BSC for the past several decades. For example, the BSC Action Plan notes that in addition to new house build in Lower Oldpark, there has also been a degree of refurbishment that has included new roofs, windows and doors, insulation, kitchens and bathrooms, boilers, rewiring, plumbing, etc., as part of the wider package aimed at boosting demand for housing in the area and

FIGURE 4.10 Metal bollards at the new Girdwood Community Hub prohibit access from the Catholic end of the site to the Protestant areas on Cliftonpark Avenue. The sign indicates "no through road" which serves as a metaphor for the design of the facility. Photograph by Tim Cunningham 2018.

stabilizing the Protestant population. One interviewee, who worked previously for the NIHE, described how an identical approach was deployed to help sustain the similarly isolated Protestant community in the Suffolk area of West Belfast in the 1980s because of pressure from Unionist politicians.

> We did work in the mid-80s in Suffolk, looking at why Suffolk was a 'difficult to let' estate, and at that time they were pumping literally millions of pounds into fixing up the houses, but there was nothing wrong with the houses, the problem with Suffolk was that it was a Protestant enclave, not that it was a problem estate, and all they did was knock down houses … it cost about 90K to build a housing executive house, so the amount of waste of the public resource, that was invested in maintaining a Protestant population in that area, at huge public expense, it was just inordinate, and it was treated as a housing problem, so we went out and put in new kitchens, we put in new bathrooms, we painted the front of the houses, did environmental improvements, everybody got a new black fence … that is not what the problem was there, so you were getting almost these sort of risible debates about what the fix was, and I think that this is the uncomfortable thing …[187]

Another interviewee also recalls how new house build was used to try and maintain a Protestant presence in the north of the city during the same period.

> I can remember speaking to the regional controller of the NIHE at the time, a guy called Billy McGivern and talking to him about this, and he said to me basically, look, the Protestant population is ageing, they're moving to smaller houses, there's nobody new coming in, and what they're actually doing is their representatives are getting new houses built and they're moving around North Belfast, but they're not moving into these areas that are empty.[188]

Bollens also identified this phenomenon in his study of Belfast in the 1980s recommending that a strategy be developed that "identifies and prioritizes Protestant neighbourhoods in terms of their potential for regeneration ..." and proposed that "difficult decisions be made regarding which neighbourhoods should be cut off from publicly funded life support"[189]. For Bollens, the solution lay with creating a consolidated, more viable Protestant community in the city that would over time feel less threatened and could assume a greater willingness to allow some normalization of Belfast's geography to meet a portion of Catholic objective needs. This approach would require negotiation around which Protestant communities would be saved and which would be sacrificed as part of a wider strategy aimed at consolidation of territory and population. In a context in which "there will not be another Torrens" is a rallying cry and where every inch of territory is replete with meaning it is unsurprising that this suggestion failed to gain traction. There is an important point however that the current challenges facing the Protestant community in Lower Oldpark reflect similar difficulties that have arisen across the city in the past several decades as the demographic balance of the two communities has shifted.

From this perspective, the Conceptual Framework approach adopted for this area of North Belfast is less of an innovative approach to addressing the legacy of deprivation and dereliction through housing-led regeneration but rather a repackaged attempt to "hold the line." Undoubtedly, the biggest challenge facing this "not an inch" approach is the demographics of the two communities themselves – it is clear that despite the efforts of Unionist politicians, and the presence of walls and barriers, housing need ensures that Catholic advances in the city continue.

> We got notification through a letter to the Department saying we're going to take down a peace wall, this is a great victory ... the young people within the area did the survey and they've agreed now that it should come down. But the reason that this wall was coming down was because there were no longer Protestants who lived on the other side it ... the demographic of the area had changed so much that the interface wall, was in fact, between two CRN communities, and therefore could come down.[190]

Notwithstanding the concerns among representatives of the Protestant communities in North Belfast about the levels of dereliction and blight that afflict their areas, there remain significant doubts about the likelihood of successfully repopulating inner city, traditionally Protestant areas by adhering to territorial boundaries when the populations that shaped those original boundaries have long since moved elsewhere. There is a strong view across a range of stakeholders that the population movements that have taken place over the past 30 years are inexorable and the level of choices available elsewhere renders efforts to rebuild working-class Protestant communities a futile exercise.

> The fact of the matter is, Unionists are leaving North Belfast, and they have been leaving in droves, this is a common result of what happens in big cities. It's the Northern Ireland version of white flight and it's precisely what happened in Chicago.[191]

At the same time, one must concede that in the short term, the Lower Oldpark area and interface has "gotten settled" and recognize the context within which urban regeneration policy takes shape in areas like North Belfast.

> We often feel that we get a bit of the blame for it, but we're prepared to work with communities, and are doing, in many locations and in places to try to break down the barriers, but we can't just simply say, we're going to make this land available, we're going to rip down the wall between Cliftonville and Lower Oldpark, we're going to build houses in there, and Catholics will apply for them. World War III would break out if we did that.[192]

The new house builds on Manor Street, which provided assurance to the community in the Lower Oldpark that they will not experience "another Torrens," also needs to be considered within the context of the last occasion that this community felt under threat at the advancement of Catholic housing onto Protestant territory.

> In the summer of 1986, there was massive rioting in Manor Street to prevent Catholics moving into the new houses, gunfire and all the rest of it … So there's absolutely massive rioting, to try to push the Catholics out of these houses, in this newly built street … and, in the end the houses had to be demolished. And that was the last real attempt by the Housing Executive to push beyond that line … and any attempts to get round that, or into Girdwood, have failed, and that's 25, 24 years ago now.[193]

The redevelopment that took place on the Girdwood site, although quite different to the "visionary" approach contained within the original Masterplan, compares more favorably when viewed in the context of the violence that Cliftonpark Avenue and Manor Street experienced during the 1970s and 1980s. In a city where housing is still "a hand grenade" and where public housing officials feel that the removal of walls around the Lower Oldpark could lead to "World War III breaking out"[194], planners, developers, and indeed everyone concerned needs to exercise caution before condemning locally agreed political compromises. Some of those most impacted by an urban policy dictated by adherence to traditional boundaries are of course the local

Catholic communities who experience longer waiting lists for much-needed housing. Representatives of that community recognize that urban planning, as with every other aspect of life in North Belfast is dictated primarily by political compromise.

> My whole instinct would be … to go up and say, look, take the walls down, there's a need for houses, put them in, etc. etc. The difficulty here is that you could end up by default, …. and I think this is what most people are afraid of, you could nearly re-create deep division on interfaces, and I suppose, I'm quite wary of doing that. … objective need says one thing, tactics nearly say another.[195]

There is no doubt that the Unionist community in the Lower Oldpark face very real and legitimate concerns, not least of which is a sense that public agencies offer their community "a dismal vision" and are content to "let their areas die." There is also little doubt that the "Conceptual Framework" redevelopment has had an immediate effect on reducing tensions and violence in the area and has gone some way to taking the heat out of a toxic situation that developed around the original Girdwood Masterplan.

In many respects, the approach to housing-led regeneration at the heart of the BSC strategy mirrors similar efforts over the past number of decades to provide life support to isolated and under-populated working-class Protestant communities in the city. Ultimately, Unionist rejection of the original Girdwood Masterplan and the political imperative to maintain territorial boundaries played an important role in the loss of a substantial private sector investment in an area of the city almost wholly dependent upon public resources. The price of containing the Catholic community within the New Lodge is a further consolidation of the socio-spatial segregation of the north end of the city that in turn perpetuates a sectarian zero-sum game in which the default position for both communities is poverty and social deprivation. In particular, the desire to retain Protestant territory must be considered in the context of how it impacts to overcrowding within the Catholic community in the city.

> Ultimately, there will be an explosion, I mean there are people who are desperate for housing, there's huge overcrowding in the New Lodge, Ardoyne and the Unity area …[196]

At the same time, patterns of housing inequality in the Catholic community need also to be considered in the wider context in which recent data has highlighted the extent to which the Catholic population has expanded across Belfast.[197] These population movements have been driven by social mobility and reflect the spatial impact of Catholic progression in the labor market.[198] In many respects, these population movements are similar to those that took place in the 1950s and early 1960s, although they are of a much greater order given added impetus by the Protestant exodus from the city from the late 1960s onward.

Demographic imbalances in the city have been further fueled by the fair employment reforms of the late 1980s which "cut with the grain" of a restructuring within the Northern Ireland labor market and the rise of a new knowledge economy. As rope makers and welders gave way to management consultants, the process of deindustrialization and the employment equality reforms of the late 1980s facilitated the large number of young Catholic graduates entering the labor market at the same time that a large proportion of older Protestant males were exiting through retirement.[199] One consequence has been that a large proportion of Catholic first-time buyers seeking a first step on the property ladder have found a willing number of older retired Protestants who were happy to avail of a boom in house prices and move further east in the city or to surrounding suburbs. The fact that Catholics were required to choose selectively in what are considered "safe" areas meant that Protestants living in areas like the Cliftonville Road in North Belfast and the Stranmillis, Lisburn Road, Ormeau, and Ravenhill Roads in the south and east of the city were able to secure a premium price for their property.

A cursory glance at the website of any property broker will reveal the degree to which Catholic homeowners in Belfast are required to pay an extra premium to live in "safe" territory. Several decades on however, the notion of what is considered "safe" territory has changed significantly with many Catholics now living in former Protestant strongholds in east Belfast as well as the "mixed" suburbs around the southeast and northwest. In some cases this has given rise to "gated communities" as private developers have sought to use urban design to further reassure prospective Catholic buyers. In many other cases Catholics simply purchase old red-brick properties and "keep a low profile." The rise in private apartment buildings has also provided an opportunity for a growing Catholic middle class to secure a foothold on the property ladder. Many of these new developments are in the city center with some, notably the developments on Sandy Row, in firmly Unionist territory. Defensive design features including gates and buzzer entry systems have ensured that these developments have remained largely impervious to sectarian tensions although there have been some notable incidents including controversy that arose when a resident of one private sector development had an Irish tricolor in a window that was visible from the street.

On occasion, these population movements precipitate more serious controversy, particularly when a Catholic presence into traditional Protestant territory is deemed to be a step too farr precipitating a violent backlash in the form of a forced eviction. In September 2017, Catholic residents were ordered out of a "Shared" housing development in Cantrell Close on the Ravenhill Road in east Belfast by Protestant paramilitaries.[200] In fact, from April 2016 until March 2017, a total of 477 people reported that paramilitaries had forced them from their homes, illustrating the extent to which the vision enshrined in the Good Friday/Belfast Agreement of "freedom from sectarian harassment" remains an aspiration rather than a reality for many people in the city. Although regrettable, the reality is that in a context in which the Protestant population of Belfast continues to decline and the Catholic population continues to grow, with the two communities reaching almost parity within the city, further tensions and instances of intimidation are inevitable. The arrival of ethnic minority communities including immigrants, into traditional Protestant territory to take advantage of the surplus housing in these areas is also likely to lead to tensions. It would be unsurprising if communities that wish to retain boundaries in order to prevent incursion by Catholics are comfortable in the long term accepting the arrival of new residents from outside Northern Ireland. This is not to suggest that Protestants are any more racist than Catholics but rather that surplus land and housing in Protestant areas, combined with increasing social economic deprivation, strained public services, a political leadership that wants to repopulate these areas exclusively by members of their own community, and paramilitary gangs that have a history of engaging in intimidation of outsiders provides a toxic recipe.

Another way in which these tensions continue to play out, particularly in the north of the city, is with increased demand for security barriers aka "Peace Lines." A study by Jarman revealed that during the 1990s, Loyalists and Unionists were more likely to argue in favor of peace walls compared to their Republican and Nationalist neighbors in an effort to solidify territorial boundaries.[201] In some instances, new peace lines were erected in the form of fences and barriers; in others, such as Girdwood, architects and planners ensure that the problem is "designed away" through strategic location of buildings and roads. The difficulty with this "finger in the dike" approach is that there are holes in the dam across the city, and the "tipping point" has long since passed when it would have been possible to keep Belfast a "Protestant" city. An important question therefore is whether the demographic transformation that Belfast is currently experiencing will pass relatively peacefully or will serve to reignite into much wider conflict.

4.3.5 Closed Systems, Core Values, and Conflict Transformation

In drawing conclusions about the architecture and planning processes shaping contemporary Belfast one might conclude that the city of the new millennium appears on the surface at least to be replicating some of the mistakes made at the tail end of the last one. The Girdwood Masterplan, despite being subject to extensive consultation, which included a range of public meetings and discussions, formal consultation exercises, including a full Equality Impact Assessment, ultimately fell victim to zero-sum identity politics. The vision of a new "city quarter" on the site of the former Gaol and army barracks gave way instead to a sterile and segregated mirror image of the kind of defensive structures that have dominated the city for the last four decades with a publicly financed leisure center built and designed to serve as a buffer halting further progression of the Catholic population into Protestant territory. One might also question whether an approach to post-conflict reconstruction of the city that directs benefits away from those areas that suffered most during the conflict and is designed to contain the growth of the growing Catholic community within discrete boundaries will allow a new generation of "conflict entrepreneurs" to precipitate a return to large-scale politically motivated violence.

Based on this case study the framework provided for in the Northern Ireland Act might be seen to have done little to reconfigure the planning system in such a way that emancipatory, rather than repressive aspects prevail. In autopoietic terms, the targeted subsystems in question have merely been "irritated" by the procedural requirements of the wider regulatory framework designed to promote equality rather than being "reconfigured." The net outcome is a city ecosystem in which redevelopment and social and economic gains are steered away from those who need it most.

This however is only part of the story. In a context in which political representation and the population of the city are moving closer toward parity between the two communities, the compromise at the heart of the Conceptual Framework ought more fairly to be viewed as a political compromise designed to prevent a return to conflict. With the inexorable growth of the Catholic population across the city political representatives of both communities are aware of the capacity of territorial tensions to precipitate violence. The Conceptual Framework is imperfect but illustrates well Ignatieff's point that in some cases it may be the case that the only realistic way for diverse populations to live together *is* to live side by side.[202]

The cross-community consociational arrangements at the heart of the Belfast Agreement have been subject to a significant degree of criticism, not least because these arrangements, it has been argued, entrench sectarianism and group divisions.[203] Part of this argument is predicated upon the notion that political frameworks that are based upon communal identity encourage and facilitate separation and segregation more generally. The evidence from this case study suggests in fact that the opposite is the case, namely, that the consociational arrangements of the Northern Ireland Act merely give de jure and legal recognition to the de facto political reality of a divided community. In other words, the legal system merely gave formal recognition to the need for cross-community political arrangements in the Northern Ireland Act of 1998 notwithstanding that the political system has operated on cross-community agreement for ensuring "horizontal governance" independently and for much longer. The primary driver for cross-community consensus within the contemporary political system is the need to focus on building and maintaining relationships between the two communities to prevent a return to violence. North Belfast is an area of the city in which problematic community relations have, since the 1970s, resulted in gun battles, rioting and civil disorder, with entire streets burnt to the ground as part of a campaign of sectarian intimidation. It ought not be a surprise therefore if politicians and many others wish to proceed with caution. To date, politicians in Northern Ireland have shown remarkable dexterity in reaching compromise on a range of difficult issues, from the redevelopment of Girdwood to the decommissioning IRA arms.

The latter process is instructive in illustrating, like the debates around the Conceptual Framework, that no manual or prescriptive formula exists for how best to deal with conflict legacy issues. The decommissioning of IRA arms took place in a secret location, with no documentary evidence provided to indicate exactly how much weaponry was put beyond use. There is widespread acceptance that the event happened, and the fact of its occurrence has been confirmed by a small number of hand-picked witnesses who are sworn to secrecy but who have sufficient credibility on all sides that their word has been good enough to convince the most skeptical critics. The agreement of the IRA to put their weapons beyond use was however predicated on a lack of detail for public consumption about the exact mechanisms involved - much like the Conceptual Framework.

In a similar vein, when asked about the negotiations that were taking place around an alternative to the original Girdwood Masterplan, one politician responded,

> North Belfast, has been if you like the microcosm of the overall conflict, the violence went on here … five years longer at higher intensity …. after 1998, the Good Friday Agreement … this [area] has the most peace walls … and, in certain ways, if you crack North Belfast, you can crack it anywhere. And also, some of the biggest steps have actually been taken here, quietly, not always quietly, but, to a great extent a lot of quiet work, there's a lot of, I think community leadership, on the ground as well, so there's a willingness there, now, to get it sorted …[204]

The quiet and "below the radar" nature of the work that Gerry Kelly refers to needs to be considered within the context of the paradoxes that are at the heart of the peace process and the compromises that are necessary to achieve a successful transition from conflict in a part of city where identity and territory are inextricably linked and are woven into the fabric of the built environment.

Efforts to adopt a more holistic approach to conflict transformation that takes account of the social and economic legacies of conflict need to recognize the volatile relationship that exists between territory, identity, and inequality in places like North Belfast. Ignatieff argues that the compromises that lie at the heart of these deliberative processes will at times require activists and scholars to acknowledge that they have no standing on the question of "whose place is this?"[205] The question of "whose place" assumes even greater salience in a context in which, to paraphrase A.T.Q. Steward, locals carry the religious geography of the city in their minds, almost from birth.[206]

Ignatieff's conclusions reflect the findings of this case study where "the less than perfect political processes" that have prevented a return to violence have indeed delivered solutions that fall outside conventional boundaries of "best practice." That these solutions also fall outside the boundaries of conventional planning, architecture, and economic development theory makes them more questionable. Ignatieff goes on to argue that "outsiders …" need to recognize "… the dangers that lurk in our own certainties and convictions and avoid them as best we can". …[207] His argument here mirrors that of Turner about the need to eschew certainty and embrace the undecidable when it comes to resolving conflicts.[208] Respect for "the local" in the Belfast context means acceptance of a de facto approach to planning that allows for continued division and segregation of the two communities in some areas of the city and allows for some isolated Protestant communities to remain on "life support" through public funding in order to help manage the wider demographic changes taking place across the city. If the lesson of the Girdwood Masterplan and the Conceptual Framework is the need for compromise and acceptance of the less-than-perfect solutions that grassroots deliberative frameworks deliver up, then the question arises as to what this means for conflict transformation and peacemaking more generally.

The answer would appear to be that a more holistic approach to conflict transformation that is focused on structural and systemic inequalities requires a more reflexive approach that embraces deliberative political, planning, and architectural processes. In practice, this means scholars and practitioners providing expertise, sharing and presenting their research findings in discussions and debates around the future redevelopment of the city but ultimately accepting that the solutions that are arrived at might be less than optimal from their own point of view. In other words, an approach that sets aside the

> utopianism that concentrates on describing (desirable) end-states and on pointing out how far away we are from attaining those end states, but that disengages from the task of providing answers to the question about how we get from here to there.[209]

Stewart offers prescient advice to those who do seek to offer their expertise in a situation where "nuance is all-important":

> The function of wise constitutions and just reforms is to help humanity achieve a future that is better than the past, but if they are not to have the opposite effect, they must take account of the grain, and not cut against it.[210]

The views of Ignatieff and Stewart, writing almost four decades apart, mirror the findings of this research namely the dangers of failing to recognize the importance of "the local." They also fit with the notion, inherent within systems theory, that "if an 'alternative' is to *compete* with an established system it must be presented in terms which cannot be disregarded as totally foreign, permanently outside and of no concern to the system itself."[211] The key lesson for scholars and practitioners from this study is, to paraphrase DeGrieff, that they ought to try and provide informed suggestions about how we get from where we are, to where we want to be, but also accept less-than-perfect *outcomes* that arrive from genuinely deliberative and grassroots processes. The findings here indicate that the main threat to political stability lies not with a failure to address the socio-economic legacy of the Northern Ireland conflict but rather to attempt to do so in a way that fails to take account of the subtle ties between identity and territory that have the capacity to reignite violence.

NOTES

1 The 'Holy Cross' incident was one of the most controversial in the recent history in Northern Ireland when a group of Protestants staged a long-running protest outside a Catholic girl's primary school. The protest was covered by both national and international media and resulted in several legal challenges relating to the way in which the protests were policed. The European Court of Human Rights ultimately rejected a claim that the police had failed to protect the children https://www.bailii.org/eu/cases/ECHR/2010/2015.html although several of the families did receive damages in a civil case relating to the failure of the state to project the rights of the children.

2 The Rev. Dunlop was assisted in his task by Mr. Roy Adams, Chief Executive Building Design Partnership (BDP) and Monsignor Tom Toner, a local Catholic priest.

3 Dunlop, J., Adams, R., and Toner, T. *North Belfast Community Action Project: Report of the Project Team*, (Department for Social Development 2002) p. 51.

4 Dunlop, J., Adams, R., and Toner, T. *North Belfast Community Action Project: Report of the Project Team*, (Department for Social Development 2002) p. 47.

5 Interview: Maurice Johnston, Area Manager for Belfast, Northern Ireland Housing Executive (2011).
6 Dunlop, J., Adams, R., and Toner, T. *North Belfast Community Action Project: Report of the Project Team*. (Department for Social Development 2002) p. 51.
7 Interview: Rev. Norman Hamilton, Protestant Clergyman, Former Moderator of the Presbyterian Church (June 2010-June 2011). (2011). Green territory refers to land occupied by the Catholic community, and orange territory is land occupied by the Protestant community.
8 Northern Ireland Housing Executive, *The North Belfast Housing Strategy: Tackling Housing Need* (NIHE 2000).
9 Northern Ireland Housing Executive, *The North Belfast Housing Strategy: Tackling Housing Need* (NIHE 2000). These figures confirm the long-standing differential level of housing need facing the Catholic community outlined in Chapter 3. The continuation of these patterns is evidenced in a recent article, see Rory Winters, 'Close to 100% of social housing need in North Belfast is concentrated in predominantly Catholic neighbourhoods', *The Detail* (Belfast, 29 February 2020).
10 Northern Ireland Housing Executive, *The North Belfast Housing Strategy: Tackling Housing Need* (NIHE 2000).
11 Interview, Emma Little Pengelly, currently DUP Deputy First Minister of Northern Ireland. At the time of interview she was a Political Adviser to the First Minister Peter Robinson (DUP) in the Office of the First Minister and Deputy First Minister. (2011).
12 Northern Ireland Housing Executive, *The North Belfast Housing Strategy: Tackling Housing Need* (NIHE 2000) 21.
13 Interview: Maurice Johnson, Area Manager for Belfast, Northern Ireland Housing Executive (2011).
14 Dunlop, J., Adams, R., and Toner, T. *North Belfast Community Action Project: Report of the Project Team,* (Department for Social Development 2002) p. 51.
15 Dunlop, J., Adams, R., and Toner, T. *North Belfast Community Action Project: Report of the Project Team*, (Department for Social Development 2002) p. 54.
16 BDP. *Crumlin Road Gaol and Girdwood Barracks Draft Masterplan* (BDP 2007) p. 8.
17 At that time the Stormont Assembly was suspended and Northern Ireland was administered by 'Direct Rule' from Westminster. David Hanson, a member of Tony Blair's Labour Government, was Minister for Social Development with responsibilities for this area.
18 BDP. *Crumlin Road Gaol and Girdwood Barracks Draft Masterplan* (BDP 2007) p. 8.
19 BDP, 'Crumlin Road Gaol and Girdwood Barracks Draft Masterplan' (BDP 2007) 8.
20 BDP. *Crumlin Road Gaol and Girdwood Barracks Draft Masterplan.* (BDP 2007) p. 4.
21 BDP. *Crumlin Road Gaol and Girdwood Barracks Draft Masterplan.* (BDP 2007) p. 30.
22 Interview: Norman Hamilton, Protestant Clergyman and former Moderator of the Presbyterian Church (2011).
23 BDP. *Crumlin Road Gaol and Girdwood Barracks Draft Masterplan* (BDP 2007) p. 4.
24 BDP. *Crumlin Road Gaol and Girdwood Barracks Draft Masterplan* (BDP 2007) p. 4.
25 Interview: Roy Adams, Chief Executive of Building Design Partnership at time of interview, lead author of the Crumlin Road Gaol/Girdwood Masterplan and Chair of the Crumlin Road Gaol/Girdwood Park Advisory Panel. (2011).
26 Interview: Norman Hamilton, Protestant Clergyman and former Moderator of the Presbyterian Church (2011).
27 Interview: Kyle Alexander, Director Maze/Long Kesh Redevelopment Project at the time of interview. Former member of the Crumlin Road Gaol/Girdwood Park Advisory Panel. (2011).
28 Interview: Prof. Frank Gaffikin, School of Natural and Built Environment, Queen's University Belfast. (2011).

29 Interview: Tim Losty, at the time of interview Director of the North Belfast Community Action Unit, Office of the First Minister and Deputy First Minister, and the public official with lead responsibility for taking forward the Crumlin Road Gaol/Girdwood Masterplan. (2011).
30 Interview: Lee Reynolds, Director of Strategy Democratic Unionist Party at the time of interview. He was subsequently appointed Special Advisor to the First Minister, Arlene Foster in November 2020. (2011).
31 Interview: Kyle Alexander, Director Maze/Long Kesh Redevelopment Project at the time of interview. Former member of the Crumlin Road Gaol/Girdwood Park Advisory Panel. (2011).
32 P Collier, *The Future of Capitalism: Facing The New Anxieties* (Harper Collins 2018).
33 Stewart, A.T.Q. *The Narrow Ground: Aspects of Ulster 1609–1969* (Blackstaff 1977) p. 182.
34 Interview: Lee Reynolds, Director of Strategy Democratic Unionist Party at the time of interview. He was subsequently appointed Special Advisor to the First Minister, Arlene Foster in November 2020. (2011).
35 Interview: Maurice Johnson, Area Manager for Belfast, Northern Ireland Housing Executive (2011).
36 BDP. *Crumlin Road Gaol and Girdwood Barracks Draft Masterplan* (BDP 2007) p. 51.
37 Interview: Norman Hamilton, Protestant Clergyman and former Moderator of the Presbyterian Church. (2011). 'Down the road' leads to the Catholic New Lodge while 'Up the road' refers to the PUL community in Lower Oldpark.
38 Jacobs, J. The Death and Life of Great American Cities. (Modern Library 1993)
39 Colin, W. *Architectural Reflections: Studies in the Philosophy and Practice of Architecture* (Manchester University Press 2000) p. 25.
40 Yiftachel, O. 'Social control, urban planning and ethno-class relations: Mizrahi Jews in Israel's 'Development Towns' (2003) 24 (2) *International Journal of Urban and Regional Research* 418–437.
41 Mihai, M. 'Architectural transitional justice? Political renewal within the scars of a violent past' (2018) 12 *International Journal of Transitional Justice* 515.
42 BDP. *Crumlin Road Gaol and Girdwood Barracks Draft Masterplan* (BDP 2007) p. 38.
43 Bollens, S. *Urban Peace-Building in Divided Societies: Belfast and Johannesburg*. (Westview Press 1999) p. 107.
44 Department for Social Development. Building Successful Communities: Lower Oldpark and Hillview Area Action Plan. (Department for Social Development 2016) p.9.
45 Interview: Lee Reynolds, Director of Strategy Democratic Unionist Party at the time of interview. He was subsequently appointed Special Advisor to the First Minister, Arlene Foster in November 2020. (2011).
46 Interview: Maurice Johnson, Area Manager for Belfast, Northern Ireland Housing Executive (2011).
47 Interview: Tim Losty, at the time of interview Director of the North Belfast Community Action Unit, Office of the First Minister and Deputy First Minister, and the public official with lead responsibility for taking forward the Crumlin Road Gaol/Girdwood Masterplan. (2011).
48 Interview: Tim Losty, at the time of interview Director of the North Belfast Community Action Unit, Office of the First Minister and Deputy First Minister, and the public official with lead responsibility for taking forward the Crumlin Road Gaol/Girdwood Masterplan. (2011).
49 Interview: Lee Reynolds, Director of Strategy Democratic Unionist Party at the time of interview. He was subsequently appointed Special Advisor to the First Minister, Arlene Foster in November 2020. (2011).

50 Interview: Lee Reynolds, Director of Strategy, Democratic Unionist Party at the time of interview. He was subsequently appointed Special Advisor to the First Minister, Arlene Foster in November 2020. (2011).
51 Teubner, G. 'Legal Irritants: Good faith in British law or how unifying law ends up in new divergencies' (1998) 61 (1) *The Modern Law Review* 11.
52 Interview: Marie-Therese McGivern, former Director of Development Belfast City Council (March 1999-October 2009) and member of the Crumlin Road Gaol/Girdwood Park Advisory Panel. Chief Executive and Principle of Belfast Metropolitan College at time of interview. (2011).
53 Murtagh, B. 'Planning for anywhere: Housing policy in Northern Ireland,' Housing studies (1998) 13 (6) 833.
54 Interview: Prof. Frank Gaffikin, School of Natural and Built Environment, Queen's University Belfast. (2011).
55 Interview: Tim Losty, at the time of interview Director of the North Belfast Community Action Unit, Office of the First Minister and Deputy First Minister, and the public official with lead responsibility for taking forward the Crumlin Road Gaol/Girdwood Masterplan. (2011)
56 Interview: Norman Hamilton, Protestant Clergyman and former Moderator of the Presbyterian Church (2011).
57 Interview: Tim Losty, at the time of interview Director of the North Belfast Community Action Unit, Office of the First Minister and Deputy First Minister, and the public official with lead responsibility for taking forward the Crumlin Road Gaol/Girdwood Masterplan. (2011).
58 Interview: Marie-Therese McGivern, former Director of Development Belfast City Council (March 1999-October 2009) and member of the Crumlin Road Gaol/Girdwood Park Advisory Panel. Chief Executive and Principle of Belfast Metropolitan College at time of interview. (2011).
59 Interview: Maurice Johnsonl, Area Manager for Belfast, Northern Ireland Housing Executive (2011).
60 BDP. *Crumlin Road Gaol and Girdwood Barracks Draft Masterplan* (BDP 2007) p. 64.
61 Interview: Prof. Duncan Morrow, Chief Executive, Community Relations Council for Northern Ireland at time of interview. Former member of the of the Crumlin Road Gaol/ Girdwood Park Advisory Panel. (2011).
62 Interview: Prof. Duncan Morrow, Chief Executive, Community Relations Council for Northern Ireland at time of interview. Former member of the of the Crumlin Road Gaol/ Girdwood Park Advisory Panel. (2011).
63 Interview: Marie-Therese McGivern, former Director of Development Belfast City Council (March 1999-October 2009) and member of the Crumlin Road Gaol/Girdwood Park Advisory Panel. Chief Executive and Principle of Belfast Metropolitan College at time of interview. (2011).
64 Interview: Maurice Johnson, Area Manager for Belfast, Northern Ireland Housing Executive (2011).
65 Interview: Prof. Duncan Morrow, Chief Executive, Community Relations Council for Northern Ireland at time of interview. Former member of the of the Crumlin Road Gaol/ Girdwood Park Advisory Panel. (2011).
66 Interview: Lee Reynolds, Director of Strategy, Democratic Unionist Party at the time of interview. He was subsequently appointed Special Advisor to the First Minister, Arlene Foster in November 2020. (2011).
67 Interview: Gerry Kelly MLA, former member of Girdwood Advisory Panel and Sinn Féin elected representative for North Belfast since 1998. He is currently a member of Sinn Féin's Ard Chomhairle (National Executive Committee). (2011).

68 BDP. *Crumlin Road Gaol and Girdwood Barracks Draft Masterplan* (BDP 2007) p. 64.
69 Interview: Maurice Johnson, Area Manager for Belfast Northern Ireland Housing Executive (2011).
70 BDP. *Crumlin Road Gaol and Girdwood Barracks Draft Masterplan* (BDP 2007) p. 65.
71 The Northern Ireland Housing Executive is required by Article 22 of the Housing (NI) Order 1981 to allocate dwellings in accordance with a scheme approved by the Department of Social Development for Northern Ireland.
72 *A Citizen's Inquiry: The Opsahl Report on Northern Ireland* (1993).
73 Boyle and Hadden argue for example that residential integration 'is not impossible with the combined support of tenants, the police and other agencies'. No mention is made how the problem is shaped by demographic imbalances and how these imbalances can be surmounted. Boyle, K. and Hadden, T. *Northern Ireland: The Choice* (Penguin 1994) p. 177.
74 Interview: Maurice Johnson, Area Manager for Belfast Northern Ireland Housing Executive (2011).
75 Interview: Gerry Kelly MLA, former member of Girdwood Advisory Panel and Sinn Féin elected representative for North Belfast since 1998. He is currently a member of Sinn Féin's Ard Chomhairle (National Executive Committee). (2011).
76 Interview: Lee Reynolds, Director of Strategy, Democratic Unionist Party at the time of interview. He was subsequently appointed Special Advisor to the First Minister, Arlene Foster in November 2020. (2011).
77 Interview: Kyle Alexander, Director Maze/Long Kesh Redevelopment Project at the time of interview. Former member of the Crumlin Road Gaol/Girdwood Park Advisory Panel. (2011).
78 Interview: Norman Hamilton, Protestant Clergyman and former Moderator of the Presbyterian Church. (2011).
79 Interview: Prof. Frank Gaffikin, School of Natural and Built Environment, Queen's University Belfast. (2011).
80 Interview: Prof. Duncan Morrow, Chief Executive, Community Relations Council for Northern Ireland at time of interview. Former member of the of the Crumlin Road Gaol/Girdwood Park Advisory Panel. (2011).
81 Interview: Brian Feeney, historian and author. Former elected representative for the Social Democratic and Labour Party (SDLP) in North Belfast.
82 Nigel Dodds eventually lost his Westminster seat to John Finucane of Sinn Fein in the general election of December 2019.
83 Interview: Emma Little Pengelly, currently DUP Deputy First Minister of Northern Ireland. At the time of interview she was a Special Adviser to the First Minister Peter Robinson (DUP) in the Office of the First Minister and Deputy First Minister. (2011).
84 Interview: Maurice Johnson, Area Manager for Belfast Northern Ireland Housing Executive (2011).
85 Interview: Gerry Kelly MLA, former member of Girdwood Advisory Panel and Sinn Féin elected representative for North Belfast since 1998. He is currently a member of Sinn Féin's Ard Chomhairle (National Executive Committee). (2011).
86 Interview: Gerry Kelly MLA, former member of Girdwood Advisory Panel and Sinn Féin elected representative for North Belfast since 1998. He is currently a member of Sinn Féin's Ard Chomhairle (National Executive Committee). (2011).
87 Interview: Tim Losty, Director of the North Belfast Community Action Unit, Office of the First Minister and Deputy First Minister, and the public official with lead responsibility for taking forward the Crumlin Road Gaol/Girdwood Masterplan. (2011).
88 Nigel Dodds lost his seat to John Finucane of Sinn Féin in the 2019 General Election. The proposed boundary changes never materialized.
89 Interview: Lee Reynolds, Director of Strategy, Democratic Unionist Party at the time of interview. He was subsequently appointed Special Advisor to the First Minister, Arlene Foster in November 2020. (2011).

90 McEldowney, O., Anderson, J., and Shuttleworth, I. Sectarian demography: Dubious discourses of ethno-national conflict. In K. Hayward and C. O'Donnell (eds.), *Political Discourse and Conflict Resolution: Debating Peace in Northern Ireland* (Routledge 2011) p. 160.
91 Interview: Jennifer Hawthorne, Director of Housing, Northern Ireland Housing Executive. (2018). See also studies that have focused for example on the efforts that have gone into maintaining the small Protestant enclave of Suffolk in the west of the city.
92 Stewart, A.T.Q. *The Narrow Ground: Aspects of Ulster 1609–1969* (Blackstaff 1977) p. 182.
93 Interview: Marie-Therese McGivern, former Director of Development Belfast City Council (March 1999-October 2009) and member of the Crumlin Road Gaol/Girdwood Park Advisory Panel. Chief Executive and Principle of Belfast Metropolitan College at time of interview. (2011).
94 Interview: Tim Losty, Director of the North Belfast Community Action Unit, Office of the First Minister and Deputy First Minister, and the public official with lead responsibility for taking forward the Crumlin Road Gaol/Girdwood Masterplan. (2011).
95 Interview: Tim Losty, Director of the North Belfast Community Action Unit, Office of the First Minister and Deputy First Minister, and the public official with lead responsibility for taking forward the Crumlin Road Gaol/Girdwood Masterplan. (2011).
96 Interview: Lee Reynolds, Director of Strategy, Democratic Unionist Party at the time of interview. He was subsequently appointed Special Advisor to the First Minister, Arlene Foster in November 2020. (2011).
97 Interview: Lee Reynolds, Director of Strategy, Democratic Unionist Party at the time of interview. He was subsequently appointed Special Advisor to the First Minister, Arlene Foster in November 2020. (2011).
98 Interview: Gerry Kelly MLA, former member of Girdwood Advisory Panel and Sinn Féin elected representative for North Belfast since 1998. He is currently a member of Sinn Féin's Ard Chomhairle (National Executive Committee). (2011).
99 Northern Ireland Act 1998, s.16.
100 McCrudden, C and O'Leary, B. *Courts and Consociations: Human Rights versus Power-Sharing (OUP 2013) p. 6.*
101 The SDLP are by no means the only party to engage in this tactic, which has become known in local political parlance as a Minister doing a 'solo run.'.
102 Wilson, R. From consociationalism to interculturalism. In R. Taylor (ed.) *Consociational Theory: McGarry and O'Leary and the Northern Ireland Conflict (Routledge 2009).*
103 Ignatieff, M. *The Ordinary Virtues: Moral Order in Divided World (Harvard University Press 2017) p. 67.*
104 Interview: Roy Adams, Chief Executive of Building Design Partnership, lead author of the Crumlin Road Gaol/Girdwood Masterplan and Chair of the Crumlin Road Gaol/Girdwood Park Advisory Panel. (2011).
105 Interview: Marie-Therese McGivern, former Director of Development Belfast City Council (March 1999-October 2009) and member of the Crumlin Road Gaol/Girdwood Park Advisory Panel. Chief Executive and Principle of Belfast Metropolitan College at time of interview. (2011).
106 Interview: Prof. Frank Gaffikin, School of Natural and Built Environment, Queen's University Belfast. (2011).
107 Interview: Kyle Alexander, Director Maze/Long Kesh Redevelopment Project at the time of interview. Former member of the Crumlin Road Gaol/Girdwood Park Advisory Panel. (2011).
108 Interview: Marie-Therese McGivern, former Director of Development Belfast City Council (March 1999-October 2009) and member of the Crumlin Road Gaol/Girdwood Park Advisory Panel. Chief Executive and Principle of Belfast Metropolitan College at time of interview. (2011).

109 Interview: Kyle Alexander, Director Maze/Long Kesh Redevelopment Project at the time of interview. Former member of the Crumlin Road Gaol/Girdwood Park Advisory Panel. (2011).

110 Interview: Gerry Kelly MLA, former member of Girdwood Advisory Panel and Sinn Féin elected representative for North Belfast since 1998. He is currently a member of Sinn Féin's Ard Chomhairle (National Executive Committee). (2011).

111 Interview: Roy Adams, Chief Executive of Building Design Partnership, lead author of the Crumlin Road Gaol/Girdwood Masterplan and Chair of the Crumlin Road Gaol/Girdwood Park Advisory Panel. (2011).

112 Interview: Roy Adams, Chief Executive of Building Design Partnership, lead author of the Crumlin Road Gaol/Girdwood Masterplan and Chair of the Crumlin Road Gaol/Girdwood Park Advisory Panel. (2011).

113 Interview: Tim Losty, Director of the North Belfast Community Action Unit, Office of the First Minister and Deputy First Minister, and the public official with lead responsibility for taking forward the Crumlin Road Gaol/Girdwood Masterplan. (2011).

114 Interview: Brian Feeney, historian and author. Former elected representative for the Social Democratic and Labour Party (SDLP) in North Belfast. (2011).

115 Ignatieff, M. The Ordinary Virtues: Moral Order in Divided World (Harvard University Press 2017) p. 136.

116 O'Leary, B. *A Treatise on Northern Ireland: Consociation and Confederation*, (Vol 3, OUP 2019) p. vii.

117 Interview: Norman Hamilton, Protestant Clergyman. Former Moderator of the Presbyterian Church (June 2010-June 2011). (2011).

118 Interview: Lee Reynolds, Director of Strategy, Democratic Unionist Party at the time of interview. He was subsequently appointed Special Advisor to the First Minister, Arlene Foster in November 2020. (2011).

119 Interview: Gerry Millar, Director of Development, Belfast City Council at time of interview. (2011).

120 Girdwood Army base proposal gets go-ahead. BBC News (21 May 2012). https://www.bbc.com/news/uk-northern-ireland-18141475.

121 Girdwood barracks development: devil in the detail? BBC News (22 May 2012). https://www.bbc.com/news/uk-northern-ireland-18162637.

122 Army Base Housing Plan Agreed, Press Release, Northern Ireland Executive May 21 2012.

123 Department for Social Development, Building Successful Communities Programme, Lower Oldpark and Hillview Area Analysis (DSD 2013) 41.

124 Interview: Kyle Alexander, Director Maze/Long Kesh Redevelopment Project at the time of interview. Former member of the Crumlin Road Gaol/Girdwood Park Advisory Panel. (2011). PUL refers to Protestant Unionist Loyalist community.

125 Interview: Lee Reynolds, Director of Strategy, Democratic Unionist Party at the time of interview. He was subsequently appointed Special Advisor to the First Minister, Arlene Foster in November 2020. (2011).

126 In the 2010 Multiple Deprivation Measure Lower Oldpark has an overall rank of 7 out of 890 Super Output Areas, Department for Social Development. *Building Successful Communities Programme, Lower Oldpark and Hillview Area Analysis* (DSD 2013) p. 15.

127 Owen, M., James, A., and Ian, S. 'Sectarian demography: Dubious discourses of ethno-national conflict'. In K. Hayward and C. O'Donnell (eds.), *Political Discourse and Conflict Resolution: Debating Peace in Northern Ireland.* (Routledge 2011) p. 160.

128 The extent of population decline in these areas due to redevelopment is illustrated by the fact that between 1971 and 1978, the total population of the Shankill area declined from 46,363 to 33,297, a fall of 13,066 people or 28.1%. This compares to an overall Belfast

average of a decline of 19%. These declines are explained mainly by the clearance of the old densely populated areas and their replacement with new housing at lower densities. In 1978, of those families who left the district, the most popular destinations were Newtownabbey, Bangor, and Antrim. Northern Ireland Housing Executive, *Draft Introduction to the Seven Districts Housing Statements* (Northern Ireland Housing Executive 1980) para 2.6 ENV/21/1/39/A.

129 Interview: Maurice Johnson, Area Manager for Belfast, Northern Ireland Housing Executive (1) (2011).

130 Interview: Maurice Johnson, Area Manager for Belfast, Northern Ireland Housing Executive (1) (2011).

131 Interview: Brian Barrington. Former Political Advisor SDLP. (2011).

132 Boal, F. 'Territoriality on the Shankill-Falls Divide, Belfast' (2008) 41 (3) *Irish Geography* 349–366.

133 Interview: Lee Reynolds, Director of Strategy Democratic Unionist Party at the time of interview. He was subsequently appointed Special Advisor to the First Minister, Arlene Foster in November 2020. (2011).

134 Government of Northern Ireland, *Future Policy on Areas of Contention, Second and Final Report of the Joint Working Party on Processions, (secret)* (Government of Northern Ireland April 1971 CAB/1634/1) p. 39.

135 Interview: Brian Feeney, historian and author. Former elected representative for the Social Democratic and Labour Party (SDLP) in North Belfast. (2011).

136 Interview: Lee Reynolds, Director of Strategy Democratic Unionist Party at the time of interview. He was subsequently appointed Special Advisor to the First Minister, Arlene Foster in November 2020. (2011).

137 Interview: Lee Reynolds, Director of Strategy Democratic Unionist Party at the time of interview. He was subsequently appointed Special Advisor to the First Minister, Arlene Foster in November 2020. (2011).

138 Interview: Lee Reynolds, Director of Strategy Democratic Unionist Party at the time of interview. He was subsequently appointed Special Advisor to the First Minister, Arlene Foster in November 2020. (2011).

139 Interview: Lee Reynolds, Director of Strategy Democratic Unionist Party at the time of interview. He was subsequently appointed Special Advisor to the First Minister, Arlene Foster in November 2020. (2011).

140 Department for Social Development, Building Successful Communities Programme, Lower Oldpark and Hillview Area Analysis (DSD 2013) 3.

141 Interview: Emma Little Pengelly, currently Deputy First Minister of Northern Ireland. At the time of the interview she was a Special Advisor to the then First Minister of Northern Ireland, Peter Robinson in the Office of the First Minister and Deputy First Minister. (2011).

142 Coyles, D., Hamber, B., and Grant, A. 'Hidden barriers and divisive architecture: The role of "everyday space" in conflict and peacebuilding in Befast'. *Journal of Urban Affairs, DOI:* 10.1080/07352166.2021.1930017.

143 Department for Social Development. *Building Successful Communities: Lower Oldpark and Hillview Action Plan.* (Department for Social Development 2016) p. 9.

144 Department for Social Development. *Building Successful Communities Programme, Lower Oldpark and Hillview Area Analysis.* (DSD 2013) p. 41.

145 Department for Social Development. *Building Successful Communities Programme, Lower Oldpark and Hillview Area Analysis.* (DSD 2013) p. 42.

146 Department for Social Development. *Building Successful Communities: Lower Oldpark and Hillview Action Plan.* (Department for Social Development 2016) p. 20.

147 Department for Social Development. *Building Successful Communities: Lower Oldpark and Hillview Action Plan.* (Department for Social Development 2016) p. 13.

148 Department for Social Development. *Building Successful Communities: Lower Oldpark and Hillview Action Plan.* (Department for Social Development 2016) p. 21.
149 Interview: Norman Hamilton, Protestant Clergyman (2011).
150 Department for Social Development. *Building Successful Communities Programme, Lower Oldpark and Hillview Area Analysis.* (DSD 2013) p. 43.
151 Department for Social Development. *Building Successful Communities Programme, Lower Oldpark and Hillview Area Analysis.* (DSD 2013) p. 44.
152 Twenty units of affordable housing on either side of Cliftonpark Avenue.
153 Department for Social Development. *Building Successful Communities: Lower Oldpark and Hillview Action Plan.* (Department for Social Development 2016) p. 22.
154 Interview: Maurice Johnson, Area Manager for Belfast, Northern Ireland Housing Executive (2011).
155 Interview: Tim Losty, Director of the North Belfast Community Action Unit, Office of the First Minister and Deputy First Minister, and the public official with lead responsibility for taking forward the Crumlin Road Gaol/Girdwood Masterplan. (2011).
156 Interview: Gerry Kelly MLA, former member of Girdwood Advisory Panel and Sinn Féin elected representative for North Belfast since 1998. He is currently a member of Sinn Féin's Ard Chomhairle (National Executive Committee). (2011).
157 Interview: Lee Reynolds, Director of Strategy, Democratic Unionist Party at the time of interview. He was subsequently appointed Special Advisor to the First Minister, Arlene Foster in November 2020. (2011).
158 Interview: Prof. Duncan Morrow, Chief Executive, Community Relations Council for Northern Ireland at time of interview. Former member of the of the Crumlin Road Gaol/Girdwood Park Advisory Panel. (2011).
159 Interview: Maurice Johnson, Area Manager for Belfast, Northern Ireland Housing Executive (2011).
160 Department for Social Development. *Building Successful Communities Programme, Lower Oldpark and Hillview Area Analysis.* (DSD 2013) p. 3.
161 Short, E. 'Housing bosses admit two Belfast multi-million regeneration schemes are in areas of "low to medium need".' (November 9, 2015). Belfastmedia.com https://belfastmediagroup.com/housing-bosses-admit-two-belfast-multi-million-regeneration-schemes-are-in-areas-of-low-to-medium-need/.
162 Interview: Jennifer Hawthorne, Director of Housing, Northern Ireland Housing Executive (2018).
163 Department for Social Development. *Building Successful Communities: Lower Oldpark and Hillview Action Plan.* (Department for Social Development 2016) p. 1.
164 Interview: Jennifer Hawthorne, Director of Housing, Northern Ireland Housing Executive (2018).
165 Three of the pilot areas were in West Belfast, Lower Shankill/Brown Square; Lenadoon/Glencolin; and Lower Falls; two were in North Belfast, Lower Oldpark/Hillview and Tiger's Bay/Mountcollyer; and Doury Road, Ballymena was the one pilot area selected outside Belfast.
166 Department for Social Development. *Building Successful Communities Programme, Lower Oldpark and Hillview Area Analysis* (DSD 2013) p. 3.
167 Interview: Jennifer Hawthorne, Director of Housing, Northern Ireland Housing Executive (2018).
168 'Girdwood development: Sixty new homes to be built on site.' BBC News (6 June 2013). https://www.bbc.com/news/uk-northern-ireland-22797984.
169 Erwin, A. 'Nelson McCausland's blocking of homes plan is sectarian, court told.' (February 17, 2013) Belfast Telegraph. https://www.belfasttelegraph.co.uk/news/northern-ireland/nelson-mccauslands-blocking-of-homes-plan-is-sectarian-court-told-29075280.html.

170 As part of the settlement a statement was read out by senior counsel on behalf of the department and minister clarifying and updating the position on implementation of a Masterplan Conceptual Framework (MCF) for the Girdwood Park site agreed by elected representatives the previous May. 'Girdwood development: Sixty new homes to be built on site.' BBC News (6 June 2013). https://www.bbc.com/news/uk-northern-ireland-22797984.
171 Interview: Lee Reynolds, Director of Strategy, Democratic Unionist Party at the time of interview. He was subsequently appointed Special Advisor to the First Minister, Arlene Foster in November 2020. (2018).
172 Manus Maguire and those working through the Lower Oldpark Interface Programme and the International Fund for Ireland were singled out for particular praise.
173 Interview: Jennifer Hawthorne, Director of Housing, Northern Ireland Housing Executive (2018).
174 Interview: Lee Reynolds, Director of Strategy Democratic Unionist Party at the time of interview. He was subsequently appointed Special Advisor to the First Minister, Arlene Foster in November 2020. (2018).
175 Interview: Jennifer Hawthorne, Director of Housing, Northern Ireland Housing Executive (2018).
176 Interview: Jennifer Hawthorne, Director of Housing, Northern Ireland Housing Executive (2018).
177 Interview: Lee Reynolds, Director of Strategy, Democratic Unionist Party at the time of interview. He was subsequently appointed Special Advisor to the First Minister, Arlene Foster in November 2020. (2018).
178 Interview: Roy Adams, Chief Executive of Building Design Partnership, lead author of the Crumlin Road Gaol/Girdwood Masterplan and Chair of the Crumlin Road Gaol/ Girdwood Park Advisory Panel. (2011).
179 Interview: Marie-Therese McGivern, former Director of Development Belfast City Council (March 1999-October 2009) and member of the Crumlin Road Gaol/Girdwood Park Advisory Panel. Chief Executive and Principle of Belfast Metropolitan College at time of interview. (2011).
180 Interview: Prof. Duncan Morrow, Chief Executive, Community Relations Council for Northern Ireland at time of interview. Former member of the of the Crumlin Road Gaol/ Girdwood Park Advisory Panel. (2011).
181 Interview: Marie-Therese McGivern, former Director of Development Belfast City Council (March 1999-October 2009) and member of the Crumlin Road Gaol/Girdwood Park Advisory Panel. Chief Executive and Principle of Belfast Metropolitan College at time of interview. (2011).
182 BDP. *Crumlin Road Gaol and Girdwood Barracks Draft Masterplan.* (BDP 2007) p. 4.
183 Interview: Tim Losty, Director of the North Belfast Community Action Unit, Office of the First Minister and Deputy First Minister, and the public official with lead responsibility for taking forward the Crumlin Road Gaol/Girdwood Masterplan. (2011).
184 BDP. *Crumlin Road Gaol and Girdwood Barracks Draft Masterplan.* (BDP 2007) p. 4.
185 Interview: Roy Adams, Chief Executive of Building Design Partnership, lead author of the Crumlin Road Gaol/Girdwood Masterplan and Chair of the Crumlin Road Gaol/ Girdwood Park Advisory Panel. (2011).
186 Interview: Marie-Therese McGivern, former Director of Development Belfast City Council (March 1999-October 2009) and member of the Crumlin Road Gaol/Girdwood Park Advisory Panel. Chief Executive and Principle of Belfast Metropolitan College at time of interview. (2011)
187 Interview: Prof. Brendan Murtagh, Prof. Brendan Murtagh, School of Natural and Built Environment, Queen's University Belfast. (2011).
188 Interview: Brian Feeney, historian and author. Former elected representative for the Social Democratic and Labour Party (SDLP) in North Belfast. (2011).

189 Bollens, S. *Urban Peace-Building in Divided Societies: Belfast and Johannesburg (Westview Press 1999) p. 148.*

190 Interview: Emma Little Pengelly, currently Deputy First Minister of Northern Ireland. At the time of the interview she was a Special Advisor to the then First Minister of Northern Ireland, Peter Robinson in the Office of the First Minister and Deputy First Minister. (2011).

191 Interview: Brian Feeney, historian and author. Former elected representative for the Social Democratic and Labour Party (SDLP) in North Belfast. (2011).

192 Interview: Maurice Johnson, Area Manager for Belfast, Northern Ireland Housing Executive (1) (2011).

193 Interview: Brian Feeney, historian and author. Former elected representative for the Social Democratic and Labour Party (SDLP) in North Belfast. (2011).

194 Interview: Maurice Johnson, Area Manager for Belfast Northern Ireland Housing Executive (2011).

195 Interview: Gerry Kelly MLA, former member of Girdwood Advisory Panel and Sinn Féin elected representative for North Belfast since 1998. He is currently a member of Sinn Féin's Ard Chomhairle (National Executive Committee). (2011).

196 Interview: Brian Feeney, historian and author. Former elected representative for the Social Democratic and Labour Party (SDLP) in North Belfast. (2011).

197 Nolan, P. *Northern Ireland Peace Monitoring Report Number Three* (Community Relations Council 2014) pp. 115–117.

198 Shuttleworth, I., Barr, P., and Gould, M., 'Does internal migration in Northern Ireland increase religious and social segregation? Perspectives from the Northern Ireland Longitudinal Study (NILS) 2001–2007' (2013) 19 *Population, Space and Place 84.*

199 Osborne, B. and Shuttleworth, I. *Fair Employment in Northern Ireland: A Generation On (Blackstaff 2004).*

200 Hughes, B. 'Catholics ordered out of Belfast shared housing in sectarian threat.' (September 28, 2017) The Irish News. https://www.irishnews.com/news/northernirelandnews/2017/09/28/news/catholics-ordered-out-of-belfast-shared-housing-in-sectarian-threat-1148290/.

201 Jarman, N. *Towards Sustainable Security: Interface Barriers and the Legacy of Segregation in Belfast* (Community Relations Council 2008) cited in Byrne, J. (2011) *The Belfast Peace Walls: The Problems, Policies, and Politics of the Troubles Architecture* (unpublished DPhil Thesis, University of Ulster).

202 Ignatieff, M. *The Ordinary Virtues: Moral Order in Divided World* (Harvard University Press 2017) p. 45.

203 See, for example, Wilson, R. From consociationalism to interculturalism. In R Taylor (ed.), *Consociational Theory: McGarry and O'Leary and the Northern Ireland Conflict* (Routledge 2009); see also O'Leary, B. Debating consociational politics: Normative and explanatory arguments. In S. Noel (ed.) *From Power-Sharing to Democracy: Post-Conflict Institutions in Ethnically Divided Societies* (McGill-Queen's University Press 2005); Anderson, J. and Shuttleworth, I. 'Sectarian Demography, territoriality and political development in Northern Ireland' (1998) 17 (2) *Political Geography 187.*

204 Interview: Gerry Kelly MLA, former member of Girdwood Advisory Panel and Sinn Féin elected representative for North Belfast since 1998. He is currently a member of Sinn Féin's Ard Chomhairle (National Executive Committee).

205 Ignatieff, M. *The Ordinary Virtues: Moral Order in Divided World.* (Harvard University Press 2017) p. 136.

206 Stewart, A.T.Q. *The Narrow Ground: Aspects of Ulster 1609–1969.* (Blackstaff 1977) p. 182.

207 Ignatieff, M *The Ordinary Virtues: Moral Order in Divided World.* (Harvard University Press 2017) p. 137.

208 Turner, C. *Violence, Law, and the Impossibility of Transitional Justice* (Routledge 2016) pp. 165–173.
209 De Greiff, P., 'The future of the past: Reflections on the present state and prospects of transitional justice' (2020) 14 *International Journal of Transitional Justice* 258.
210 Stewart, A.T.Q. *The Narrow Ground: Aspects of Ulster 1609–1969* (Blackstaff 1977) p. 185.
211 Carson, W. (1981) *The Other Price of Britain's Oil: Safety and Control in the North Sea,* (Martin Robertson 1981) p. 301.

5 The Legacy of Coercive Architecture and Planning in the U.S.

5.1 SUBSIDIZING SEGREGATION AND THE SUBURBAN DREAM

As outlined in Chapter 1, the utopian promise that lay at the heart of American suburbia was always, at most, "a partial paradise" predicated upon the principle of exclusion.[1] This quintessentially U.S. vision, captured in popular culture by a host of comedy shows such as "The Brady Bunch" or "The Wonder Years," was in fact shaped and influenced by colonial practices in South Asia that "provided a perfect basis for the construction of a modern bourgeois identity".[2] As outlined in Chapter 2, the HOLC area descriptions of the 1930s illustrate the key components that shaped metropolitan differentiation in the U.S. through the twentieth century, notably the deployment of legal and architectural/design instruments to preserve racial and social homogeneity.

This section explores how these techniques came to shape patterns of settlement in the decades following World War II. By and large the postwar years followed the model laid down by the earlier HOLC assessors in the 1930s with one important and notable exception. The pre-war area descriptions were predicted upon a racial binary that privileged White Anglo-Saxon Protestants from other racial and immigrant groups. As a result, African Americans, Jews, Irish, Italian, Polish, and other ethnic groups were treated as a homogenous and equally undesirable presence. Following World War II this binary shifted and immigrant groups came to be "Whitened," allowing them access to the burgeoning suburbs across the nation.[3] Notwithstanding their contributions to the war effort African Americans remained beyond the pale and continued to be regarded as an existential threat to the survival of the suburban dream. Of crucial importance in securing racial exclusivity across the country was the Federal Housing Administration (FHA). The organization played a crucial role in expanding postwar homeownership in the U.S. From 1934 to 1972, the percentage of American families living in owner-occupied dwellings rose from 44% to 63%.

During that period, the middle-class suburban family living in their own home courtesy of a long-term, fixed-rate mortgage, became synonymous with the American Dream. This lifestyle was largely bankrolled by the FHA which revolutionized the postwar real estate market by insuring and acting as guarantor for mortgages made by private lenders for home construction.[4] For the first time in U.S. history the favorable lending terms underwritten by the U.S. Treasury made it cheaper to buy a home than to rent one.

The FHA Underwriting Manual, as outlined in Chapter 2, was the bedrock upon which U.S. cities and suburbs were financed, designed, populated, and segregated. Predicated upon the view that any "change in social or racial occupancy generally leads to instability and a reduction in values"[5] racial and social exclusion were

DOI: 10.1201/9781003208129-5

inextricably linked to economic value. This principle underpinned access to mortgages, street design, location, and transportation to the new suburban developments in the decades immediately following World War II as architects, planners, real estate developers and civic leaders developed strategies for sifting and sorting communities by race and class that has had a toxic and lasting impact on social connection and civic engagement across the U.S. Some of the more explicit aspects of racial bias inherent in FHA practices were formally outlawed by the Fair Housing Act 1968 This chapter will show that many of the more exclusionary aspects of suburbia have morphed into other economic and design practices that continue to shape residential patterns in the country with huge consequences for the political, social, economic, and environmental health of the nation.

From its inception, only segregated subdivisions "enforced by deed restrictions, and sometimes separated by walls from neighborhoods where people of color resided, could qualify for FHA support."[6] The FHA not only refused to guarantee mortgages for African Americans but also to whites who might lease or subsequently sell property to African Americans regardless of the applicant's creditworthiness.[7] Some developers and real estate professionals either outright rejected the "infiltration" theory upon which this dubious practice was based or sought to profit from it. Mohl for example notes that a considerable body of real estate literature from the early decades of the twentieth century argued that African Americans made reliable mortgage customers, and that residential property values rose after Blacks moved into transitional neighborhoods.[8] He also highlights how the postwar economic boom created significant employment opportunities for Americans of all races. In Miami, for example, the postwar service and tourist economy provided steady employment opportunities for Black workers, giving Black families in the city the financial resources to purchase homes.[9] It is also worth noting that given widespread racial prejudice African Americans were often of a higher social class than the white residents of the neighborhoods that they wished to move into. Left to their own devices, and without government intervention, it is likely that those developers and lenders who wished to supply African Americans with the housing they needed would have done so and those who supported an alternative vision would have followed exclusionary practices. The crucial tipping point at this moment in history was that the federal government did not remain neutral but took an explicitly segregationist line and embraced the views of the most extreme and prejudiced sections of the marketplace.

The purpose of constitutional guarantees of equality is to uphold the rights and freedoms of all citizens, including those who would otherwise be treated unfairly, especially racial and religious minorities. In this case, the government acted contrary to its own laws and values providing official justification for prejudice, lawlessness, and racial violence. The majority of those who purchased suburban homes in the postwar period were first-time buyers who gained a significant financial asset that they could pass on to their children, use as collateral, or sell, using the equity that was accumulated to invest in another bigger and more desirable property. This hugely increased opportunities for wealth creation among working and lower middle-class Americans who historically had not been able to gain entry to the property market. In a context in which official government policy stipulated that mortgage guarantees, which underpinned this entire financial edifice were contingent upon racial exclusion the arrival of *any* African American presence in a suburban development was engineered as an economic threat.

In other words, official government policy provided a powerful financial incentive for white homeowners to police the racial composition of their neighborhoods and encouraged them to refuse to sell their homes to African Americans or veto proposals for integrated developments. This resulted in many local municipalities using their powers to issue building permits and other measures to prevent developers constructing residential units if they believed that they would be inhabited by African Americans. This also encouraged and conferred legitimacy upon those who perpetrated extreme acts of harassment and intimidation upon those African Americans who were keen to avail of the suburban lifestyle. Across the country, many of those African American families who sought the suburban dream found themselves firebombed, shot at, and subject to all manner of racial violence. In many cases local municipalities and law enforcement officials willfully ignored or actively participated in these actions. Sometimes victims of racial violence were arrested and charged with instigating the trouble in the first place. In other instances those who sold property to African Americans in white neighborhoods found themselves before the courts. Racial violence was further legitimized by measures such as racial zoning ordinances and racially restrictive covenants that provided additional legal barriers to African American entry to suburbia and meant that firebombers, cross-burners, and stone throwers, could claim they were upholding local laws by promoting segregation.

This policy had a huge impact in shaping not just individual properties but entire suburbs as racially exclusive white enclaves. Perhaps the most notorious example is Levittown in Pennsylvania, a massive development of 17,500 mass-produced two-bedroom homes financed by the FHA on condition that it be all white with a variety of measures in place to ensure "protection from adverse influences." That this happened in the decades after World War II, when many of the African Americans who were denied access to mortgages were either returning servicemen who had risked their lives for the country, or who had worked in war industries and munitions factories, was especially egregious. One contemporary observer noted noted how the

> FHA adopted a racial policy that could well have been culled from the Nuremberg laws.[10]

Like Goethe's sorcerer's apprentice, the FHA predilection for social and racial homogeneity established a chain of recursive planning and architecture processes that led developers, real estate professionals, and municipal planning authorities to perpetuate an ever-increasing degree of differentiation by demography that has shaped postwar settlement in the U.S. White flight is a generic term that belies this much more complex form of "metropolitan differentiation" that evolved in the decades following World War II, leading to "highly stratified suburbs" ranging from upper-middle class to working class, which were also segregated by race, and increasingly also divided according to age and family status.[11] One consequence of this is the rise of what geographer Wei Li refers to as the "ethnoburb," i.e. suburban ethnic clusters where one ethnic minority group has a significant concentration but does not necessarily comprise a majority." Upon arriving in Los Angeles Li was instructed to look for housing in Monterey Park, which was described to her by one of her academic colleagues as "a Chinese area."[12]

The cluster of affluent suburbs along the North Shore of Lake Michigan, where the genteel upper-middle-class, white, and largely Christian Lake Forest, with its lily-white steepled places of worship and Christmas trees give way to the menorahs of the equally prosperous but more Jewish suburbs of Highland Park and Deerfield illustrates how this form of differentiation plays out in the Upper Midwest. In 1959 a developer in Deerfield building a new subdivision was hit with a stop-work order when town officials became aware of his plans to set aside 20% of housing on the site for African Americans. The incident became something of a cause célèbre attracting the attention of James Baldwin, Martin Luther King Jr., and Eleanor Roosevelt. Deerfield featured in the national media and was labeled the "Little Rock of the North" with cross burnings and other incidents of racially motivated violence providing the backdrop to a heated debate about the impact of African Americans on property values and "the character" of the town. The property on which the redevelopment was to take place was condemned and the developer, supported by civil rights activists, sued the town but lost. The developer's appeal to the U.S. Supreme Court was also denied and Deerfield turned the land into a park. The site earned the moniker "Deerfield's Confederate Statute," given its role in upholding racial exclusion.[13]

Amy Roost, a former resident of Deerfield has described how the municipality managed to preserve its predominantly white and upper-middle-class status over subsequent decades.[14] Following the passage of the Fair Housing Act of 1968, Illinois, like several other states across the country during the era of the "Great Society", adopted regulations designed to promote residential diversity to help atone for the exclusionary practices of the previous decades. One way the state sought to achieve this was by promoting affordable housing, one of the few means by which racial and social diversity can be promoted in existing developments. The Illinois ordinance required that 10% of all housing be designated for families and individuals earning no more than 60% of the area's median income. This would indirectly have benefited African Americans who disproportionately experience lower incomes.

In the 1970s, Deerfield, like many other municipalities, applied for and received home rule status from the state exempting it from these provisions illustrating the various mechanisms that allow local municipalities to resist more progressive statewide ordinances designed to enhance residential diversity. In 2015, a proposal for a 48-unit affordable housing project in the city was again rejected on grounds that it would lead to overcrowding in the public schools, traffic congestion, declining home values, and a loss of "community character." These arguments, along with concerns about rising crime, have been used in suburban communities across the country to limit development over recent decades, especially for affordable housing. Developers in Deerfield returned in 2018 with a more modest proposal to build 25 units and secured a vote of approval from the town. Supporters of the project, who advocate for greater social and economic diversity as well as racial integration welcomed the vote as evidence of a change in attitudes on the part of the local community.

One can but speculate about the degree to which African American social and economic mobility would have been boosted had the original development in Deerfield gone ahead in 1959. Without doubt, any families who secured homes there would have benefited not just from a significant increase in homeownership wealth but also from the educational, social, and employment opportunities that the area offers. The suburb of Deerfield is by no means unique and provides a model that was repeated across the U.S. as municipalities and homeowners over the course of several decades vetoed similar developments. The original vote in 1959 was by no means unanimous; there was, even at that time, a substantial number of town residents who wished for a more integrated residential community. Faced with the threat of government withdrawal of subsidized mortgages, it is easy to see why most homeowners in Deerfield, and elsewhere prioritized their economic interests over racial integration during this period of suburban and wealth expansion.

5.1.1 Off the Grid: Why the Streets Have No Name

Much of the criticism of the FHA over the years has focused on the way in which their *lending* practices disadvantaged African American communities. Much less has been written about the way in which the FHA shaped the architecture of suburban neighborhoods where design features frequently augmented legal and financial barriers established to ensure social and racial homogeneity. At a micro-level the FHA promoted architectural homogeneity by setting up "minimum requirements for lot size, setback from the street, separation from adjacent structures, and even for the width of the house itself."[15] One consequence of this was the elimination of regional design features with whole categories of dwellings such as the traditional "shotgun" homes of the South replaced with a bland suburban design that rendered the developments of Boston indistinguishable from those of the Midwest, the Sunbelt, or the East Coast. The design of shotgun homes facilitated air flow, creating a natural cooling effect – hence their popularity in the South. The loss of this type of vernacular design that promoted sustainability and relied upon locally sourced materials, to be replaced by standard cookie-cutter models that were much less energy efficient highlights how FHA requirements delivered adverse environmental as well as social and racial consequences over this period.

As outlined in Chapter 1, Julie Obert notes that one of the defining characteristics of eighteenth- and nineteenth-century colonial urbanism is the banality or iterability of power that promotes architecture that is "steady, careful, legible". This requires eschewing vernacular design and difference in favor of bland "boring" designs and spaces. Thomas Heatherwick has castigated the global "blandemic" in contemporary architecture that eschews "interesting" building design in favor of rectilinear glass and steel uniformity.[16] Heatherwick lays the blame for the current blandemic with Le Corbusier's modernist ideology and the commodification of building that prioritizes simple and less expensive architecture. The reality is that postwar U.S suburbs combined all these elements to create spaces that shaped a bourgeois identity predicated upon racial, social, and design homogeneity exclusion.

The "Planned Violence" that lay at the heart of FHA's strategy is illustrated by the way in which racial exclusivity was shaped by street designs that sought to render legibility and control over residential subdivisions and the exclusion of non-residents.[17] FHA guidance and local municipal politics combined to ensure that measures ostensibly designed to promote pedestrian safety served also to consolidate residential segregation. Traditional grid patterns were replaced by loops, lollipops, and curves that eliminated through traffic and promoted traffic safety through "a residential panopticon that ensured constant visibility and internalized discipline." Loewen has described how many "sundown suburbs" are characterized as defended neighborhoods where sidewalks and bike paths are rare and do not connect to those in other communities inhabited by residents of lower social and racial status.[18] He also points out how many white suburbs and smaller towns and cities across the country deliberately refused to allow street signs to discourage outsiders. Southworth has documented how developers have sealed off suburban neighborhoods by altering the street design to make "more self-contained, self-focused, and unconnected subdivisions that made it easier for residents to control their own space … an intentional device … acting as a deterrent to all nonresidents- casual visitors as well as criminals."[19] Planners and architects in Atlanta, Georgia, routinely manipulated street and road design at the local level to restrict African American mobility and access. Some tactics included designing road networks to prevent African Americans from using white neighborhoods as thoroughfares, closing pre-existing access routes as neighborhood demographics changed, and limiting north-south streets to block expansion of African American neighborhoods particularly on Atlanta's west side. Samuel Adams, former research director for the Southern Regional Council, highlighted the impact of racially motivated street design on connectivity in the city:

> Drive from downtown Atlanta six miles southwest through Atlanta's bulging westside Negro community on out into the unincorporated areas of Fulton County, and you will find exactly three streets that provide fairly direct access between the Negro areas to the northwest and the white areas to the southwest.[20]

These racially motivated street design choices created lasting issues. The 1980 Atlanta Comprehensive Plan cited major traffic congestion concerns on the west side due to the "limited street network" and "lack of adequate north-south arterial capacity."[21] The huge volume of carbon dioxide emitted by this traffic congestion is another consequence of planning and design decisions that were driven by racial exclusion rather than economic or environmental efficiency. The limited access routes identified by Loewen, Southworth, and others are virtually identical to the techniques deployed by the British Army in Belfast that inhibited movement through manipulation of road and street networks (Chapter 3). Residential segregation and chronic traffic congestion experienced by contemporary Atlanta and Belfast is evidence of how this exchange of ideas delivered adverse social, economic, and environmental consequences on both sides of the Atlantic.

In the Upper Midwest traffic calming and road safety measures were also designed with a dual purpose. In Cleveland, Ohio, in the early 1990s a banner that read "Entering Apartheid Shaker: Home of Barricades" was set up by a group of protestors who had been fighting for 15 years to remove physical barriers between the suburb of Shaker Heights and the city of Cleveland.[22] According to Shaker Heights officials the barricades were merely "traffic diverters." For residents on the Cleveland side the barricades were known as "The Berlin Wall for black people."[23] Katherine Bowman has highlighted how similar barriers ostensibly designed to improve traffic safety segregate the prosperous white suburb of Grosse Point from nearby Detroit.[24] In this instance a mixture of walls, wooden fences and even a wooden shed used for a farmers' market forms a mixture of architectural design features that discourages African American residents of Detroit from entering the affluent white suburb. Both Grosse Point and Shaker Heights have a long history of using racially restrictive covenants and other legal measures to restrict access to their communities by non-whites. Grosse Point also had an elaborate points system for rating suitability as a future resident which, unsurprisingly, was heavily weighted to maintain racial homogeneity.[25]

Perhaps the most well-known physical barrier erected to prevent residential diversity is the Eight Mile Wall in Detroit. In the late 1930s, as the city expanded, white families began to settle near an African American enclave adjacent to Eight Mile Road. By 1940 the African American community was surrounded, but neither they nor the white group could get insurance for mortgages because the FHA viewed this as proximity of inharmonious racial groupings. In 1941 a white developer built a concrete wall between the white and Black areas which secured FHA approval and mortgages were granted for the white properties.[26] The Detroit Eight Mile Wall subsequently gained notoriety and has been described by Kenneth Jackson, the preeminent historian of U.S. suburbs, as "bizarre and capricious."[27] The wall is often viewed as an aberration or an oddity but is in fact merely one example of a series of similar structures across the country.[28] The Eight Mile Wall was constructed in line with the specifications outlined in the FHA Underwriting Manual and illustrates the lengths that developers had to go in order to comply with official segregationist doctrine and how these measures became embedded into the urban fabric of metropolitan areas across the country.

As with highway and road construction the use of walls to segregate and divide communities in the U.S. goes back to at least the early part of the twentieth century. In Baltimore, Maryland, in the late 1930s, white residents banded together to erect a wall in response to the growing number of Black people in the area attending Morgan State, a historically Black educational institution. Known locally as the "Spite Wall" the structure was intended to keep Black students from venturing into a once predominantly white neighborhood. White residents had opposed the building of the College in 1917 and racial tensions increased following the state's decision to change what was then Morgan College from a private institution to a public one in 1939 to promote African American education. Construction of the wall began in 1942 at a time when Baltimore also used racially restrictive covenants to limite areas where Black people could live in the city. According to Dale Green, a professor and architectural historian at Morgan State, "For the white community, this spite wall was to send a signal and to physically create a divider that would symbolize the segregation that they stood for."[29] The wall was torn down as recently as April 2023 illustrating the longevity of many of these divisive architectural features.

Further south, in 1915 white Miamians attempted to build a wall along the edges of what was then called "Colored Town" the African American district to the northwest of downtown. The plan was tabled in the wake of several court decisions in other states that struck down racial segregation laws. Although the plan was not acted upon at this time, it illustrates again how walls and physical barriers, like highways, came to be viewed as an alternative to legal measures for supporting racial segregation as courts took a more enlightened and progressive view towards racial zoning. The use of walls to support residential segregation subsequently became a prominent feature of the architecture of Miami from the 1930s. Following construction of the Liberty Square Black public housing project the Miami Housing Authority (MHA) opened an all-white housing project, Edison Courts, a few blocks away.[30] Because of their proximity, the two developments were deliberately designed to stand apart and to expand in opposite directions as the population of the two developments grew. Because there were significantly fewer housing options available to African Americans elsewhere in the city the non-white demand for public housing increased at a much greater rate than it did for whites. This expansion prompted Dade County Officials and the project's architects to erect a four-foot-high concrete wall backed by a line of large Banyan trees on the African American side of Twelfth Avenue which was the dividing line between the two developments.[31] According to Marvin Dunn, Emeritus Professor at the Department of Psychology at Florida International University "Some of those trees still remain … The intent was not even to have visual access into the white community by having the wall, as well as these trees built along it."[32] Known locally as the "race wall" this structure would serve as a template for other builders who, by the end of the 1940s, began including concrete walls as standard features of larger African American housing projects in Miami and other cities across the country.

By the end of World War II, housing shortages in Miami were so extreme that many African Americans risked their lives moving into traditionally "white" territory. In response Dade County Commission decided to zone the Brownsville area a "whites only" community almost three decades after the Supreme Court had ruled racial zoning unconstitutional. The legal ordinance was augmented by a cement wall, paid for with federal funds, between Brownsville's "white" and African American sections. As confirmation of the wall's purpose a broad red line was painted along the length of the barrier.[33] Connolly points out how Brownsville's wall "gave the redlining HOLC map a new vivid, concrete form." Concrete was by no means the only material used to provide racial barriers in Miami with plants and trees also deputized to help shape exclusionary design. In May 1944 the Dade County Commission voted to change the zoning of an agricultural area of Opa-locka declaring it "open to Negro occupancy." However, the Commission also stipulated that a planted "buffer strip at least sixty feet in width" be established along the eastern border of the property walling off the new Black residential area from nearby white commercial and residential property. Subsequently, Opa-locka experienced a rapidly rising Black population in the years after World War II.[34]

5.1.2 Coconut Grove and Segregation by Design

The use of walls and street design features to block access from predominantly African American communities to more affluent and predominantly white neighborhoods in Miami also featured prominently in the planning and architecture of the Coconut Grove area. Fieldwork by this author illustrates both the longevity of this form of segregation by design and the invidious way in which barriers, whether made by concrete, wood, chain-link fences topped with barbed wire, or lush greenery can be augmented by severed streets[35] and a disconnected grid[36] to keep communities, literally and figuratively, in their place.

Segregationist architecture in this area of Miami goes back to the early 1940s when a private developer asked for a zoning change to allow construction of two homes for African Americans in the traditionally white section of the Grove. After complaints from white homeowners about the adverse impact on property values the company promised that if the area was rezoned to allow African Americans to live there they would build a dividing wall like the one constructed by the federal government in Liberty Square. Following protests from white residents the Miami City Commission revoked the request by the two Black families for building permits killing the proposal.[37] In June 1944, white residents of Coconut Grove again successfully resisted a proposal by the FHA for the construction of 50 duplex dwellings for Black war workers in the area.

In the years following World War II the planning board recommended that a 24-acre site in Coconut Grove be used by the MHA to provide low-rent housing for African Americans to address the chronic level of housing need facing that community. Following a lengthy battle between developers and white homeowners a compromise was reached allowing the development to proceed along a 50-foot street (Charles Terrace) with a row of single-family houses separated by a 30-foot buffer strip and a six-foot wall topped with wire.[38] In exchange the white residents would drop their objections clearing the way for developers to access FHA mortgages.[39] This incident illustrates the complex dynamics at play during this period in which the federal government, through the FHA, was unwilling to insure mortgages in mixed residential areas and white homeowners were reluctant to allow construction of housing developments for African Americans in proximity to their area due to the perceived adverse impact on property values and the potential loss of mortgage insurance. At the same time there was huge pressure from within the African American community for additional housing to meet the chronic level of need and, as with any other similar examples across the country, African American demands for additional housing were actively supported by property developers who recognized the huge profits to be gained by addressing this need.

The concrete wall that was erected on Charles Terrace remains in place and still separates the predominantly low-income African American community from the more affluent Kumquat Avenue. In addition to the wall the street pattern was also designed to promote racial segregation. Figure 5.1 illustrates how the disjointed grid pattern ensures that Charles Terrace and Kumquat Avenue run parallel to each other but without any connecting streets. Pamona Lane and Linden Lane, both located within the historic "White Grove" end at Kumquat Avenue. Similarly, Brooker Street on Charles Terrace, located within the historic "Black Grove," ends abruptly with no linkage through to Kumquat Avenue. The only way to get directly from Charles Terrace to Kumquat Avenue is via the more circuitous 37th Avenue/Douglass Road (Figure 5.2).

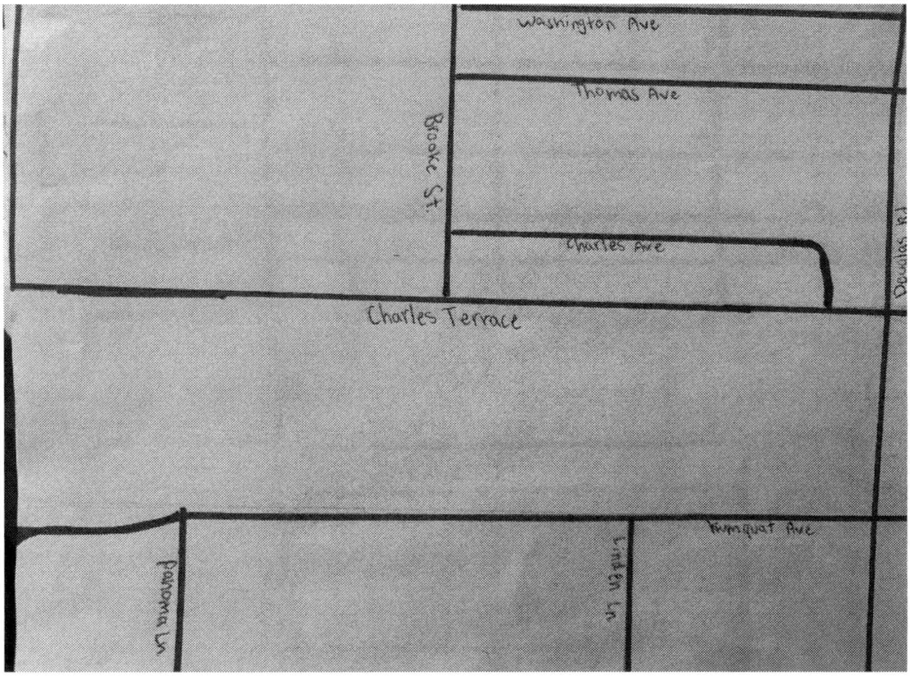

FIGURE 5.1 Map showing how street design was used to ensure the separation of Charles Terrace and Kumquat Avenue in Coconut Grove, Miami. The disjointed grid pattern and lack of connectivity between the two areas is apparent from this drawing, illustrating how racially motivated designs still affect connectivity in the area. The racial barrier, in the form of the concrete wall, still runs along the rear of Charles Terrace.

Drawing by Cara Cunningham, October 2024.

A site visit in October 2024 highlights the different ways in which the wall and the street design impact on connectivity and relationships between communities in the area. Most of the houses on Kumquat Avenue are large enough that they now tower over the wall which more closely resembles a garden fence rather than a racial barrier. One of the residents on Charles Terrace, an African American male, described to this author how the wall serves as a useful barrier for keeping out animals and potential intruders.[40] His view was that Charles Terrace is now a fairly mixed neighborhood with no tensions between the communities. Walking Charles Terrace on a Saturday evening there was a relaxed and friendly vibe around the neighborhood. One group of residents was having a cook-out and watching college football on a large screen in their front yard; others were casually going about their business. The low-income houses on Charles Terrace are unusual in this part of Coconut Grove, much smaller and distinct from the high-end properties on Kumquat Avenue and many other parts of Coconut Grove that have undergone significant gentrification in recent decades. In addition to being smaller, the properties on Charles Terrace are also much more open and accessible, in sharp contrast to many of the more affluent properties in the

FIGURE 5.2 Charles Terrace Wall, Coconut Grove, Miami. This wall still marks the boundary of the historic color line that divided the backyards on Charles Terrace from the white community along Kumquat Avenue but is now regarded more as a garden fence than a racial barrier. There is no evidence of the barbed wire that was part of the original construction.

Photograph by Tim Cunningham, October 2024.

area that are surrounded by fencing, closed off and isolated from each other (Figures 5.3 and 5.4).

The old racial barrier, nestled at the rear of the property line, genuinely appears to represent a curious relic of the past, hidden from sight, and a sad reminder of the way in which the federal government conspired with municipal authorities to erect barriers between communities based on skin color. The tranquil atmosphere around Charles Terrace, coupled with the fact that this section of the wall is now so visibly insignificant and blends so seamlessly into the surroundings, appears to have defused the toxic history of the structure.

Unlike the wall, the street design has more long-lasting and impactful consequences. The only way to access Kumquat Avenue from Charles Terrace is via 37th Avenue and I suspect that I am somewhat unusual in wandering along this route. The social and economic divisions in the area mean that residents of Charles Terrace don't have any reason to walk back into Kumquat Avenue and vice versa. Both communities are clearly distinct and separate, one wealthy, the other lower income. Like many U.S. suburbs, there was no sign of any kind of street life on Kumquat Avenue, with single-lot homes separated from each other as much as from the lower-income

FIGURE 5.3 Charles Terrace, Coconut Grove, Miami.

Photograph by Tim Cunningham, October 2024. Low-income housing on this street was separated from nearby Kumquat Avenue by a wall to the left of this picture to placate white residents and ensure access to FHA-backed mortgages.

community on the next street. One home had an array of signs indicating support for Donald Trump, while the home directly opposite had signs indicating support for Kamala Harris (see Figure 5.5). Kumquat Avenue is a beautiful, narrow street, surrounded by lush greenery but like many affluent neighborhoods across the U.S., there is a notable absence of people of any color.

Similar design practices were deployed a few blocks east of Charles Terrace on Marler Avenue.[41] Tensions came to the fore in that area in the 1990s when additional fences, some of which were topped with barbed wire, appeared. "Shades of apartheid" was the view of one resident at the time, Will Johnson, who returned home to the "Black Grove" in the mid-1990s after 18 years in the U.S. Army. In a feature article for the Miami New Times in 1998, David White, then President of the West Coconut Grove Homeowners and Tenants Association, described how he routinely walked the three blocks from his home on Marler Avenue to the Plymouth Congregational Church back in the 1930s via a footpath that led to Hibiscus Street, which was then a dirt road. For decades, the church, which is still standing, was the only racially integrated house of worship in Dade County. By the late 1990s, David White

FIGURE 5.4 Low-income housing on Charles Terrace, Coconut Grove, Miami.

Photograph by Tim Cunningham, October 2024.

> would have had to climb a ten-foot chain link fence topped with strands of barbed wire to take the first right onto Hibiscus Street from his home on Marler Avenue to go to his place of worship.[42]

Figure 5.6 illustrates how Hibiscus Street is severed at Marler Avenue, beginning again at Loquat Avenue on the white section. White recounted how when the street came to be paved in the 1950s he had asked the workers if they would pave Hibiscus Street all the way to the white area so that he could walk down to church but was told that the city would not allow it.

Instead of running continuously from the Black Grove to the white section Hibiscus Street was severed when it reached Marler Avenue (see Figure 5.6). In his article, Nielsen describes how a trail through the thicket remained until the early 1970s, when private developers built a row of condominiums on that lot and sealed them off from White's neighborhood with a ten-foot chain-link fence topped with barbed wire. The other end of Marler Avenue, Plaza Street, could have provided an alternative route for David White and his neighbors through the White Grove, but again, through the 1980s and early 1990s, the city refused to pave the street despite numerous requests.

In 1998 the West Coconut Grove Homeowners and Tenants Association demanded that the city remove the chain-link fence at the intersection of Marler and Plaza

FIGURE 5.5 Photo of Kumquat Avenue, Coconut Grove, Miami. This area was historically the "White Grove." A sign indicating support for Kamala Harris in the upcoming Presidential election is visible on the right. A sign indicating support for Donald Trump is visible on the left.

Photo by Tim Cunningham, October 2024.

Avenue on the grounds that it was blocking a public right of way.[43] Among white residents of Loquat Avenue there was strong support for keeping the fences which were viewed as necessary to prevent crime. Some members of the white community cited robberies and other intrusions perpetrated "by black males" in their area and some residents had been robbed at gunpoint.

David White's journey to church changed during his lifetime from being able to walk just over three blocks to his place of worship in the 1930s to having to traverse a much more circuitous route by the mid-1990s, illustrating how a series of architecture and planning decisions over several decades continue to contain and confine residents of the old Black Grove. Selective street paving, chain-link fences, an illegal land grab by white homeowners, and private development all played a role in cutting off streets like Marler Avenue from their natural hinterland to the south.

Walking Marler Avenue on a Saturday evening in October 2024 one can see and feel how coercive architecture continues to shape relationships and affect race relations in the area. One resident of Marler Avenue with whom I chatted, an African American woman who was tending her yard, described the lush trees and the wooden fencing that faces directly onto her property (see Figure 5.7) as a "physical, social,

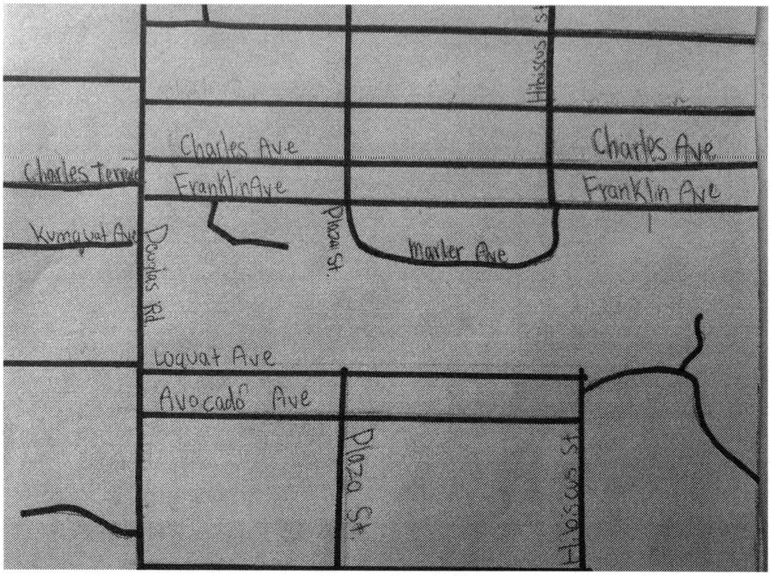

FIGURE 5.6 Map showing how connectivity between Marler Avenue in the historic "Black Grove" was severed from "White Town" and the Plymouth Congregational Church by street design. Hibiscus Avenue in the historic "Black Grove" ends abruptly at Marler Avenue, preventing direct access to Locquat Avenue in what was considered "White Town." Plaza Street is also severed at Marler and begins again at Loquat. Drawing by Cara Cunningham, October 2024.

racial and psychological barrier." She described how the architecture of the houses on Locquat Avenue, which face north and back directly onto her property, illustrates how people in the white community feel about those living on Marler Avenue, i.e., turning their backs on them. Her comments echo those made to me by a resident of the New Lodge area in Belfast who described how the architecture of the new Ulster University campus ensured that those in the building were "turning their backsides to the local community" (Figures 5.8 and 5.9).

The resident of Marler Avenue with whom I spoke had moved into the area in the mid-1990s and described how little interaction there was between residents of her street and those living in "White Town" as she pointed across the street. She recounted how some years ago she had become embroiled in a dispute with municipal authorities over trees at the rear of the property on Locquat Avenue directly facing her home. The trees had become a nuisance, overhanging Marler Avenue and impeding her ability to drive in and out of her own home. Repeated attempts to have the municipal authority trim the trees were met with indifference until she cut some of them herself. A complaint about her was made by the property owners, town officials got involved, and ultimately the trees were trimmed.

Her account was one of anger at what was perceived to be indifference toward the residents of Marler Avenue by city officials and the residents of "White Town." She also described how there were complaints from white members of the community on the other side of the barrier about theft from a property on Locquat Avenue and the

FIGURE 5.7 Marler Avenue, Coconut Grove, Miami. The fence and barrier to the right of the image on Marler Avenue remain and still mark the barrier between the two areas although the old "Black Grove" has now undergone significant gentrification and many of the residents in this area are now white. Tensions remain about the presence of the barrier. The arrow indicates to drivers that there is no access to the right onto Hibiscus Avenue.

Photograph by Tim Cunningham, October 2024.

involvement of the police. Apparently, footage from the house in question showed that some white tenants on Marler Avenue, who had rented briefly on the street, were responsible. She also described how another African American resident of Marler Avenue had complained about accusations from the white residents of Locquat Avenue that African Americans were the culprits. There is clearly quite a degree of anger within the African American community of Marler Avenue at what was seen as official indifference to their needs and a willingness to point the finger when there was a problem, whether that related to theft or interfering with trees. The resident with whom I spoke also joked about how some of the parents of children in the nearby private school could occasionally be viewed mistaking Hibiscus Street as a thoroughfare and were caught out, expecting a connection to Locquat Avenue that doesn't exist.

In many respects, the incidents that I heard about resemble the kinds of disputes that characterize many neighborhoods, especially ones in which wealthier new arrivals co-exist with lower-income residents, except that in this case there was a strong racial edge, heightened by the presence of the physical barrier to this dispute. Unlike the wall at Charles Terrace a few blocks away, the coercive architecture and

FIGURE 5.8 Hibiscus Avenue, Coconut Grove, Miami, looking south towards Marler Avenue. This is the historic "Black Grove" looking towards the old "White Town." The abrupt severing of the street up ahead at Marler Avenue and lack of connectivity along Hibiscus Avenue between the two areas is clearly visible with the arrow indicating drivers must take a right turn.

Photograph by Tim Cunningham, October 2024.

street designs that remain on Marler Avenue present a substantial impediment to improved relationships between the races. In 2021 City workers rebuilt Plaza Path, the brick walkway that connects Marler Avenue, which marks the southern boundary of the Black Grove, with Loquat Avenue to the south. The outcome is that I was able to walk unimpeded to Plymouth Congregational Church from the Black Grove, just as David White did as a child in the 1930s. Rather than evidence of improved relationships this new access route reflects the reality that much of the property in the historic Black Grove is now occupied by white residents, illustrating how gentrification has precipitated the softening of some older racial barriers in the area (Figure 5.10).

The Plymouth Congregational Church, on the Saturday evening that I meandered around the area, had just held an event in a Church Hall at the rear of the campus. Affluent white parents and their children were leaving to drive home as I arrived. The overall feel of the area around the Church was one of security and comfort, with historic architecture blending in with lush greenery and narrow walkable streets. The walk from the Church, along Devon Road, onto Hibiscus Street and around to Locquat Avenue could be described as idyllic if one wasn't aware of the reasons why

FIGURE 5.9 Hibiscus Avenue, Coconut Grove, Miami, looking north. This is Hibiscus Avenue from the historic "White Town" looking toward the old "Black Grove." The arrow indicates that drivers must turn left onto Locquat Avenue with no through access onto Marler Avenue illustrating the legacy of architectures of exclusion in the area.

Photograph by Tim Cunningham, October 2024.

the road suddenly turns left when you would expect it to go straight. The luxuriant vegetation that runs along Locquat and Marler Avenues is quite unlike the concrete and steel barriers that characterize the security barriers that divide communities in other parts of the U.S. and in Belfast. At the same time, this more aesthetically pleasing architecture of segregation is arguably more invidious, conferring a façade of respectability and even beauty upon an environment that is designed to exclude.

5.1.3 Crime Prevention & Neighborhood Safety

Another striking feature about many of the racial barriers that characterize U.S. cities and suburbs is that although they were often (but not exclusively) built to comply with FHA mortgage insurance requirements, many of these structures subsequently evolved into crime prevention measures. This evolution is highlighted above by the way in which racial exclusion in Coconut Grove led to increased fencing to deal with complaints about crime in the 1980s and 1990s. Similar dynamics shaped the architecture of the "Berlin Wall of Connecticut."

In 2014, residents of Hamden, Connecticut, finally lost their fight to keep a controversial fence, approximately 10 feet high and 1,500 feet long, that had separated

FIGURE 5.10 Plymouth Congregational Church, Coconut Grove, Miami. Located in the old "White Grove," for decades this church was the only racially integrated house of worship in Dade County. Access to the church from the "Black Grove" was gradually severed by a series of planning and architecture decisions, including street paving and the erection of a fence.

Photograph by Tim Cunningham, October 2024.

their community from New Haven for decades. Hamden residents cited fear of crime and concerns about traffic as justification for keeping the fence, which was erected by the city of Hamden in the 1950s in response to complaints about crimes committed by people living in a public housing complex in nearby New Haven. The fence was a flashpoint between the two municipalities and served as a symbol of racial as well as urban-suburban divisions. The wall had isolated residents in public housing projects from the surrounding area, meaning they had to travel almost 8 miles to buy groceries at Hamden shopping center three miles away. The trip took two buses and up to two hours to complete.

The issue came to a head in early 2010s following New Haven's demolition of the decaying public housing complexes and the construction of less dense, new urbanist-style, mixed-income, rental, and homeowner developments at the cost of $200 million.[44] The success of the new mixed-income development required the removal of the barrier however in 2012 New Haven's mayor suspended efforts to tear down the wall after homeowners in Hamden "exploded in anger over the prospect." A two-year process to try to build trust on both sides ensued but to little avail. The New Haven Housing Association (NHHA) organized a series of trust-building events, like cook outs, to try and build relationships between the two sides but reports indicate that only families from the New Haven side turned up.

Two developments ultimately precipitated the removal of the wall. First, New Haven city officials approached the Department of Housing and Urban Development (HUD) about the possibility of suing the town of Hamden for violating the New Haveners' civil rights, arguing that the decision to erect and maintain the fence was discriminatory. The U.S. Attorney's Office began an investigation that could have resulted in a lawsuit against the town of Hamden, preventing the town from receiving any federal money. At the same time, New Haven officials did a survey of the property that revealed they actually owned the land on which the fence stood and therefore they had the legal right to tear it down. For decades it had been assumed by everyone that the fence was on property owned by the city of Hamden and they had legal control over whether or not it remained. The realization by New Haven officials that they in effect owned the wall negated further challenge on the part of HUD.

The election of a new mayor of New Haven, Toni Harp, led to swift action to remove the wall once it became apparent that they had the authority to do so. Harp called the wall a "morally reprehensible" barrier that put elderly people's lives in jeopardy. There were concerns that in the event of flooding, or other storm-related disasters elderly residents could have been trapped by the fence. In an interview with the local press, one resident of Rockview expressed delight that removal of the fence would mean she no longer had to drive all the way through New Haven to bring her children to a Hamden clothing store. The reaction was quite different among Hamden residents, where Councilman Mike Colaiacovo claimed, "The people are going crazy." To placate these concerns, Hamden promised residents that traffic calming measures and increased police patrols would be established to ease concerns about crime.

The history of the fence is chronicled in a report commissioned by Hamden officials following its controversial removal.[45] Written by attorney Steve Mednick, the study found that the fence resulted in "remote and isolated conditions for the New Haven residents, with a series of dead-end streets interwoven with the projects that left open access only to New Haven, where businesses were miles away." The Woodin Street fence cut off access to bus routes that provided much-needed shopping and employment opportunities.

There are several striking features about this story. First is the extent to which Hamden residents opposed the removal of the fence and their fear that this would leave them vulnerable to crime even after the public housing projects were removed and replaced by a $200 million mixed-use development. The series of events organized by New Haven to secure agreement on taking the fence down, and the lack of engagement on the part of residents illustrates the extent to which mindsets within Hamden were closed regardless of the changes that had taken place on the other side of the barrier. Second is the extent to which residents in the New Haven housing projects acquiesced to the presence and, indeed, the extension of the wall over several decades. Compared to Charles Terrace residents in Coconut Grove, who are not geographically isolated by an old racial barrier that serves as a glorified garden wall, the residents of Rockview had to travel several additional miles to access essential services, including grocery stores. The dangers posed to elderly residents by lack of access for emergency vehicles further compounded the spatial indignities that the fence facilitated. Despite these incumbrances, the report by Steve Mednick noted,

> One of the fascinating strains in this history is the lack of any meaningful protest from the New Haven side regarding the existence of the fence ... By the late 1960s, our history shows that the demand for the fence was generated by Hamden residents and their elected representatives. On the New Haven side there was nothing but silent acquiescence.

The lack of opposition to the fence from the New Haven residents is less surprising when one considers that this is a common feature of communities across the U.S., and beyond, including Belfast, when faced with these kinds of structures. One explanation for this is the capacity of coercive architecture to shape human behavior in subconscious but detrimental ways. Sarah Williams Goldhagen has pointed out that even when we pay no conscious attention to the built environment it functions in our lived experience through environmental stimuli that can influence a person's thoughts and behaviors by "activating memories, emotions, and other kinds of cognitive associations."[46] If memories and other cognitive associations are of being put in one's place, of being excluded, and considered less worthy, or of fearing what is one the other side, these walls and barriers will play an active and detrimental role in continuing to shape people's lives and behaviors regardless of the objective level of threat. Research from Belfast has highlighted the negative impact that the more extreme physical barriers in that city (i.e., 'Peace Walls') have on mental health.[47] The power of architecture to shape behavior – for better or worse, has also been highlighted recently by Thomas Heatherwick who has drawn attention to the way in which certain types of architecture can impact adversely on physical, social, and environmental health.[48] His findings are supported by additional research from psychologists and neuroscientists about the adverse impact that building facades and design have on the physical and mental people around them.[49]

Most of these studies have focused on the adverse consequences of bland concrete and steel structures that characterize urban centers. Clearly, additional works is needed to measure the wider impact of the kind of physical barriers and defensive architecture identified here but there is certainly enough prima facie evidence to show that, like gated communities, walls and fences serve to increase isolation, fear and anxiety rather than alleviate it and that the long term psychological consequences of these structures has been significantly underappreciated. In addition, these kinds of structures clearly played a not insignificant role in perpetuating patterns of spatial disadvantage and, similar to the prison panopticon, promoted a form of self-regulation that not only inhibited, but actively discouraged free movement and association – especially with those on the other side of the barrier. Writing about the South Miami segregation wall, Marvin Dunn claimed,

> There was a sense among some white people that a physical barrier would stop Blacks from wandering into the white area and would restrict the presence of Blacks in sections that were designated for whites ... It was also a way for the police to be able to contain Black people and [for Black people] to know the boundaries that they were supposed to respect. And therefore, they'd

> be arrested if they disrespected those boundaries. So these walls had psychological and physical components to them that had the basic impact of having Black's know their place: it is on the other side of that wall.[50]

As outlined in Chapter 1, Charles Wilson has observed that architecture is by its nature assertive, creating an order that either affronts or enhances quality of life.[51] Evidence from local communities corralled behind walls and fences would suggest that architecture is actually more powerful and assertive in creating an order that affronts than has typically been appreciated and that much of that power comes from its hidden and subconscious impacts that shape acquiescence and comfort in exclusionary designs.

Despite their detrimental impact on local communities federal enthusiasm for walls continued through the 1990s, as evidenced in Baltimore, where an eight-foot-high spiked fence was installed in 1998 around Hollander Ridge, a 1,000-unit public housing development built in the mid-1970s by the Housing Authority of Baltimore City. The fence was constructed by the local housing authority with funding from the HUD and blocked access to and through Rosedale, "a contiguous and mostly white neighborhood."[52] The wall was part of a pilot scheme by HUD to explore measures to reduce crime in public housing projects. The Rosedale residents ostensibly wanted the fence to prevent crime and maintain property values, although "there was a not insubstantial vocal segment of Rosedale whose racist views were made readily apparent."[53] The Hollander Ridge development was demolished in 2000, by which time it was virtually 100% African American. In this instance there was active opposition to the structure – residents of Hollander Ridge took a civil rights lawsuit against HUD and the City of Baltimore to challenge the construction of the wall but were unsuccessful.[54] The judge ruled that there had been mismanagement of the housing complex but refused the claim that the construction of the wall was racist even though some of the residents who supported the building of the structure had expressed racist sentiments. The argument from residents of Rosedale that the wall would be effective to reduce crime and preserve property values highlights again how rationales enshrined in the FHA Underwriting Manual continue to shape contemporary debates about housing and residential diversity. Hollander Ridge also illustrates Sarah Schindler's point about the limitations of traditional legal frameworks designed to eliminate discrimination for dealing with architectural and urban design practices that segregate and exclude.

5.1.4 From Race Walls to Gated Communities

In recent decades, the individual walled structures championed by the FHA have followed their natural conclusion and morphed inexorably into gated communities where the entire neighborhood is enclosed and access requires passing through barriers that are often manned by security guards. This phenomenon has been linked to increases in immigration and a rising fear of crime but research shows that many gated communities are no safer than other suburban neighborhoods "where crime rates are already low."[55] Setha Low highlighted the interesting paradox that barriers and gates to protect against crime increase social isolation that may result in greater perceived vulnerability and more fear – which residents deal with by taking even greater precautions and thus becoming more isolated socially. Essentially an isolationist/fear doom loop is established which explains why "gating may create more fear of crime than it alleviates" with residents of these communities perceiving "the world outside their development as full of dangerous 'others.'"[56] Low also found significant levels of anxiety among those living within gated communities that the barriers, surveillance, and elaborate security systems were inadequate with some residents expressing concerns that workers coming in and out without being checked might commit crimes. Low's findings mirror the observations from Jane Jacobs over seventy years ago about 'the curse of border vacuums' in which she identified how roads and blank walls inhibit the safety and natural surveillance provided by 'eyes on the street.'

Gated communities are marketed as "traditional" neighborhoods which in the U.S. frequently serves as a racist dog whistle. Residents are encouraged to keep an eye out for those who 'look out of place' another highly racialized trope. The consequences of this practice are evidenced in incidents like the killing of Travyon Martin, a teenager who was fatally shot by George Zimmerman, a neighborhood watch coordinator in a gated community in Sanford, Florida. Zimmerman had a history of reporting "suspicious" outsiders to the police. Trayvon Martin's killing illustrates how the highly racialized nature of crime control and access to space in the U.S. led to the killing of a young unarmed African American male who was visiting his father and had gone out to the local 7–11 to buy a packet of Skittles. Community efforts to promote neighborhood safety, like complete streets, which encourage more active public spaces and social connections have proven more effective than physical barriers in reducing crime and are ultimately healthier for the national sense of well-being and the physical safety of those who 'look out of place' in many of the more affluent communities i.e. non-whites.

Following the same logic as the FHA Underwriting Manual there is also a widely held perception that gated communities result in greater stability in property values through their forced exclusivity which is another reason for their popularity. In practice, there is no evidence to support this view although gated communities have been shown to have a detrimental effect on property values in neighboring areas. Government policy that promoted physical barriers to ensure neighborhood success provided the historical foundation for many suburbs in the U.S. that now incorporate design features that make them look and feel like gated communities without the actual gates.

Other communities across the country have gated themselves off incrementally through the implementation of defensive design features over an extended period. This phenomenon is evident in Coral Gates in Miami, where in 2010 residents insisted that the City build a 10-foot-high wall, stretching nearly half a mile, and costing city taxpayers $1.5 million, around their community to reduce crime.[57] A year later, Miami-Dade County officials, following a complaint from a member of the public, claimed the wall was illegal because it blocked pedestrian access to public streets and ordered that the wall be removed. The City refused and the wall became a permanent feature of the landscape augmented by street barriers that had the effect of limiting vehicle access to homes. In 2019, further controversy arose when some residents paid for surveillance technology that tracked the license plate of every vehicle entering the community.[58] The area experienced very little crime, although a carjacking incident had taken place two years previously. The surveillance technology or "virtual fence" drew protests from several residents who felt that this was an invasion of privacy. The technology deployed was unregulated and unmonitored by state or local laws and following "heated debates within the community about the relative merits of privacy versus security", the cameras were eventually moved from public to private properties.

Several decades of FHA policy predicated upon promoting social as well as racial homogeneity means that many gated communities in the U.S. are now further differentiated on demographic grounds, some of which restrict occupancy to those aged 55 and over with physical barriers and armed guards ensuring that empty nesters and retirees live apart from those with children and young families. The outcome is a "numbing homogeneity" where Girl Scouts are prevented from selling cookies outside their own discrete residential enclave and children resort to "trunk or treating" in sterile parking lots because they do not have access to sidewalks or public spaces that would allow them to knock on front doors. Canvassers and campaigners who form a central part of the democratic process are also excluded from these developments.

Contemporary social and legacy media, at both local and national levels, espouse the virtues of over-55 lifestyle communities with glamorous empty nesters relaxing poolside. The apotheosis of this kind of monocultural urban design and architecture is found in communities like "The Villages," located about 60 miles north of Orlando, Florida. The Villages, like many other such developments, prohibits anyone under 19 from residency.[59] The exclusion of children and young people (those in their thirties or forties may live there but only if co-habiting with at least one person aged 55 or older) is allied to the exclusion of any semblance of local democracy. Many gated communities have their own governance systems, typically a homeowner association that performs many of the functions normally associated with local municipalities. This has led some scholars to argue that homeowner associations ought to be considered "state actors" given that their authority and functions are a de facto form of local government.[60] The Village Center Community Development District for example has no mayor or city council and no local elections. Taxes, in the form of amenity and utility fees, which are used to cover emergency services, maintenance, and many other municipal functions, are raised by an unelected Board of Supervisors chosen by the developer-owner.

The proverb "It takes a village to raise a child" widely attributed to the Igbo people of Nigeria, is now practically possible in many contemporary U.S. suburbs exemplifying the extent to which values and practices in American planning and architecture deviate from established norms developed over several millennia about what makes a successful community. Separation and differentiation by age exacerbate intergenerational conflicts over wealth, politics, foreign wars, climate change, and a host of other issues by limiting physical space in which these issues can be debated. The typical U.S. suburban development may not yet incorporate armed guards and exclude local democratic governance structures like The Villages, but all too often neighborhoods are planned and designed to achieve social disconnection enshrined in legal ordinances and supported by atomized architecture.

5.1.5 Full Circle: 9/11, Military Urbanism and the Social Life of the Street

Allied to the rise of gated residential communities is the emergence of a new wave of defensive architecture and urban design features as the streetscapes of U.S. cities were "deputized" as part of the 'war on terror' following the attacks of 9/11 (Chapter 1).[61] In New York City, Wall Street, which derives its name from the defensive well erected by Dutch settlers in the mid-seventeenth century to defend against invasion by the British or native Americans, a fortified barrier was put in place to prevent attack on potential high-profile targets like the New York Stock Exchange (Figure 5.11).

As a white 50-something male I felt reasonably comfortable taking the photograph and was not aware of any reaction from the police officers in the image to my presence. I am not sure how a young Muslim male might feel in similar circumstances. My own experience here is informed by growing up in Northern Ireland where it was unlawful to capture such images. The Northern Ireland (Emergency Provisions) Act 1973 created the offence of "gathering information likely to be used by terrorists." On one occasion, when I was a student, I was hitchhiking outside a police station with a friend and we were told by police officers to move away from their installation. We did not have any photographic equipment but our physical presence was deemed to be potentially threatening lest we were observing the unmarked vehicles entering and leaving the facitility. Current anti-terrorism laws in

FIGURE 5.11 Defensive barrier erected on Wall Street, New York City, to prevent car bomb attacks on key buildings including the stock exchange and Trump Tower. Similar structures secured the urban fabric of many other buildings in New York City following 9/11. Several other buildings in the city owned by Donald Trump were provided with similar protection following his election in 2016.

Photograph by Tim Cunningham, July 2022.

the U.S. and the U.K. also contain this offence but my experience taking photographs in New York was not impacted. In D.C., while taking photographs of defensive architecture near the Capitol, I was asked on one occasion what I was doing. When I explained that I was working on this book I was allowed to continue. My white skin and Irish accent probably went some way to reassuring the security personnel that I did not pose a threat. The same accent and skin color in Northern Ireland in the 1970s and 1980s would have had the opposite effect.

These structures reflect an increased militarization of urban space since 9/11 and the way in which coercive architecture can create a chilling effect, especially for "suspect communities." These barriers form part of a wider network of defensive architecture established in U.S. cities and installations across the world. The U.S. consulate in Belfast was moved from the city center to a much more inaccessible location in the south of the city, surrounded by shrubs, fending, and with a disguised entrance. Key buildings in Washington, D.C., were also heavily fortified with similar structures specifically designed to prevent vehicle bombings by increasing pedestrianization and reducing traffic access (see Figure 5.12).

Concerns that the aesthetic of the U.S. capital was being disrupted by "unsightly and makeshift security barriers" led to the development of the National Capital Urban Design and Security Plan in October 2002.[62] The plan aimed to integrate security considerations with a more welcoming streetscape that preserves the city's historic urban design character. The proposals, which were prepared by the Interagency

FIGURE 5.12 Defensive street design at Pennsylvania Avenue preventing vehicle access to the White House to limit the potential for attack by car bomb. Similar motives precipitated pedestrianization in Belfast in the 1970s.

Photograph by Tim Cunningham, Washington D.C., June 2021.

Security Task Force of the National Capital Planning Commission brought together public agencies, the private sector, security experts, the professional design community (including landscape architects) and other stakeholders. Members of the technical advisory group were listed in the final report and include several figures from the Secret Service as well as Donald Rumsfeld, the then U.S. Secretary of Defense. Among the findings and recommendations were that street furniture, including benches, planters, bollards, and lights, should be hardened to increase their barrier effect. The details contained within the Security Plan, and the colorful drawings and photographs, contrasts with the secret Joint Working Party document formulated for Belfast in the early 1970s (Chapter 3). The willingness of the U.S. to publicly illustrate how the design of the public realm can enhance security means that scholars and public commentators can at least be spared debate about whether this type of thing takes place and focus more on the consequences and wider implications of this approach to urban design.

Evidence of these more subtle and aesthetically desirable forms of defensive architecture can also be found beyond the capital, notably in New York City, where heavy stone bollards serve several functions including preventing car bombs, enhancing walkability, and providing casual seating space (Figure 5.13). This illustrates another

FIGURE 5.13 This image illustrates how "Broadway Vision," a design framework established by NYC's Department of Transportation to reclaim and transform public spaces from Union Square to Columbus Circle in Manhattan, incorporates several design elements, including the heavy defensive stone bollards on the left of the image, street painting, and plant pots. The aim is to create a secure but aesthetically pleasing urban realm.

Photograph by Tim Cunningham, New York City, 2024.

aspect of this more aesthetically conscious but somewhat invidious form of defensive design where efforts to reduce the threat posed by vehicles that may be used for attack "lean in" to campaigns to reduce auto-centricity, enhance walkability, and promote the social life of the street.[63] Like many similar initiatives, Broadway Vision enjoys support from a wide range of stakeholders, ranging from advocates for greater pedestrianization and walkability to local businesses who recognize the value of turning road space devoted to moving vehicles into seating space reserved for paying customers.

Having walked along Broadway and experienced the new public realm first hand, with significantly reduced traffic and the attendant noise, pollution, and danger to pedestrians, I can personally attest to the benefits of the changes. Efforts to promote walkability and create more complete streets ought to be applauded and encouraged, and there is no inherent problem with incorporating security features into these new designs provided they are openly discussed and subject to democratic approval and oversight. The difficulty is that such measures can, as illustrated in this book, perform other less desirable functions.

One problem is that enhanced walkability and a quality public realm in many contemporary urban centers also tends to accompany and assist gentrification. Broadway Vision, The High Line, and Central Park are prime examples of quality urban design significantly enhancing the ambulatory experience. The reality however is that less affluent New Yorkers do not typically avail of these spaces but rather find their movement and life experiences constrained by hulking concrete structures like the Cross Bronx Expressway. The life experiences of those communities who have yet to benefit from the joys of street furniture and an enhanced public realm will be considered in the next section.

5.2 A HOUSE DIVIDED: THE ARCHITECTURE OF ISOLATION

5.2.1 INTRODUCTION

A recent report by the U.S. Surgeon General Vivek Murthy found that the country is presently experiencing an "epidemic of loneliness and isolation" with a shocking one in two adults of all ages and socioeconomic backgrounds across the U.S. reporting that they feel lonely, isolated, invisible, and insignificant.[64] The report also highlighted the adverse health effects of this epidemic on individuals notably greater levels of cardiovascular disease, dementia, stroke, depression, anxiety, and premature death. These problems are compounded by severe societal consequences including higher healthcare costs, worse performance at work, lower academic achievement, lower levels of economic prosperity, and, crucially, less representative government. The Surgeon General has warned that "we need a more connected society … [or] we will continue to splinter and divide until we can no longer stand as a community or a country."

The national consequences of this epidemic have also been identified by Thompson, who argues that solitude is having a profound effect on individual well-being and on America's "civic and psychic identity."[65] Problems associated with a lack of social connection in the U.S. have also been identified by Seth Kaplan who has expressed concerns about the way in which many Americans neglect their local community to the detriment of themselves, their neighbors, and the nation.

> Beyond the home, we don't belong to place-based mutual aid societies, ethnic clubs, civic organizations, or religious congregations the way that our grandparents did … We don't have the kind of close-knit relationships with our extended family and friends that we used to have, or interact much with our neighbors, leaving many without a support system to call upon in a crisis.[66]

Part of the problem is that feelings of loneliness brought about by social disconnection feed distrust which in turn impacts on national well-being. A recent report by the Rand Foundation found that the U.S. lacks a shared notion of the common good and is experiencing political polarization and loss of public confidence in institutions of governance.[67] Onora O'Neill, in her series of Reith lectures for the BBC pointed out that Confucius warned his disciple Tzu-kung that there are three conditions necessary for government to function: weapons, food, and trust.[68] If a ruler cannot hold onto all three, Confucius advised that a ruler should give up weapons first, food next, and hold onto trust until the end. His contention was based on the observation that weapons are useless when armies lose trust in their leaders and switch allegiances. O'Neill noted how governments throughout history have also withstood food shortages when citizens retained trust in their institutions, as evidenced by the way in which rationing was accepted by many nations during and immediately following World War II. Trust however is the sine qua non for successful governance, which is why the historically low levels of trust felt in the U.S. are of such concern. In 1972, polling data showed that roughly "45% of Americans felt they could reliably trust other Americans, however that shrank to roughly 30% in 2016."[69] This figure corresponds with near-historic high levels of polarization. Significantly, lack of trust in institutions is linked to a serious decline in social connections with each other. In 2018, only 16% of Americans reported that they felt very attached to their local community.[70]

One way in which trust and confidence in each other, and in our institutions of governance can be nurtured is through healthy and diverse social networks. Michael Sandel noted that democracy does not require perfect equality,

> but it does require that people of different backgrounds and social positions encounter and bump into one another during everyday life. By physically sharing common spaces we learn to negotiate and abide our differences and care for the common good.[71]

Cass Sunstein, coming at the problem from a different direction, highlighted the tendency for people who cluster in like-minded groups to move to more extremist and absolutist positions thereby promoting and sustaining polarization.[72] More socially connected communities are also typically associated with increased levels of civic engagement and a more representative government. Part of the reason for this is that political tolerance can be cultivated by "core discussion networks," where people disagree over difficult and controversial issues but work out ways to co-exist together.[73] Evidence shows that the size and diversity of these networks in the U.S. have diminished over recent decades, as core discussion networks become more politically homogeneous feeding political polarization and an "us and them" mentality.

Laurie Santos, psychology professor at Yale University, has pointed out that one simple but highly effective way in which greater trust and happiness can be supported and nurtured is through casual social interactions – including conversations at the post office, the deli, or with the barista in the coffee shop.[74] Other benefits of greater social connection include increased economic prosperity and upward mobility by providing opportunities for less well-off individuals to gain a foothold in the labor market or obtain higher-paying jobs. Chetty for example found that low-income residents of a city benefit dramatically from contacts with new neighbors who have greater social connectedness, usually in more affluent areas.[75] The corollary of this is "lack of sustained contact or social interaction with individuals and institutions that represent the mainstream society … makes it much more difficult for those who are looking for jobs to be tied to the job network."[76] This is not counterintuitive. If you are poor and without a job your chances of finding employment are significantly more limited if most of the people you know are also poor and without a job. Similarly, Thayler and Sunstein have pointed out that controlling for other risk factors, teenage girls who see other teenagers are having children are more likely to become pregnant themselves.[77] This is one reason why corralling huge numbers of African Americans into high-rise public housing, spatially and socially isolated from the economic mainstream of cities produced such devastating patterns of concentrated poverty.

Communities with higher levels of social connection also typically enjoy significantly better health outcomes. Setha Low pointed out how "the depression and anxiety of physical distancing and not seeing people during lockdown reflected how much a feeling of belonging is based on face-to-face interaction in familiar places."[78] Joanna Lombard, Professor of Architecture at the University of Miami, has shown that the greatest predictor of longevity and health is our level of social interaction.[79] High levels of social connection also result in increased resilience to natural disasters. A study by Klinenberg has highlighted how social infrastructure played a key role in saving lives during a heat wave in Chicago in July 1995.[80] The strength of emerging data, and the effects of the COVID-19 pandemic led the Biden Administration to pay more attention to this problem. A National Strategy to Advance Social Connection was established with six pillars, the first of which focuses on the design of the built environment:

> Decisions about the layout of our cities, from the usability and reach of public transportation to the design of housing and green spaces, have a direct effect on social interaction in a community. This is why strengthening social infrastructure that promotes social connection is critical to advancing key aspects of community health, resilience, safety and prosperity.

The Surgeon General's report and related research points to an obvious conclusion, namely that fostering social connection is imperative for the physical, social, economic, and political health of the nation. The problem for the U.S. is that over the course of the past seven decades the built environment was specifically designed to inhibit association and discourage social interaction. Our brains are adapted to expect proximity to others, yet across the country cities, towns, and suburbs are designed in large part for social distancing and isolation.

5.2.2 Social Capital and Social Exclusion

It is a maxim of life in the U.S. that any difficulties experienced by the general population are felt more acutely among people who are Black or Brown and those who are poor. That includes lack of social capital, civic decline, and the attendant problems, including worse physical and mental health, as well as social, political, and economic well-being. Data shows that the demographic groups that face the most serious levels of economic disadvantage also face the highest levels of social poverty.[81] This is partly a function of the planning and design of our built environment and disparities in access to civic space. Americans with only a high school education or less "are now more likely to live in civic deserts, lacking commercial places (e.g. coffee shops) and public spaces (e.g. community centers, parks, and libraries) that are hubs of community connection." Partly as a result, these Americans are more likely to be socially isolated. The same report concluded,

> The density of civic infrastructure and the robustness of American friendship networks correspond remarkably closely. Americans with less access to civic infrastructure – such as parks, coffee shops, and libraries – have many fewer friends and report greater difficulty making social connections. Americans with no access to public or commercial spaces are more than three times more likely than those with the most access to report having no close friends (32 percent vs. 9 percent).[82]

As a result, "associational life has apparently become a "high end good" that most people can't access."[83] Corinth et al. highlight how government has tended to focus on economic poverty rather than social poverty largely because the former lends itself to easier data collection, giving politicians and policy makers more information to formulate interventions. "We lack good data on individual social wellbeing, particularly regarding social capital – the strength and number of relationships with people and institutions that provide us value. It's hard to solve a problem you can't measure."[84]

Fewer resources are targeted at addressing this "poverty of social disconnection," and knowledge and expertise in this area are still evolving. Tackling social poverty requires finding ways to build and strengthen relationships – within families, communities, and civic institutions, and it is difficult to see how this can be achieved without a comprehensive program to address the legacy of segregationist architecture and planning that blights the lives of poorer communities and leaves large sections of the population in "civic deserts."

As with many other problems in the U.S., the origins of the current crisis of social isolation and loneliness can be traced to the deliberate policy of promoting metropolitan differentiation and segregation by the federal government and the real estate industry in the decades immediately following World War II. Although American cities were traditionally ethnically homogeneous at the neighborhood level, often containing separate blocks of African American, Irish, Italian, Polish, and Jewish communities, these blocks were typically located in proximity or even directly adjacent to each other, allowing friendships, business connections, and romances to develop across racial and ethnic lines.[85] The boundaries between neighborhoods were often quite porous and subject to changing patterns of settlement as various ethnic groups sought to navigate and negotiate territory. This often led to tensions as evidenced by the flare-ups and occurrences of violence that developed as different racial and ethnic groups vied to protect boundaries. Some of these social and ethnic tensions around community, territory, and identity were captured in popular culture by movies like West Side Story, a modern retelling of Shakespeare's Romeo and Juliet. Historically, it was also common in U.S. cities, and, indeed cities across the globe, for domestic workers and gardeners as well as the lower-income residents to live near or indeed within the upper-middle-class homes in which they worked given the unsociable hours there were required to report for duty and the lack of motorized transportation. This explains why racial zoning ordinances when they were first introduced typically included exemptions for domestic staff and servants. Putnam has observed that the outcome of this historical ethnic patchwork quilt was racially, ethnically, and economically homogeneous blocks located within municipalities that were often pluralistic, requiring a degree of cooperation and interaction. This dynamic changed significantly with the advent of modernist planning, the interstate highway program, and the rise of suburbia when the physical and social fabric of city neighborhoods was shattered and entire municipalities became increasingly homogeneous in ethnic and class terms as porous boundaries were frozen in concrete.

Although African American communities were disproportionately targeted by the planners and architects during urban renewal and the highway program they were by no means the only victims. Poor white ethnic communities experienced equally devastating consequences as their neighborhoods were torn apart. Robert Caro, author of the seminal work on Robert Moses "The Power Broker," uncovered numerous tales of human tragedy, isolation, and social disconnection while researching the impact of the Cross Bronx Expressway in the Tremont area of the Bronx. At the time the neighborhood was a mixed Jewish and Irish community with some African Americans.

> … I'm interviewing them about what their life was like before and what their life was like now. And I remember it hitting me … I type up the interview … And then I realized I was typing over and over the same word: "lonely." They were saying they had friends. They had family. But now they didn't know anybody. "Lonely is a word … In my opinion, you don't use the word "lonely" about yourself unless it's very, very overwhelming in your life … Sometimes you interview an elderly couple … They don't know anybody. They used to have this wonderful life with friends around - a sense of community. Now they have nothing … a huge, forced migration. A lot of them wound up in places they didn't want to be because they couldn't afford anything. So, you're talking about a human tragedy here.[86]

It is estimated that in New York City alone Robert Moses forcibly evicted 250,000 people for his highway projects and a further 250,000 for urban renewal, highlighting the vast scale of the devastation inflicted during this period. Given severe housing shortages in the city, and with many of those affected living in rent-controlled apartments, the social and economic consequences of the forced removals were devastating. In urban communities in the U.S., lack of material wealth had been somewhat alleviated by historic architectural and design features that served to enhance social life and community cohesion.

> "… it didn't matter that they were not very well-off people because they had neighbors. They had stores where everyone knew your name and your kid's name, and you could send them out for milk and all. The old men would sit around benches on Southern Parkway and play chess. And the women … had their baby carriages, and they'd sit there on Southern Boulevard and talk. So although they didn't have much money, they had a lot … And all of a sudden this thing came along … and they had nothing … The community – the people were scattered."

Caro's description of the Tremont neighborhood pre-urban renewal and the Cross Bronx Expressway echoes similar descriptions by Jane Jacobs of neighborhoods in New York, Boston, and Philadelphia as well as accounts of thriving African American communities like Sweet Auburn in Atlanta, Overtown in Miami, Greenwood Avenue in Tulsa, Rondo in St Paul, and numerous others across the U.S (Chapter 3). The common theme provided by all these descriptions is of socially cohesive working-class neighborhoods with walkable streets and a preponderance of "eyes on the street" i.e. natural surveillance provided by the presence of other people. Old men playing chess and grandmothers with strollers formed part of the "sidewalk ballet" identified by Jacobs that left strangers and residents feeling comfortable in the knowledge that they were part of a wider social gathering.

The flipside of the vibrant street life that was characteristic of many urban neighborhoods before highways and urban renewal are the dead spaces or "shatter zones" underneath urban expressways where poor lighting, concrete or metal supports, and a lack of street life combine to create an environment that is welcoming only to crime and anti-social behavior. Section 3.1 outlined the financial loss and dislocation experienced by inner-city communities across the country as roads and urban renewal programs tore up neighborhoods and forcibly removed large sections of inner-city populations to more distant "second ghetto" areas outside the urban core.

For those left behind the long-term impact was equally bleak as many communities were cut off from their natural social and economic hinterland by hulking strips of concrete and metal infrastructure that segregated and isolated entire neighborhoods. Amy Stelly, a planner and community activist in New Orleans, described to this author the drug dealing, prostitution, and illegal selling of firearms that regularly takes place underneath the Claiborne Avenue Expressway in her own neighborhood. Similarly, residents of Sunset Park in Brooklyn described to this author feeling fearful at night when criminality and an absence of eyes on the street combine to make the areas under the Brooklyn Queens Expressway a space to be avoided. Walking part of this area in the early hours of darkness with a group of my students from the Pratt Institute, some of whom are required to traverse this path to access public transport, is a stressful and unpleasant experience. Huge volumes of traffic, with its attendant noise and impatient commuters, combined with the dark and desolate environment under the highway render this a hostile and eerie area to be avoided. On our walk we perhaps witnessed some low-level criminality but avoided looking too closely lest we found out (Figure 5.14).

The field of neuro-architecture, which examines how architectural spaces and urban designs influence human cognition, emotions, and behaviors has shed further light on the way in which buildings and other structures affect our mood and well-being.[87] Using wristband readings to measure arousal and on-the-spot emotion surveys in Lower Manhattan, Ellard found that when subjects walked past a Whole Foods store with a long, smoked-glass frontage, "their arousal and mood states decreased, and they quickened their pace. In contrast, their mood improved when they reached

FIGURE 5.14 A classic shatter zone located underneath the BQE at Hall Street in Brooklyn. This space is eerie, inducing feelings of discomfort compounded by the noise and large volume of traffic. The image shows someone pushing an infant in a stroller. Although there are pedestrian crossings along this stretch, navigating the road is an unpleasant experience with a hazardous feel. Ones bodily sensations change entering this space, with an increase in adrenaline and a nervous system on the look out for potential dangers – not least from the vehicles that sometimes serve off the main route into the spaces beneath the BQE. The experience is stressful and to be avoided.

Photograph by Tim Cunningham, New York City, November 2024.

areas with more varied storefronts such as restaurants and smaller shops." The same study also found that "people are strongly affected by building façades, with complex and interesting façades having a positive effect, while simple and monotonous ones have a negative effect." The work by Ellard and others provides measurable and scientific data to support what Jane Jacobs, Holly Whyte, Jan Gehl, Fred Kent, and successive generations of architects and urbanists have been arguing since the mid-1960s - namely the importance of diverse, mixed-use neighborhoods with active street life that create safer, more vibrant, and happier communities. Additional insight into the psychological impact of these structures is evident from recent comments by Rep. Alexandria Ocasio-Cortez who described the insidious and powerful psychological impact of the Cross Bronx Expressway in her community

> One of the things that I really did not appreciate until I went to college and lived in a different city is how much the civic and urban planning really affects the psyche of the communities that occupy and have to endure these decisions … there is a psychic weight to living in communities that are designed to be disconnected. It affects your social life … we have some of the longest commutes in all of New York City. And it is a commute not just to work, it is a commute to do anything. It is a commute to connect socially. It is a commute to connect spiritually if you are part of a faith community. It is a commute to get your groceries. And so, these decisions are designed to disconnect, disempower, and isolate people … you can really see how it actually builds in organizing challenges to communities who actually want to empower themselves politically.[88]

Anthony Foxx, Transportation Secretary in the Obama administration, expressed similar feelings about growing up in Charlotte, North Carolina, in an area devoid of grocery stores, doctors' offices, and other necessities where "freeways were carrying people through my area but not really to it."[89] His community was surrounded by highways and fences but he had no sense as a child that the prominent physical barriers were compounded by more invidious economic and psychological ones. Like Ocasio-Cortez, it was only later in life when he began to understand the range of decisions that had led to his community and countless others', being *constrained* rather than connected by infrastructure, that he fully appreciated how transportation and city planning impact and shape lives. "*Transportation gets us places … but transportation also makes places … and it can make good places or it can make places that are hard to get out of ….*"[90]

The impact of urban highways on social disconnection and increased feelings of isolation and loneliness are also confirmed by recent quantitative research based on large social media datasets. One study which covered 50 U.S. cities showed that the barrier effect of urban highways was "especially strong for short distances and consistent with historical cases of highways that were built to purposefully disrupt or isolate Black neighborhoods."[91]

Another outcome of auto centric planning practices is that U.S citizens have been required to spend measurably more time every day alone in their vehicles - at least until the pandemic. The proportion of Americans travelling to work in a private vehicle rose from "61 percent in 1960 to 91 percent in 1995, while all other forms of commuting – public transit, walking, etc. declined."[92] Not only has commuting increased but so too has the *time* spent commuting, which in turn has contributed to a detrimental impact on community life, civic engagement, and social interaction. Putnam calculated that "each additional ten minutes in daily commuting time cuts involvement in community affairs by 10 percent." He also found that increased commuting time among residents of a community has negative externalities lowering average levels of civic involvement *even among non-commuters.* Part of the problem is that high-commute communities compete with place-based ties so that peple have to choose between spending time with their work colleagues or their neighbors, but lack the ability to do both. More frequently, the time and energy expended commuting means that residents of high-commute communities do neither. The detrimental physical, social, economic, and civic costs of commuting is part of the reason many employees so willingly embraced remote working during the pandemic and wish to continue with that practice in some form. The fact that many CEOs and senior executives can typically afford to live in closer proximity to their workplace doubtless explains the insistence from those higher up the corporate ladder that employees need to 'get back to the office'. The lack of decent, reliable and affordable transit in the U.S. is another factor that significantly diminishes the commuting experience. Sprawl has created an additional problem for parents who are required to spend considerable time and money driving to and from organized sporting activities or recreational activities. The adverse consequences of dispersed development impacts every aspect of life and are confirmed by a recent study which found "relying on a car for more than 50 per cent of the time for out-of-home activities is associated with a decrease in life satisfaction.[93] This is a quite shocking statistic that is directly related to the shape and design of our living spaces. The authors of this report recommend that planners and decision-makers ought to promote multimodal transportation and land use policies that may help reduce car dependance and its potential negative impact on subjective wellbeing.

There is a tendency to view commuting as a problem that afflicts only suburban and rural communities however as Rep. Ocasio-Cortez highlighted inner-city neighborhoods also experience this phenomenon due to the way in which these areas were dismembered by concrete and steel with small local businesses replaced by flyovers and wide intersections. The "psychic weight" that she alludes to that disconnects people socially, spiritually, and politically, leaving them isolated and disconnected mirrors the comments by numerous others about the reconfiguration of urban space since World War II. The problem of "spiritual disconnection" is worthy of comment given that churches, especially ones that served African Americans, were a target for destruction during the urban renewal era of the 1950s and 1960s. In some cases this was because of their connection to the civil rights movement but more often, churches were destroyed because they were embedded within the fabric of communities earmarked for land clearances to make way for highways, hospitals, universities, middle-class developments, and other urban renewal projects.

For those who moved out to the postwar suburbs social disconnection was severe but played out in slightly different ways from those who were left in the urban core. As young families were enticed with the prospect of buying a home cheaper than they could rent one in the city, many children grew up in isolated nuclear families largely separate from grandparents and extended relatives. Recent research indicates that the psychological impacts of the design features that the FHA helped promote are greater than previously recognized. Goldhagen notes that the suburban housing boom was predicated upon "an architecture of isolation" creating "inward looking, and inward feeling, living spaces, where residents typically enter homes through the garage while "front" doors "gaze sorrowfully onto the street."[94] William Whyte, Betty Friedman and other social scientists described over the years how the sterile architecture of post-war suburbs led to excessive conformity and a high degree of conservatism that had a stultifying effect on the "corporate men" commuting to workplaces each day and the young women they left behind who were isolated for long hours by designs that inhibited social interaction. Scholars like Lewis Mumford castigated the low-grade uniform environment of U.S. suburbia "from which escape is impossible."[95] Baumgartner's ethnographic study of a suburban New Jersey community in the 1980s found "a culture of atomized isolation, self-restraint, and moral minimalism."[96] Kenneth Jackson meanwhile observed how suburban auto centricity led to the demise of the sidewalk and the front yard so that social intercourse, the main characteristic of urban street life, vanished. He concluded that "there are few places as desolate and lonely as a suburban street on a hot afternoon."[97] Fieldwork for this book included walking ethnography carried out in several suburban developments in various parts of the U.S. One common theme of the sites visited is the way in which the act of being a pedestrian is viewed with, at best, mild curiosity by residents. Unless one is wearing athletic gear and 'power-walking,' or accompanied by a dog or an infant in stroller, most contemporary suburbanites do not expect to see people on foot. My white skin, age, and apparel mean that I am viewed as an oddity rather than a threat by residents who pass by in their vehicles. In many respects this curiosity is justified given the lack of public spaces, retail outlets, or indeed anything remotely worth visiting other than a dull, repetitive landscape of interest only to social scientists.

There is also a wider political consequence to these design and planning practices that serve to differentiate and segregate by demography. Oliver found that by creating communities of homogeneous political interests suburbanization reduced the local conflicts that engage and draw the citizenry into the public realm in the first place resulting in low civic engagement and neighborliness.[98] It is a paradox of civic life that people often become actively engaged in local politics when something happens that they disagree with. Greater social homogeneity and stringent design rules and regulations written into housing codes reduced engagement in local politics because many of the disputes that were debated in that realm have been contracted out to third parties like home owners associations. These detrimental impacts of architecture and planning on social interaction and civic engagement are compounded by additional problems relating to physical safety and mental health that arise from deliberately designing streets and roads as moats and racial buffers to promote segregation and exclusion.

5.2.3 Closed Off: The Anxious Generation and Accessing Space

Another legacy of the isolationist architecture that characterizes the contemporary U.S. is evident in debates about other social ills that afflict the country notably anxiety among young people. Smartphones and social media are often cited as the primary reasons why Americans are "hanging out" with each other less. As Tanya Panova highlighted however, for many teenagers, the smartphone is merely "the solution to a range of other problems" that include neighborhoods that are designed in ways that leave friends too far apart to spend time together and which have a dearth of interesting locations for social interaction.[99] Several other observers have also acknowledged the role of architecture and urban planning for explaining why people spend less time together.[100]

It may sound trite, but the reality is that hanging out and spending time with friends outside the digital world requires a physical location in which the act can take place.

Over the past decades, the number of accessible and free "third spaces" (i.e., non-home or non-work venues) that young people can utilize has decreased significantly. Parks, libraries, green spaces, and playgrounds are the type of public places that support the development of social connection.[101] And yet these very places have struggled in recent decades as public spaces have been increasingly privatized and under-resourced. This problem is not unique to the U.S. A feature article on the dilemma currently facing teenagers in the U.K. highlighting that "There's nowhere to hang out for free anymore and it's making us lonely" epitomizes the current problem in many towns, cities, and suburbs, where physical distance and lack of affordable social spaces are the key reasons cited by young people why they were unable to spend more time with their friends.[102]

Clearly a range of factors has contributed to social disconnection, loneliness, isolation, political polarization, and mental health issues, including anxiety and depression, especially among young people. Much criticism has been leveled at the role of digital communication and the need for greater cooperation from tech companies. Former Secretary of State, Hilary Clinton, expressed concern about the impact of big tech on the mental health of young people and the adverse effects of social media. Jonathan Haidt's recent work "The Anxious Generation" received significant attention from parents, educators, and child psychologists for identifying the addictive and adverse consequences of smartphones and social media.[103]

Much analysis of this problem focuses primarily on the role of social media in developing and promoting algorithms that drive addictive online behaviors, yet many of the solutions proposed by the same analysts lie within the realm of the architects and planners who shape local neighborhoods. Haidt for example, recommends that one method by which parents can address the problems facing "the anxious generation" is by "sending their kids to the grocery store for a clove of garlic while they are cooking dinner and encouraging them to walk to school with a group of friends."[104] These are desirable and worthy objectives but practically impossible for most parents across the U.S. When I put Haidt's suggestions to a group of college students in suburban New Jersey there was loud derision about the prospect that they could walk from their homes to a grocery store selling garlic or indeed anything else. Growing up in the shadows of U.S. Route 9, the students with whom I interacted bemoaned the lack of gathering spaces and the plethora of liquor stores and retail outlets selling auto supplies scattered around their immediate neighborhoods. The students also complained how they needed a car to get to class, and indeed, to get anywhere.

The non-profit "Strong Towns," has pointed out that in the past 55 years, the number of children who walk to school has declined from 40.7% to 11% because the built environment has transformed in ways that make walking to school unsafe and practically impossible.[105] The Strong Towns Director of Community Action Edward Erfurt experiences this problem firsthand with his own son attending middle school less than a mile from home but physically unable to walk there because of the danger of traffic. The street to the school has no sidewalk, "14-foot-wide travel lanes (two feet wider than highway lanes) and a double yellow line, which is common on high-speed roads and makes drivers feel comfortable driving fast."[106] His organization, which focuses on designing safer streets, proposes a simple, three-step response to the problem: (1) remove the double yellow line to visually communicate that the street is a slow, residential area, (2) narrow the travel lanes to 10 feet to "further emphasize that drivers must maintain slow speeds," and (3) use the extra 8 feet to create a walking area, physically separated from the travel lanes with curbing or bollards. Although such change is relatively inexpensive and would be a game changer for parents and children alike in facilitating walking and even biking to school the reality is that measures like this are needed across the U.S. and not every community, especially ones that need them most, have an urban design professional on hand to advise and advocate for safer streets let alone a municipality who might be interested or willing to put these ideas into practice.

Although data is still emerging highlighting the impact of social disconnection on the health of individuals and communities awareness of the importance of interactions between people of different racial groups has been around for some time. Gordon Allport's seminal work "The Nature of Prejudice" published in 1954 proposed that under appropriate conditions, interpersonal contact between different racial groups could effectively reduce prejudice. Crucially, however, one of Allport's conditions was that intergroup contact should involve equal status, common goals, intergroup cooperation, and support from authorities or customs. As outline in Chapter 2, this approach was anathema to the FHA, which insisted that racial homogeneity and strict segregation was a sine qua non for ensuring successful residential developments.

Subsequent studies have adopted a more nuanced approach to Allport's findings. His essential requirements are now viewed more as facilitating factors while other studies have confirmed that intergroup contact can reduce prejudice even in suboptimal conditions. The intergroup contact theory is widely accepted within the field of social psychology, with support from organizations such as the American Psychological Association and provides the basis for many Diversity, Equity, and Inclusion (DEI) initiatives. Meta-analytic tests by Pettigrew and Tropp in 2006 strongly supported the wider contact theory finding that positive results extended beyond racial and ethnic encounters to include other groups.[107]

Given that intergroup contact is so important in promoting tolerance and reducing prejudice across all groups, one must seriously question the wisdom of planning, designing, financing, and constructing the plethora of "lifestyle communities" that restrict membership to one specific demographic. One might also question the impact on the country as a whole of generations of Americans forced by FHA lending policies to grow up in racially homogeneous suburbs and attend racially homogeneous schools with little or no contact with other racial or ethnic groups. In a lecture at Brown University, Richard Rothstein asked how the U.S. can ever develop a common national identity if so many African Americans and whites live so distant from each other and are unable to understand each other's life experiences. As Heather McGhee has noted, segregation is often viewed erroneously as merely a problem that affects African Americans but the reality is that "white people are the most segregated people in America".[108] The curious paradox is that although segregation is aimed at limiting the choices of people of color, "it's white people who are ultimately isolated."[109]

McGhee points out that in a survey taken after the uprisings in Ferguson, Missouri, following the killing of Michael Brown, an unarmed Black teenager, the majority of white American said they regularly came into contact with only "a few" African Americans. In 2016, three-quarters of white people reported that their social network was entirely white.[110]

The lack of common national identity that Rothstein alludes to, like the lack of common purpose identified by the Rand Foundation, will remain elusive so long as architects, planners, real estate professionals, and lawyers conspire to make diverse neighborhoods a practical impossibility. Since World War II, segregation and social sifting have increased rather than decreased in most parts of the U.S. The consequent lack of social connection, empathy, and understanding impacts political and policy choices creating more expensive and inefficient healthcare, education, housing, and transportation. Dr. Melba Sullivan, who has a doctorate degree in clinical community psychology and was an uncommon person of color to grow up in the predominantly white suburb of Deerfield, near Chicago, argues that diversity is important because it creates "the opportunity for people to build emotional ties that enhance their ability to empathize and show up as a compassionate friend with someone who doesn't look like them."[111]

Derald Wing Sue has pointed out that a balanced picture of another racial group requires spending time with healthy and successful individuals of that group. One outcome of historical segregationist policies is that, currently white Americans are most likely to have contact with people of color who represent only a narrow spectrum i.e. those who need special help or who have gotten into trouble. Sue notes that police officers, court officials, social workers, and healthcare professionals typically interact professionally with people who are poor, have mental health issues, and come from dysfunctional families. White professionals in the caring and law enforcement professions meet white people who have these kinds of issues but their views of the white population as a whole are balanced by the wide number of functioning and successful white people they interact with in their daily lives. For many white people in the U.S. an "absence of contact with the wide range of healthy, successful, and well-adjusted individuals and families in African American communities provides a distorted perspective of people of color."[112] Sue outlines a series of recommendations that white people should take in order to overcome personal racial fears and discomfort, including inviting neighbors, colleagues, coworkers, or students of color home for dinner or for a holiday. He also suggests that people ought to frequent minority-owned businesses and live in an integrated or culturally diverse neighborhood, attend neighborhood organizational meetings, and attend or help to organize block parties.

Asking white Americans to become more friendly with their African American neighbors, frequent minority-owned businesses, and live in an integrated or culturally diverse neighborhood is a bit like asking parents to allow their children to walk to school with their friends or go to the grocery store to pick up a clove of garlic – extremely difficult, thanks to the planning and design practices that shape our built environment. Moreover, efforts to promote diversity are undermined by the plethora of legal, economic, as well as physical barriers to trying to build more cohesive and integrated spaces where people can rub shoulders with those who look different. These barriers include exclusionary zoning laws that prohibit multifamily dwellings and local businesses, mortgage tax relief, transportation funding, and other "ostensibly race-neutral measures that frustrate efforts to create more socially connected spaces."[113] Efforts to promote "upzoning" and increase the amount of "missing middle housing" that would allow neighborhoods to become more economically and racially diverse are often met with fears that greater economic and racial diversity will lower property values, increase crime, and/or have a detrimental impact on "neighborhood character." Rothstein has outlined how recursive planning practices build upon existing patterns of disadvantage to further concentrate poverty and segregation

> it's much easier to build affordable housing in low income segregated neighborhoods … so that's what we do … we reinforce segregation by building affordable housing in low income segregated neighborhoods where its easiest to build …. land is cheaper, you don't need to hold 100 community meetings for approval … it's easier to rent apartments because there are lots of people who can walk by and see a sign in the window … it's much easier to reinforce the segregation of the past.[114]

Archer has argued that the specific configuration of a given space plays a "crucial role in the formation and sustenance of the consciousness of all those who exist there". He contends that "space is integral to the delineation and facilitation of the whole spectrum of relations among individuals, institutions, and social fractions (classes, races, castes, etc.) … the very terms of human praxis – economic, political, moral, religious, artistic, etc. – are embedded in, and sustained by, the configuration of surrounding spaces."[115] In a context in which spaces, identity, and human praxis has been embedded in and sustained by spaces where Black and Brown bodies were viewed as alien or a threat, it is unsurprising that many of those who grew up in those spaces baulk at the presence of racial difference in "their" spaces.

Alexandra Lange has described how shopping malls, both in cities and suburbs in recent years have introduced ordinances banning young people unless they are accompanied by an adult. As with many coercive ordinances limiting access to space, there are strong racial dimensions to these practices. Lange references the segment "Shopping While Black" from the sitcom *Everybody Hates Chris* to highlight how security guards routinely follow African Americans in shopping malls. Research from mixed-income developments in Chicago shows how "the enduring power of an urban underclass narrative" informed by "embedded assumptions about race, class, and culture" led some residents, both African American and white, to complain about young African American males engaging in behaviors deemed unacceptable.[116] These behaviors included "hanging around," loitering, barbequing in public, and playing chess. These issues arose in developments that were deliberately designed to limit the amount of public space showing how little room there is for Jane Jacobs' sidewalk ballet of assorted activities and neighborhood characters in developments designed to ensure that integrated housing does not mean integrated living.

The "enduring power" of the urban underclass narrative and the legacy of generations of white Americans raised in lily-white suburbs also comes to the fore with incidents like the arrest of Donte Robinson and Rashon Nelson at a Philadelphia Starbucks store in 2018. The two men, both African American, had gone to the store for a business meeting to discuss a real estate project and asked to use the bathroom but were told it was for paying customers only. Like many people at Starbucks they sat at a table without making a purchase. Unlike the rest of the white clientele who frequented the store they were asked to leave and when they refused the police were called and they were arrested. The arrest went viral, and the incident provoked a national outcry over racial profiling. The two men reached a non-disclosed financial settlement with Starbucks and the chain took the unusual step of closing all its stores for one day to provide training on racial bias.[117]

Christian Cooper, a Black, male, birdwatcher, asked a woman in Central Park to leash her dog because she was in an area of the park that was designed to protect wildlife. The woman, who was white, refused and called the police, falsely claiming that an African American male was threatening her.[118] Cooper has written a memoir, with one chapter devoted to the Central Park incident.[119] This case is illustrative of several other incidents that have arisen in the same location which show that Central Park functions as "white public space." Young Black male researchers for example report that "police officers trail behind them when they are photographing or mapping social behavior, and they find it difficult if not impossible to interview in this setting."[120] Lolade Siyonbola, a Black graduate student in African studies at Yale University, was interrogated by campus police officers after she fell asleep in the common room of her dorm and a white student called the police.[121] Kelly Fyffe-Marshall and her three friends, two of whom were also African American, were questioned by police in Rialto, California, as they checked out of an Airbnb they had rented.[122]

This phenomenon is also evident in the way racial tensions play out around the issue of gentrification. For some of those who move to an integrated or culturally diverse neighborhood, the legacy of decades of social and racial differentiation, with behavior strictly regulated by zoning codes and homeowner association rules, means that they are more likely to call 311 than invite an African American neighbor to dinner. Unlike 911, 311 calls relate to what are considered nuisance but unthreatening behavior. The 311 system exists to ensure that those experiencing excessive noise, etc., have an outlet to make their complaint. Several long-term African American residents of Bedford-Stuyvesant have recounted to this author the extent to which this system is used to regulate and alter the character of the neighborhood. I have heard complaints about 311 calls being made not just about loud music and parties but also about excessive noise complaints relating to church bells and in one case a Mr Softee ice cream truck. The complaints I listened to are in line with other studies that show a significant rise in 311 calls following the arrival of new white residents in historically Black neighborhoods. Videos on social media show people calling the police on African Americans engaged in all sorts of mundane activities – babysitting, eating lunch, going for a swim.[123] The residents with whom I have spoken express anger not just at what they consider to be ludicrous and bizarre behavior on the part of certain individuals but also what they see as a wider effort to stop "Brooklyn being Brooklyn". These long-term residents typically feel that their behavior and lifestyle is under threat not just from rising property prices (which is undoubtedly a concern) but also at the way in which their quotidian actions are being targeted and criminalized. Other long-term residents complain about the fact that new neighbors (i.e. young, affluent, and predominantly white gentrifiers) refuse to make eye contact or say hello – findings that have been confirmed by other studies.

Setha Low has pointed out that public culture, like public space, is "socially produced by the history, political economy, and materiality of a specific time and place." She argues that "public culture mediates the relationship of public space and the public sphere by communicating ideas and feelings that shape structures of social relations and interactions."[124] In a context in which the materiality of the built environment has, for several decades, communicated ideas and feelings that certain people, notably those of color, do not belong in a wide number of spaces, it is hardly a surprise that African American college students, businessmen, or those renting vacation property find themselves questioned by security or law enforcement officials. From Yale University to Central Park in New York City, and from Rialto, California, to the suburbs of Philadelphia, from main street to the suburban shopping malls, these incidents, and countless others, many of which do not receive national attention, reflect decades of laws and public policy that created racially and socially homogeneous neighborhoods denying generations of white Americans "the humanizing and socializing effects of being active in a diverse public sphere."[125]

5.2.4 Barriers, Buffers, and Dangerous by Design

The promotion of roads, highways, and street design to safeguard white neighborhoods from encroachment by "inharmonious racial groups" has also had serious consequences for the physical health of those communities who were severed from the social and economic life of metropolitan centers. Black and Brown neighborhoods in the U.S "tend to have more high-speed roads, poor visibility, and heavy traffic volume and a lack of facilities for people walking."[126] "In many cities communities of color house a disproportionately high share of the most deadly roads, devoid of pedestrian infrastructure."[127] For example, in Philadelphia, 46% of the most dangerous roads are in poor areas mostly populated by people of color.[128] Deborah Archer traces the current lack of pedestrian infrastructure in Black neighborhoods to the etiquette of Jim Crow and slavery that required African Americans to keep on the street and prohibited them from sidewalks.[129]

The outcome of decades of this form of infrastructural racism is that Black and Brown people in the U.S. are significantly more likely to experience a pedestrian fatality or injury. Research shows that Black, American Indian, and Alaska Native populations account for almost "22% of all pedestrian deaths in metro areas despite accounting for just under 13 per cent of the population."[130] Black pedestrians are killed at more than twice the rate of their white counterparts.[131] A study in Florida found that even after accounting for differences in income, Black populations were "strongly related to increased pedestrian death and injury."[132] Significantly, the same study also found that in lower-income areas, pedestrian death and injury was "associated with difficulties accessing household-supporting destinations", for example workplaces and grocery stores. For affluent areas, the factors associated with increased pedestrian death and injury "were those related to recreation and nightlife – specifically, bars and clubs, hotels and restaurants."[133] In other words, African Americans are typically killed and injured in the course of accessing essential services while for more affluent white Americans the problem was more strongly associated with going out at night when one might assume alcohol consumption contributes to impairment of senses.

As with many other areas of life, a substantial body of work has emerged in recent years connecting road deaths, road design, and racial background, challenging the notion that pedestrian fatalities are a function of individual culpability and human error.[134] Activists and experts are increasingly demanding that the road and street design manuals used by traffic engineers and planners be altered to produce safer travelling experiences for pedestrians and cyclists.[135] Urban planner Veronica O. Davis offers a manifesto for repairing divided communities and identifies the link between race and pedestrian death and injury through for example lack of sidewalks and pedestrian crossings.[136]

> You show me any city in the US … show me where there's no sidewalks, and it is highly likely those communities are going to be Black, brown, low-income, or some combination of the three. It's the same thing for a high-injury network in any city.[137]

The previous section (Section 5.1) highlighted how the design of sidewalks and paths that disconnect many affluent suburbs from non-white and poorer communities and a lack of street signage formed part of a wider design strategy aimed at consolidating racial segregation.[138] Anthony Foxx, Transportation Secretary in the Obama administration, pointed out that only 49% of low-income neighborhoods have sidewalks, but 90% of high-income neighborhoods have sidewalks.[139] He drew attention to the human cost of this disparity through the case of Raquel Nelson, a 30-year-old single mother from Cobb County, outside Atlanta. Nelson's four-year-old son was struck by a car when the family were crossing the 85-foot-wide four-lane street to their apartment building across the road from their bus stop.[140] Nelson, like most people in her neighborhood, crossed the busy road rather than walk one-third of a mile to the nearest crosswalk and another one-third of a mile back to her apartment.[141] She was convicted of vehicular homicide in the second degree for not crossing at the sidewalk but received a suspended sentence. The case became a cause célèbres for advocates for better transportation with campaigners pointing out that Cobb County was built for car use, but some less affluent residents who have subsequently moved in are dependent upon public transportation. Nelson, like many newer residents in the area, was unable to afford a car but many members of the all-white, middle-class jury that convicted Nelson admitted to never using public transit themselves. The NAACP argued that the jurors failed to understand the plight of Nelson, an employee, mother, and student at Kennesaw State University.[142] As with other examples provided in Chapter 3, the planning and design of the streets and roads in Cobb County adhered to racist practices that sought to keep African Americans, physically and figuratively, in their place.

The racial dimensions of road design and the legacy of the FHA Underwriting Manual are also evident in the death of Cynthia Wiggins who was hit and killed by a dump truck while she was attempting to cross a seven-lane highway to get to the mall where she worked. Like Nelson, Wiggins couldn't afford a car and rode the bus from her inner-city home to the suburban mall where she worked. Despite numerous requests, the mall owners had actively resisted requests to have a bus stop on the mall property. Instead, the bus stopped outside the mall on the other side of the highway.[143] Documents produced in subsequent legal proceedings revealed that the decision-making process was motivated at least in part by a desire to prevent people of color from accessing the mall. Reports emerged that mall managers had stated they were opposed to a bus stop on the premises because they didn't want to attract "the kind of people" who use public transit to their facility. Members of Buffalo's Black community claimed that the mall was "trying to use the highway as a moat." More than a century after the original FHA Underwriting Manual extolled the benefits of roads and expressways as "buffers" to prevent access by "inharmonious racial groups" the planning and design of the built environment continues to reflect that reality often with fatal consequences.

Architecture and planning decisions, especially lack of sidewalks, create additional problems by providing more opportunities for confrontations between young people of color and law enforcement officials.[144] One Chicago study found that 93% of tickets for bicycle infractions were issued on streets with no bikeways. These areas were disproportionately Black and Latino.[145] These road designs force pedestrians, cyclists, and e-scooters to break the law simply because there are no better alternatives by riding on the sidewalk to avoid a busy street or walking on the side of a road without a sidewalk.[146]

A recent study in Washington state found that Black pedestrians are stopped by police at an average rate over four times that of their share of the population for jaywalking.[147] Statewide records show that most stops occur at least a 3.5-minute walk, round-trip, from the nearest marked crosswalk, illustrating how jaywalking laws criminalize behavior that is a natural response to lack of pedestrian infrastructure. In L.A., a city with a 9% Black population, nearly one-third of pedestrians issued jaywalking tickets over the last decade were Black. In Atlanta, African Americans make up almost 52% of the population, yet 90% of those ticketed for jaywalking and 81% of those cited for not using the sidewalk were African American.[148] In the same city, 8% of streets are responsible for 88% of traffic fatalities and more than half of serious injuries. Those areas with the highest levels of death and injury have a larger share of Black residents, lower rates of vehicle ownership, and lowest level of sidewalk coverage. This illustrates again how a traffic violence doom loop ensures lack of pedestrian crossings not only increases injuries and death but also contributes to problematic interactions between communities of color and law enforcement that can lead to more serious consequences. The racialized nature of transit planning, provision, and policing is illustrated in a short film by Vox-ProPublica entitled "Walking while Black" which shows how young African American men in Jacksonville Florida are targeted by law enforcement for walking across a designated crosswalk while the pedestrian signal is colored 'red'.[149] The footage of numerous white police officers doing the same thing illustrates how racial harassment, rather than concerns for pedestrian safety, appears to be primary motive behind enforcement of jaywalking laws in many municipalities.

5.2.5 Inequality in the Air: Pollution and Road Design

Another legacy of the coercive architecture and planning of the past seven decades is the adverse health consequences experienced by minority communities caused by vehicle pollution. Understanding the extent of this phenomenon is still developing as scientists continue to research the toxic impact of living near highways. Existing data connects vehicle pollution with respiratory symptoms in children.[150] Research from the U.S. Environmental Protection Agency (EPA) concluded that communities of color, especially African Americans, faced far higher risk than white populations.[151] Researchers at the University of Minnesota found unequal exposure to nitrogen dioxide, produced by cars, construction equipment, and industrial sources, that has been linked to higher risks of asthma and heart attack.[152] Minority groups on average are exposed to 38% higher levels of outdoor nitrous dioxide than whites in the communities where they live. Nationwide, the difference in exposure is equivalent to approximately 7,000 deaths per year from heart disease. Research from the University of Washington demonstrated that exposure to traffic pollution can lead to high blood pressure, which again disproportionately impacts Black and Brown communities in the U.S.[153] Even after controlling for income, large disparities remain between white and non-white communities.

More recently, researchers from Imperial College London found that particles from the wear of vehicle tires are especially damaging due to the presence of toxic chemicals and heavy metals, including lead. This study is significant, given that research on vehicle tire wear has been historically neglected compared with work on fuel emissions.[154] As scientists continue to revise upwards the health impact of roads and highways the data on health inequalities illustrates the physical consequences of driving "white men's road through Black men's homes." Black children "are nearly three times as likely to have high levels of lead in their blood, which stunts intelligence and leads to greater violence in adulthood." Compared with white children, Black children "are almost one and a half times as likely to have asthma – and five times likelier to die from it." *The Guardian* newspaper recently carried a feature on "asthma alley" in the South Bronx highlighting environmental injustices in that neighborhood.[155] *The Economist* concluded in 2020,

> Greater exposure to fine particulate matter – the sort of pollution which most damages lungs – and delays to treatment brought on by a lack of good health insurance may explain why covid-19 was killing African Americans at twice the rate of white Americans.[156]

At a very basic level these disparities highlight the long-term consequences of infrastructural racism in terms of the physical health of Black and Brown communities. Scholars have noted however that poor health, especially among children also contributes to other disparities, notably educational outcomes. Rothstein for example has highlighted how conditions like asthma lead to missed days at school and adversely impacts the cognitive abilities of children. Other studies have found a link between lead exposure and later, delinquent, criminal and antisocial behavior. Although lead is no longer used in vehicle fuel in the U.S. children living closer to busy roads were found to be "more likely to have high blood lead levels, because the soil near those roads was still contaminated due to the use of leaded gasoline decades ago."[157] The physical, mental and social health of communities that were segregated and divided by roads and highways has been further compounded by the adverse environmental consequences to which we now turn.

5.2.6 Climate Change, Heat Islands, and the Highway Effect

Notwithstanding hurricanes, flash flooding, and other consequences of climate change extreme heat is currently the leading cause of weather-related deaths in the U.S.,[158] impacting disproportionately those with "pre-existing health conditions (e.g. chronic obstructive pulmonary disease, asthma, cardiovascular disease, etc.), limited access to resources, and older people."[159] As global temperatures rise "heat extremes in urban areas are becoming more common and the number of deaths and attendant illnesses are expected to increase in the US."[160] The urban heat island effect i.e. the phenomenon whereby the built environment absorbs and traps heat, can cause temperatures to vary "as much as 10C within a single urban area."[161] A major predictor for the surface urban heat island effect is the density of total impervious surface area (ISA), typically found in concrete, brick, and asphalt (i.e., parking lots, roads, and pavements).[162]

Many of the hottest urban areas in the U.S., unsurprisingly, tend to be inhabited by poorer residents and communities of color. A recent study for example found that deaths during heatwaves were more likely to occur in Black (non-Hispanic) individuals than other race/ethnicities, "more likely to occur at home than in institutions or hospital settings, and more likely among those living in census tracts that received greater public assistance."[163] Deaths during heat waves were also "more likely among residents of the city with higher relative daytime summer surface temperature and less likely among residents living in areas with more green space."[164] Another study found that residents living in neighborhoods with higher racial diversity, extreme poverty, and lower levels of formal education were statistically more likely to be exposed to the neighborhood heat island effect.[165] Many roadways and large building complexes found in contemporary U.S. cities were constructed in the 1940s and 1950s using high-density materials such as cinder block and brick. These materials gain heat during the day and release it back into surrounding neighborhoods in the evening, preventing a natural cooling down of the ambient environment. These resultant higher evening temperatures "are precisely the factors that can exacerbate excess mortality and morbidity." [166] Higher intra-urban land surface temperatures (LSTs) also correlate with higher summertime energy use which impose greater financial burdens on poorer communities. Increased healthcare costs further exacerbate inequities.

Greenspace, trees, or water features within cities have a cooling effect on LST.[167] Unfortunately, urban highways and urban renewal destroyed large sections of greenspace and trees that had historically provided a cooling effect and health benefits for local communities. Perhaps the most celebrated example of this was the Tremé neighborhood in New Orleans, where over 400 mature trees that provided shaded walkways and neighborhood gathering spots in the wide median of North Claiborne Avenue were destroyed in the 1960s to make way for I-10. More than any other structure, I-10 exemplifies the devastating social, economic, environmental, and health consequences of the highway program. The Claiborne Avenue Alliance has partnered with Thriving Earth Exchange and LSU School of Public Health to conduct a survey of I-10 neighborhoods to determine the extent and impact of the highway's pollution as part of a series of measures to secure removal of the structure.

The consequences of climate change, and the detrimental impacts of hulking concrete structures on the environmental health of the communities in which they were located is one outcome that the highway planners of the 1950s could not reasonably have foreseen. What is clear however is that the environmental costs of the "highway robbery" perpetrated by the interstate system is significantly greater than has thus far been appreciated and is an important focus of the contemporary environmental justice movement in the U.S.[168] Moreover, continued highway expansions (many of which defy economic as well as environmental logic),[169] the targeting of the Environmental Protection Agency by the Trump administration, and ongoing resistance on the part of some state authorities to support environmental protection measures,[170] show that despite scientific evidence, these injustices continue in many cases largely unabated. Ignorance is often said to be no defense. Unlike their counterparts of the 1950s and 1960s, contemporary politicians, planners, and public officials cannot even resort to that dubious justification for actions that continue to perpetuate acute environmental harms.

5.2.7 Spatial Inequalities and Transportation Injustice

The spatial reconfiguration of U.S. metropolitan areas in the decades following World War II, the destruction of urban neighborhoods by interstate highways, the use of roads, highways, and walls as racial barriers, the relocation of public housing developments to spatially isolated locations far from downtown areas, and a funding model that excluded African Americans from growing suburbs, does not entirely explain how patterns of racial inequality and concentrated poverty were shaped in U.S. cities through the twentieth century. Another important part of the jigsaw in ensuring widespread racial segregation is the role of transportation planning that left many African American communities disconnected from jobs, businesses, better housing, education opportunities, and white people.

In theory, the fundamental purpose of transportation planning is to ensure the efficient movement of people and goods as economically and safely as possible. It is also a truism that physical mobility serves to advance social and economic mobility. This explains why many of those stuck in poverty are also physically stuck in places that are far removed from jobs, education, and in many cases, access to affordable and healthy food. Sherrilyn Ifill, the president and director-counsel of the National Association for the Advancement of Colored People (NAACP) Legal Defense and Education Fund, has pointed out,

> I think people tend not to notice or to pay attention to issues of transportation and transportation infrastructure as a civil rights issue and it's one of the matters I've been pressing. Because access to jobs, access to mobility, reducing housing segregation, all of these things are deeply dependent on transportation decisions that happen in many ways out of sight out of mind.[171]

Ifills' a point is well made and illustrates why demands for racial equality have been inextricably linked with transportation planning and policy since the end of the Civil War. *Plessy v Ferguson*, the landmark case that legitimized "separate but equal" Supreme Court doctrine originated when Homer Plessy violated Louisiana state law by boarding a whites-only rail car. Rosa Park's refusal to give up her seat and the Montgomery

Bus Boycott also highlight the important role that transportation policy has played in the civil rights struggle by African Americans. As with segregated lunch counters and restroom facilities, growing public and judicial opposition to "whites only" railcars and sections of buses led white elites to devise more elaborate, invidious, and less obvious ways to maintain division of the races. Once again, planners and architects played a key role in ensuring that the physical, social, and economic mobility of African Americans were curtailed through ostensibly neutral and technocratic policies and practices.

Given that urban renewal and highway routing resulted in many African American communities being relocated to more remote sections of metropolitan areas mass transit could have gone some way to mitigating the impact of this dislocation by facilitating access to jobs and services. The problem was that transit planning in the U.S., like highway and road planning, was highly racialized serving to consolidate white supremacy. Just as highway route selections and street networks were deliberately designed to maximize severance and serve as physical barriers by impeding residential mobility, so too, transit routes were often deliberately designed to impede, not facilitate, the physical movement of African Americans around metropolitan areas. To fully appreciate the impact of transportation planning on patterns of racial disadvantage, one needs to acknowledge the way in which auto centricity and the growth of suburbia reshaped the spatial dynamics of the U.S. metropolis. Robert Fishman of Rutgers University accurately summarizes the relationship between the two:

> The [highway] system was supposed to save central cities by rescuing them from automobile congestion … the new peripheral "beltways," originally designed to enable long-distance travelers to bypass crowded central cities, turned into the "Main Streets" of postwar suburbia. Cheap rural land along the beltways became the favored site for new suburban housing, shopping malls, industrial parks, and office parks that drew people and businesses out of central cities … More than any other measure the 1956 Interstate Highway Act created the decentralized, automobile-dependent metropolis we know today.[172]

Urban geographer Peter O. Muller draws a similar but more succinct conclusion that auto centric planning turned the metropolis "inside out."[173] As outlined in Chapter 2, the federal government through the offices of the FHA used its financial muscle to facilitate this inversion of traditional patterns of settlement, in part driven by Cold War anxieties about nuclear conflagration. The imperative for dispersal of essential populations did not conflict with pre-war policies designed to promote suburbs as safe, white spaces. In addition to home loans, transportation funding also played a key role in promoting racial segregation and exclusion. By funding 90% of the cost of the new road infrastructure from a highway trust fund that was supported by revenue from gasoline a powerful incentive was created for municipalities to focus transportation investments on roads, with mass transit neglected and going into decline.[174] As many poorer, non-white city residents did not own cars, they lost out.

It would however be an error to view the decline of mass transit in the U.S. as an unfortunate consequence of inadequately thought out economic incentives. The transition from rail to road in the U.S. is part of a systemic process designed to enhance racial exclusion through "secessionist automobility." Gioelli has highlighted how

> white suburbanites opposed the extension of metropolitan public transit into their communities at every turn, using a new racial language of crime, which successfully connected older anxieties with new fears of the crack epidemic and war on drugs in the 1980s.[175]

Sarah Schindler has also noted how white communities routinely engage in "architectural exclusion" through the design and siting of public transit and transportation infrastructure.[176] This includes local community lobbying to prevent transit stops in neighborhoods to block access for people who typically ride transit i.e. communities of color and poor people.[177]

In Atlanta, both African American and white communities supported an effective widespread transit system in the years immediately following World War II. The problem was that bus ridership "dropped by double digits when the buses were desegregated."[178] Residents of the white suburbs of Atlanta spent decades opposing the extension of the MARTA (Metropolitan Atlanta Regional Transit Authority) system to their neighborhoods.[179] Like many other U.S. cities, Atlanta's dense network of streetcars that served many of the poorer and historically redlined communities in the city was ripped up in the 1950s.[180] One county official reported that at public meetings in her Atlanta suburb residents loudly protested extending the MARTA bus service because "Blacks would steal TVs."[181] This opposition made it difficult for transit-dependent residents – primarily poor people and people of color – to access job opportunities in the suburbs. As a result, MARTA became a city-only service that did little to relieve commuter traffic.[182] A former Republican legislator explained the underlying racial motivations:

> They will come up with 12 different ways of saying they are not racist in public … But you get them alone, behind a closed door, and you see this old blatant racism that we have had here for quite some time.[183]

As recently as 2019 Gwinnett County voted down MARTA expansion for a third time despite increasing racial diversity in the county.[184] Kevin Kruse has noted the link between racist planning and road congestion: "Atlanta's traffic is at a standstill because its attitude about transit is at a standstill too."[185]

In some cases municipalities openly refused large financial incentives to improve connectivity in order to limit racial mobility. In the upper Midwest the northern suburbs of Detroit "fought for decades to not take $600 million in federal money to build a subway because there was a lot of fear of bringing Black people into White neighborhoods."[186] Communities in suburban San Francisco and Washington, D.C. also successfully organized to oppose transit stops in their areas to prevent access by people who lived in poorer areas of the city. This has not only significantly impeded the ability of people living in the center of cities to access jobs in more prosperous white suburban communities but also had devastating consequences for climate change illustrating again Heather McGhee's research showing how racially motivated choices tend to leave everyone worse off.[187]

This form of architectural exclusion is by no means a historic phenomenon but rather continues to shape patterns of settlement, employability, time spent communiting, and climate change. Baltimore, Maryland, is a major U.S. city without a major transportation network. In June 2015, Larry Hogan, the then Governor of Maryland canceled the Red Line. This was a planned light rail that would have improved east-west mass transit in Baltimore. He refused to authorize state funding for the project that aimed to boost the local economy by creating up to 10,000 construction jobs and provide career training at the predominantly African American high schools on the west side of Baltimore where Freddie Gray was killed by police officers in 2015.[188]

Hogan's decision redirected funding that had been earmarked for the Red Line to a transportation initiative in rural and suburban areas outside of the city, one outcome of which was to disproportionately benefit white suburbanites at the expense of inner-city African Americans. His actions led the Obama Administration to carry out a civil rights investigation by the Justice Department. Anthony Foxx, Obama's transportation secretary personally expressed disappointment at the decision given how it would impact on prospects for social and economic mobility for African Americans in Baltimore but the DOJ inquiry was subsequently dropped by the Trump Administration. Wes Moore, the first Black governor of Maryland, subsequently vowed to jump-start the Red Line to provide access for residents to jobs, healthcare, schools, grocery stores, recreation, and entertainment. There are however now serious funding challenges to doing this given that the original pot of money that had been earmarked for the project is no longer available and Moore will have a difficult job trying to secure an equivalent fund again, especially given the reelection of Donald Trump in 2024.

In many respects the Baltimore Red Line example epitomizes the ongoing racialized nature of transit and transit funding in the U.S., and highlights one of the fundamental weaknesses of the transportation funding model that is still skewed in favor of roads. Perhaps unsurprisingly there have been calls to rethink the way in which transportation funding is allocated and move the dial so that transit gets a bigger share of the pie.[189] As Department of Transportation (DOT) Secretary, Anthony Foxx challenged the assumption that states are the best decision-makers for deciding how federal funding dollars ought to be distributed and called for a review of the transit funding model to explore how Metropolitan Planning Organizations could play a bigger role.[190] Following the election of Donald Trump in 2024 transit funding once again has become mired in the struggle between what his supporters view as "anti-car" and "woke" policies and efforts by city officials to promote more efficient and sustainable public transportation. One of the first acts of the Trump administration was targeting New York City's congestion pricing which was introduced to help provide additional funding to boost transit in the city. As with highway routing, debates about transit in the U.S. serve to highlight the fact that very often, transportation planning serves as a proxy for wider political and racial issues.

In fact, perhaps more than any other area of public policy, funding for mass transit illustrates how racist ideology is inherently inefficient in practice, establishing a zero-sum game in which everyone loses out, including white Americans.[191] Jake Berman has pointed out that a comprehensive, efficient, and successful public transportation system requires buy-in from all communities; otherwise, economies of scale are lost, and the system is unable to provide fast, frequent, and reliable services that take people where they want to go.[192] White, suburban secession from mass transit therefore "shakes the very foundations" of the entire system resulting in a much less efficient and scaled-back network with limited service.[193] The net outcome of decades of underinvestment in transit and the promotion of auto-centricity in the U.S. is that it is both difficult and unsafe to access jobs, polling stations, places of education, places of worship, as well as many other daily necessities such as food and health care without a vehicle. Geographically dispersed and sprawling development "makes public transportation inefficient to operate, producing infrequent, inconvenient and unreliable service… while wide, heavily trafficked roads with subpar or non-existent sidewalks and few safe crossings make walking or biking unpleasant at best and deadly at worst."[194]

In many metropolitan areas across the U.S., the inequalities resulting from this approach to transportation planning are quite stark. Fewer than 10% of Americans currently live within walking distance of frequent transit, like buses or metro trains which adversely impacts employment and education opportunities stifling social and economic mobility. A recent report highlighted that a huge proportion of U.S. community college campuses (which are disproportionately favored by non-whites and first-generation college students) are located miles from the nearest transit stop, and the distance "puts critical educational opportunities firmly out of reach for students who can't or don't drive."[195] Despite decades of auto centric development and policy approximately 28 million Americans (about 9% of the population) do not have access to a car.[196] Again, as with other areas of U.S. life, racial disparities are significant with almost one in five Black households lacking access to a car – three times the rate of white households.[197] Black people are significantly more likely to take public transit, representing 24% of all transit riders, nearly double their proportion of the population.[198] They are also more likely to face poor traffic infrastructure, police stops and tickets, and injury and death while traveling than their White peers.[199]

The neglect of transit in the U.S. has produced detrimental outcomes for the entire country, not just those it was designed to exclude.[200] Despite African Americans being disproportionately disadvantaged by poor transportation planning the majority of counties in the U.S. with high rates of zero-car households are rural and predominantly white. Older Americans who are no longer able to drive are also significantly disadvantaged by a lack of affordable and accessible public transportation. Increased traffic pollution, and employers who are unable to fill positions because willing employees cannot access workplaces are some of the other results of a design process that leaves large swathes of the country, and the people living in them, disconnected and separated from where they need to go. A recent study by the Urban Institute found that more than half of low-wage workers are now people of color, up from 20% in 1971.[201] Black, Latino, and women workers are also more likely to be segregated into lower-quality and lower-paying jobs. These are precisely the groups of people who need access to safe, regular, and well-connected transit networks that can be relied upon to get them to their places of employment.

Crime is a reality for many transit users, especially women, but again this can partly be attributed to the way in which many transit stops are poorly designed, the lack of investment in transit facilities that can help create more welcoming and safer spaces, and the lack of volume. It is a truism that busy and crowded transit networks feel, and are, safer than quiet and desolate ones. Katie Causey recently contrasted the dangerous design features of Palo Alto transit station with the much safer design she encountered in San Carlos.[202] She describes the fear she felt travelling on the train as she realized she would be arriving in San Carlos after dark and how these fears were allayed as soon as she arrived in a station that was very different from the one she was used to:

> When I realized the sun had set and I had lost track of time, my stomach dropped … The idea of being alone at a train station at night – especially … those [with] enclosed tunnel underpasses – terrified me. When I walked to the San Carlos train station however, it was designed completely differently than those in Palo Alto. The train is safely elevated above pedestrian and cyclist paths. Unlike the tunnels of Palo Alto, the design is incredibly open with high visibility for pedestrians: In the middle of the night standing anywhere in the station, I could see the entire area. If anyone else had been there, I would've seen them, and if something happened to me, it would have been impossible not to see … It is amazing how at ease you feel when these essential locations that thousands of community members use every day are designed with safety in mind.

Causey's description of how her "stomach dropped" when she anticipated entering a poorly designed station after dark illustrates again the psychological impact of architecture and its capacity to encourage or discourage certain forms of behavior – in this case use of transit. The well-designed station she describes in San Carlos is probably unfamiliar to many people in the U.S., who are more accustomed to transit spaces that are poorly designed and feel threatening like the one in Palo Alto – if indeed mass transit exists at all in their neighborhood.

Another outcome of white secession from public transit is that more and more people miss the socializing aspects of public transit. Yingling Fan, Professor of Urban and Regional Planning at the University of Minnesota has highlighted the important social benefit of public transit as a core public space where people of different racial backgrounds and social class can come together and interact.[203] This is especially important in a context where such public spaces are increasingly under threat and where patterns of settlement mean many people have few interactions with people who look different. The other consequence of course is increased climate change from the externalities caused by vehicle emissions. Traffic pollution has moved from a phenomenon that mostly affects those living near highways to one that affects the entire planet. In the 1950s a plan for a monorail system for Los Angeles to deal with smog and ease congestion was killed following public protests. Current data shows significant levels of health disparities in the city including "higher rates of asthma and unnatural cognitive decline." [204] The long-recognized problems of breathing in air contaminated by vehicle emissions have been compounded by wider environmental concerns. A Los Angeles County Climate Vulnerability Assessment found that over half the residents of the city, almost 5.7 million people, "live in a neighborhood that has a wide exposure to at least one of five severe climate threats – extreme heat, wildfire, drought, extreme precipitation and sea-level rise."[205] The city does have a light rail metro system, but as with many other urban centers across the country, the L.A. network is limited with too few lines connecting too few people with where they want to go. As a result, 73% of Los Angelenos drive to work alone. Although the L.A. metro rail system is expanding local opposition to transit stops, fear of crime, and lack of funding mean that progress is painfully slow. The pathology with which cities like L.A. are viewed by the current Trump administration means that the default position involves deploying troops from the National Guard and the U.S. Marine Corp rather than federal dollars to support funding for mass transit.

Writing this, in the summer of 2025 as L.A. is currently experiencing protests about the actions of I.C.E. officers following a winter in which the city was engulfed in flames caused by wildfires the words of Ta-Nehisi Coates are especially apposite.

> Something more fierce than Marcus Garvey is riding on the whirlwind … the flight from us sent them sprawling into the subdivided woods. And the method of transport through these new subdivisions, across the sprawl, is the automobile, the noose around the neck of the earth, and ultimately, the Dreamers.[206]

5.2.8 Hobson's Choice

Turning the metropolis "inside out" in the decades following World War II led commentators to highlight the numerous social, economic, and environmental problems associated with this lifestyle change including increased segregation, loneliness, isolation, inconvenience, and environmental damage. Some scholars argue that these problems were a result of choices freely made by the American people. There are several problems with this assertion, not least of which is that "the suburban dream" was a choice that was denied to large sections of the population, notably, communities of color. Another difficulty with this proposition is establishing the degree to which turning the metropolis inside out was a choice that Americans entered freely. Useful insight here is provided by considering the practices of a transportation provider called Thomas Hobson, a seventeenth-century stable owner in Cambridge, England, who ran a livery stable where he rented out horses. Hobson implemented a strict policy that meant customers were required to take the horse nearest the stable door – other stable owners allowed customers to freely choose an animal from those available. This policy gave rise to the term "Hobson's Choice" given that his customers were required to take the animal nearest the stable door or take none. Something akin to Hobson's Choice operated in the growth of suburbia and the decline of U.S. cities in the decades following World War II. First, federal government policy, supported by the real estate market, in the decades preceding and immediately following the war starved urban centers of finance and rendered entire neighborhoods ineligible for mortgages and repairs. This, as Jacobs noted, led to "inexorable decline" as communities were refused credit to buy property or improve the property they already owned. Although some communities found ways around this policy its impact was severe. As a result the inner-city housing market was devastated and deprived of a whole generation of purchasers who could have supplied the demand that would have reinvigorated declining areas.

Second, the suburban housing boom was of such vast proportions, and the lending terms were so favorable, to make a single-lot home accessible to most white, middle-income families. With mortgages underwritten by the U.S. Treasury it was often cheaper to purchase a suburban home on a long-term low-interest basis than it was to rent an apartment in a city. Third, survey data, such as that contained within the DDB Needham LifeStyle has shown that in the last quarter of the twentieth century, i.e. the years of rapid suburbanization, suburban living gradually became less attractive compared to residence in either the central city or smaller towns. To this day, when it comes to asking people about their lifestyle choices walkable communities, access to amenities, especially shopping, and connectivity are consistently at the top of most people's wish list. The preference for walkable and connected communities with mixed-use development is further evidenced by that most American of indicators – the market itself. The most expensive properties in the U.S. tend to be located within walkable, dense and mixed-use developments, an indication of the extent to which people are prepared to pay a premium to escape the "numbing homogeneity" of suburban enclaves.

Fourth, Putnam contrasts the greater sifting of communities by race and class with marketing campaigns in the 1990s by developers who continued to sell suburban life as a return to the hometown lifestyle of the past where people knew each other and were socially connected with their neighbors. In other words, developers marketed suburbs as homely neighborhoods where people had a strong sense of community, even though data showed the opposite was the case. Developers recognized the value of community as an ideal, even though that is not what they were selling.

Perhaps the strongest point challenging the notion that vast numbers of Americans "choose" the suburban lifestyle is the huge number of people who protested strongly at being forced to leave their communities because of urban renewal projects and, failing in their efforts, were forced to move. In some cases, in mostly white neighborhoods, highway routing and urban renewal programs were halted by the political mobilization of elites, including historic preservationists, journalists, and other campaigners. In the early 1940s, the personal intervention of the President, Franklin Delano Roosevelt saved the Battery Park district in lower Manhattan from the devastation that would have been wreaked by Robert Moses' proposed Brooklyn-Battery Park bridge project. Jane Jacobs' successful campaign against the Lower Manhattan Expressway in the late 1950s and 1960s prevented another Robert Moses infrastructure project from devastating Soho, Washington Square Park, and Little Italy. In this case Jacobs was assisted by numerous well-connected civic leaders and white elites as well as the inhabitants of local communities all of whom rejected the promise of greener pastures of suburbia for the urban lifestyle. On other occasions, residents wrote to President Kennedy to object to the racist nature of urban renewal and highway plans, securing the support of organizations like the NAACP but mostly these interventions were unsuccessful.

The process by which the community of Tremont learned about their evictions in New York City reveals much about the decision-making process at the time. Robert Caro reveals how letters were sent to residents from Moses informing them that they had 90 days to leave their homes. The letters were deliberately and erroneously designed and worded to look as if they were official court documents, which was not the case. Moses' goal was to instill fear into the residents and ward off potential resistance underscoring the degree to which many of those living in urban communities wished to remain where they were. To conclude, that the problematic aspects of suburbanization ought to be offset because Americans choose that lifestyle grossly neglects the extent to which many people, like Thomas Hobson's customers, choose the only option that was realistically available to them.

NOTES

1 Fishman Robert, Bourgeois Utopias: The Rise and Fall of Suburbia (Basic Books, 1987). p. 4.
2 John Archer, "Colonial Suburbs in South Asia, 1700-1850, and Spaces of Modernity" in Visions of Suburbia, (ed.) Roger Silverstone (Routledge 1997), p. 29.
3 R. Roediger, Working Towards Whiteness: How America's Immigrants Became White: The Strange Journey from Ellis Island to the Suburbs. (Basic Books 2018). See also N Ignatiev, How the Irish Became White (Routledge 2008).
4 The FHA was later supplemented by the Servicemen's Readjustment Act of 1944 (the "GI" bill), which created a Veterans Administration (VA) program to help the 16 million soldiers and sailors of WWII purchase a home after the war. Because the VA very largely followed the policy and procedures of the FHA, the impact of the two organizations can be considered together.
5 Federal Housing Administration. Underwriting Manual: Underwriting and Valuation Procedure Under Title II of the National Housing Act. U.S. Government Printing Office. (Washington D.C. 1938). para. 937.
6 Hayden D (2003) *Building Suburbia: Green Fields and Urban Growth 1820-2000*, Pantheon.
7 Hayden D (2003) *Building Suburbia: Green Fields and Urban Growth 1820-2000*, Pantheon.
8 Raymond Mohl "Making the Second Ghetto in Metropolitan Miami, 1940–1960", Journal of Urban History 21 (March 1995). p. 420.
9 Raymond Mohl "Making the Second Ghetto in Metropolitan Miami, 1940–1960", Journal of Urban History 21 (March 1995).
10 Charles, A. *Forbidden Neighbors: A Study of Prejudice in Housing* (1955) pp. 229–35.
11 Putnam, R. *Bowling Alone: The Collapse and Revival of American Community.* (Simon and Schuster 2000) p. 209.
12 Lange, A. *Meet Me at The Fountain: An Inside History of the Mall.* (Bloomsbury 2022) p. 241.
13 LeeAnn, T. *Deerfield District Agrees to Rename Park Tied to Decades-Old Housing Inequality.* (NBC Chicago June 19, 2020).
14 Amy, R. Illinois Public Media. Illinois Issues: Why Is Deerfield (Still) So White? November 2018. NPR.
15 Jackson, K. *Crabgrass Frontier: The Suburbanization of the United States.* (Oxford University Press 1985). p. 208.
16 Building Soul – with Thomas Heatherwick. 'The Cult of Modernist Architecture'. BBC Radio 4. https://www.bbc.co.uk/programmes/m001r7x6.
17 Boehmer E and Davies D, eds. (2018) Planned Violence: Post/Colonial Urban Infrastructure, Literature and Culture. Palgrave Macmillan.
18 Loewen, J. *Sundown Towns: A Hidden Dimension of American Racism.* (The New Press 2018) p. 255. Loewen defines a sundown town as any organized jurisdiction that deliberately excluded African Americans and was intentionally all-white. The term originates from historic signage that warned African Americans not to be caught in the municipality after sundown. Contravention of this informal policy of racial exclusion could result in physical harm, including death.
19 Blakely and Snyder. *Fortress America: Gated Communities in the United States.* (Brookings Institution 1997).
20 Bayor, R. *Roads to Racial Segregation: Atlanta in the Twentieth Century*, 15 J. URB. HIST. 3 (1988) p. 16.
21 Bayor, R. *Roads to Racial Segregation: Atlanta in the Twentieth Century*, 15 J. URB. HIST. 3 (1988) p. 16.
22 Liz Sisley, "Shaker Barricades," *Cleveland Historical*, accessed June 13, 2025, https://clevelandhistorical.org/items/show/824.
23 'Fence Is Not Neighborly in a Suburb of Cleveland June 27 1987 *New York Times.*

24 Gowman K. (2014) The Other America: 50 Year Case Study. https://the-other-america.com.
25 Wu Nicholas. Princeton Correspondents On Undergraduate Research. 'The Uncomfortable Truths of Home.' November 27, 2017. https://pcur.princeton.edu/2017/11/the-uncomfortable-truths-of-home/?subscribe=pending#subscribe-blog-blog_subscription-3.
26 David, A. L. *Internal Combustion: The Races in Detroit, 1915–1926* (Westport Conn, 1976).
27 Jackson, K. *Crabgrass Frontier: The Suburbanization of the United States.* (Oxford University Press 1985). p. 209.
28 Travieso C. "A Nation of Walls," Places Journal, September 2020. Accessed 13 June 2025. https://placesjournal.org/article/a-nation-of-walls/?cn-reloaded=1&cn-reloaded=1.
29 Adams C. '80-year-old segregation wall finally comes down in Baltimore.' April 12, 2023 NBC News. https://www.nbcnews.com/news/nbcblk/80-year-old-segregation-wall-finally-comes-baltimore-rcna79148.
30 Connolly, N.D.B. *A World More Concrete.* (University of Chicago Press 2014).
31 Dorothy Jenkins Fields. 'Tracing Overtown's vernacular architecture,' Florida theme issue 23 (1998) *Journal of Decorative and Propaganda Arts.*
32 Grace, D. 'The Walls that divided Miami' 2023 *The New Tropic: Life Like You Live Here.*
33 Grace, D. 'The Walls that divided Miami' 2023 *The New Tropic: Life Like You Live Here* 134.
34 Mohl, R. 'Whitening Miami: Race, housing, and government policy in twentieth-century Dade County' (2001) 79 (3) *The Florida Historical Quarterly* (Winter, 2001) 319–345, 331.
35 Segregation Walls Are A Sad Reminder Of South Florida's Jim Crow Past. February 16, 2022. CBS News Miami. https://www.cbsnews.com/miami/news/miami-jim-crow-segregation-wall/.
36 Gurley K. Metro Atlantic: Architecture, cities, transit, and urbanism. "The (Segregation) Walls That Divide Us. May 6 2013. https://metroatlantic.wordpress.com/2013/05/06/the-segregation-walls-that-divide-us/?blogid=40005200&blogsub=spammed#subscribe-blog.
37 Dunn, M. *Black Miami in the Twentieth Century.* (University Press of Florida 1997) p. 207.
38 Travieso C. "A Nation of Walls," Places Journal, September 2020. Accessed 13 June 2025. https://placesjournal.org/article/a-nation-of-walls/?cn-reloaded=1&cn-reloaded=1.
39 'Where The Sidewalk Ends, Racism Begins' (December 6, 2013) *Not Now Silly.* https://notnowsilly.com Accessed June 13, 2025.
40 Interview with local resident, October 2024.
41 Nielsen, K. "The Wall". February 5, 1998. Miami New Times. https://www.miaminewtimes.com/news/the-wall-6360290.
42 Nielsen, K. "The Wall". February 5, 1998. Miami New Times. https://www.miaminewtimes.com/news/the-wall-6360290.
43 Nielsen, K. "The Wall". February 5, 1998. Miami New Times. https://www.miaminewtimes.com/news/the-wall-6360290.
44 Paul, B. 'New Haven-Hamden "Berlin Wall" coming down (May 4, 2014) *New Haven Independent.* https://www.newhavenindependent.org/article/new_haven-hamden_berlin_wall_coming_down.
45 Steven, M. 'There Are No Angels In The Room: A History of the Woodin Street Fence", Report Compiled for the Legislative Council of the Town of Hamden, December 2014.
46 Goldhagen, S. *Welcome to Your World: How the Built Environment Shapes Our Lives* (Harper 2017) p. 59.

47 Maguire A., D French, D O'Reilly, 'Residential Segregation, dividing walls and mental health: a population – based record linkage study. J Epidemiol Community Health. 2016 Sep;70(9):845-54. doi: 10.1136/jech-2015-206888. Epub 2016 Feb 8. PMID: 26858342; PMCID: PMC5013154.
48 Building Soul – with Thomas Heatherwick: Why Boring Buildings Are Bad for US. BBC Radio 4. https://www.bbc.co.uk/programmes/m001r1b2.
49 Eleanor Young. 'Boring facades can seriously damage your health. The RIBA Journal. 2 May 2025. https://www.ribaj.com/products/facade-design-psychology-neuroscience.
50 The Walls that Divided Miami. The New Tropic. February 2022.
51 C, *Architectural Reflections: Studies in the Philosophy and Practice of Architecture*, (Manchester University Press 2000) p. 25.
52 D'Oca D. The Arsenal of Exclusion and Inclusion. August 8, 2010. https://arsenalofexclusion.blogspot.com/2010/08/perimeter-fence.html Accessed June 13, 2025.
53 D'Oca D. The Arsenal of Exclusion and Inclusion. August 8, 2010. https://arsenalofexclusion.blogspot.com/2010/08/perimeter-fence.html Accessed June 13, 2025.
54 Thompson v U.S. Department of Housing and Urban Development., 348 F. Supp. 2d 398, 404,427, 428, (D, Md 2005).
55 Low, S. *Behind the Gates: Life, Security, and the Pursuit of Happiness in Fortress America* (2003) p. 11.
56 Low, S., *Why Public Space Matters* (Oxford University Press 2023).
57 County, City Spar Over Coral Gate Wall. June 14, 2011. CBS News Miami. Accessed June 13, 2025 https://www.cbsnews.com/miami/news/county-residents-spar-over-legality-of-coral-gate-wall/.
58 Rivero, D. A Miami Neighborhood Divided Over Unregulated Surveillance Technology. WLRN Public Media. August 26, 2019. https://www.wlrn.org/news/2019-08-26/a-miami-neighborhood-divided-over-unregulated-surveillance-technology Accessed June 13, 2025.
59 Allman, T. D. *Finding Florida: The True History of the Sunshine State*. (Atlantic Monthly Press 2013).
60 David J. Kennedy, "Residential Associations as State Actors: Regulating the Impact of Gated Communities on Nonmembers," Yale Law Journal, vol. 105, no. 3 (December 1995), pp. 761–93.
61 Parkinson J (2012) Democracy and Public Space: The Physical Sites of Democratic Performance. Oxford University Press. p. 168.
62 National Capital Planning Commission: National Capital Urban Design and Security Plan in October 2002. p. 3.
63 See for example the work of the Social Life Project and the Placemaking Movement https://www.sociallifeproject.org.
64 Our Epidemic of Loneliness and Isolation: The U.S. Surgeon General's Advisory on the Healing Effects of Social Connection and Community. 2023.
65 Kelley, L. 'How solitude is rewiring American identity' (January 8, 2025) *The Atlantic*.
66 Religious News Service. 'Want to fix America's woes? Start by getting to know your neighbors, researcher says' (October 27, 2023) *Message*.
67 Mazarr, M., Sweijs, T., and Tapia, D. 'The sources of renewed national dynamism' (2024) *Rand Foundation*.
68 O'Neill, O. *A Question of Trust: The BBC Reith Lectures*. (Cambridge University Press 2002).
69 Our Epidemic of Loneliness and Isolation: The U.S. Surgeon General's Advisory on the Healing Effects of Social Connection and Community. 2023. p.13.
70 Our Epidemic of Loneliness and Isolation: The U.S. Surgeon General's Advisory on the Healing Effects of Social Connection and Community. 2023. p.13 p. 16.

71 Sandel, M. *What Money Can't Buy: The Moral Limits of Markets*. (Farrar, Straus and Giroux 2012) p. 203.
72 Sunstein, C. *Going to Extremes: How Like Minds Unite and Divide*. (Oxford University Press 2009).
73 Our Epidemic of Loneliness and Isolation: The U.S. Surgeon General's Advisory on the Healing Effects of Social Connection and Community. 2023.
74 Think Fast Talk Smart Podcast. No. 179 https://podcasts.apple.com/us/podcast/think-fast-talk-smart-communication-techniques/id1494989268?i=1000684854642.
75 Chetty, R., Jackson, M.O., Kuchler, T. *et al.* Social capital I: measurement and associations with economic mobility. *Nature* 608, 108–121 (2022). https://doi.org/10.1038/s41586-022-04996-4Walls.
76 Wilson, J. W. *The Truly Disadvantaged: The Inner City, the Underclass, and Public Policy*. (University of Chicago Press 1987) p. 60.
77 Thayler R and S. Sunstein. *Nudge: The Final Edition*. (Penguin 2021) p. 66.
78 Low, S. *Why Public Space Matters*. (Oxford University Press 2023) p. 3.
79 Steuteville, R. Healthy Communities Explained in Three Steps, Congress of New Urbanism, June 29, 2023. https://www.cnu.org/publicsquare/2023/06/29/better-neighborhoods-make-healthier-lives.
80 Klinenberg, E. 'Adaptation' (December 30, 2012) *The New Yorker.*
81 Corinth, K., Winship, S., and O'Rourke, T. 'Perspective: It's time to face up to our social poverty problem' (October 11, 2024) DeseretNews. https://www.deseret.com/opinion/2024/10/11/social-capital-poverty-solutions/.
82 Cox, D and Pressler, S. (2024) Disconnected: The Growing Class Divide in American Civic Life. Findings from the 2024 American Social Capital Survey. https://www.americansurveycenter.org/research/disconnected-places-and-spaces/.
83 Carney, T. *Alienated America: Why Some Places Thrive While Others Collapse* (HarperCollins 2019).
84 Corinth K, Winship S, and O'Rourke T, DeseretNews "Perspective: It's time to face up to our social poverty problem", October 11, 2024 https://www.deseret.com/opinion/2024/10/11/social-capital-poverty-solutions/.
85 Putnam, R. *Bowling Alone: The Collapse and Revival of American Community*. (Simon and Schuster 2000).
86 99% Invisible. Podcast. Episode PB01. 18 January, 2024. The Power Broker #1: Robert Caro. https://99percentinvisible.org/episode/power-broker-01-robert-caro/transcript.
87 Bond, M. 'The hidden ways that architecture affects how you feel.' (2017, June 5) *BBC Future*. https://www.bbc.com/future/article/20170605-the-psychology-behind-your-citys-design.
88 99% Invisible. Podcast. Episode PB04. 19 April, 2024. The Power Broker #4: Rep. Alexandria Ocasio-Cortez. https://99percentinvisible.org/episode/the-power-broker-04-rep-alexandria-ocasio-cortez/.
89 Center for American Progress. "Secretary Foxx Discusses A Transportation Plan for the 21st Century." Accessed June 13, 2025.
https://www.youtube.com/watch?v=QqtJgWaU0zM&t=494s.
90 CAP 20, Bridging the Divide: A Transportation Plan for the 21st Century, https://www.americanprogress.org/events/bridging-the-divide-a-transportation-plan-for-the-21st-century/.
91 Aiello L, A. Vybornova, S. Juhász, M. Scell, and E. Bokányi, "Urban highways are barriers to social ties", PNAS 2025 Vol. 122 No. 10. https://www.pnas.org/doi/10.1073/pnas.2408937122.
92 Putnam D, *Bowing Alone: The Collapse and Revival of American Community*. (Simon and Schuster 2000) p. 212.
93 Saadaoui, R, D. Salon, H. Jamme, N. Corcoran, J. Hitzeman. "Does Car Dependence Make People Unsatisfied With Life? Evidence From a U.S. National Survey." Travel

Behaviors and Society. Vol. 35. April 2025 https://www.sciencedirect.com/science/article/abs/pii/S2214367X24002175?via%3Dihub.

94 Goldhagen, S. *Welcome to Your World: How the Built Environment Shapes Our Lives* (Harper 2017), p. 25.
95 Mumford L, *The City in History: Its Origins, Its Transformations, and Its Prospects. (Mariner Books, 1968), p. 486.*
96 Putnam D, *Bowing Alone: The Collapse and Revival of American Community.* (Simon and Schuster 2000), p. 210.
97 Jackson, K. *Crabgrass Frontier: The Suburbanization of the United States.* (Oxford University Press 1985). p. 272.
98 Oliver, E. 'The effects of metropolitan economic segregation local civic participation' (January 1999) 43 *American Journal of Political Science.*
99 Panova, T. 'Don't blame your teen's smartphone, blame their neighborhood' (January 3, 2022) *Psybertronic.*
100 Thompson, D. 'Why Americans suddenly stopped hanging out' (February 14, 2024) *The Atlantic.*
101 Healthy Places by Design, Social Isolation Learning Network. Socially Connected Communities: Solutions for Social Isolation. Healthy Places by Design 2021.
102 O'Sullivan, S. 'There's nowhere to hang out for free anymore and it's making us lonely' (27 November, 2023) 29 *Refinery.*
103 Haidt, J. *The Anxious Generation.* (Penguin Press 2024).
104 Haidt, J. *The Anxious Generation.* (Penguin Press 2024) p. 272.
105 Erfurt E. 'How to Make a Street Safer Before the Kids Go Back to School'. 23 August, 2024. Strong Towns.
106 Erfurt E. 'How to Make a Street Safer Before the Kids Go Back to School'. 23 August, 2024. Strong Towns.
107 Pettigrew, T. F. and Tropp, L. R. 'A meta-analytic test of intergroup contact theory' (2006 May) 90 (5) *Journal of Personality and Social Psychology* 751–83.
108 McGhee H. *The Sum of Us: What Racism Costs Everyone and How We Can Prosper Together.* (One World 2021). p. 168.
109 McGhee H. *The Sum of Us: What Racism Costs Everyone and How We Can Prosper Together.* (One World 2021). p. 174.
110 McGhee H. *The Sum of Us: What Racism Costs Everyone and How We Can Prosper Together.* (One World 2021). p. 175.
111 Amy, R. Illinois Public Media. Illinois Issues: Why Is Deerfield (Still) So White? November 2018. Npr.
112 Sue, D. W. *Race Talk and the Conspiracy of Silence: Understanding and Facilitating Difficult Dialogues on Race.* (Wiley 2015) p. 218.
113 Poon, L. 'City life is too lonely: Urban planning can help' (December 14, 2023) *Bloomberg City Lab.*
114 Richard Rothstein lecture, Center for the Study of Race and Ethnicity in America, Brown University, August 14, 2018. Available at https://www.youtube.com/watch?v=r9UqnQC7jY4&t=95s.
115 Archer, "Colonial Suburbs in South Asia, 1700-1850, and Spaces of Modernity" in Visions of Suburbia, ed. Roger Silverstone (Routledge 1997). p. 27.
116 Chaskin, R. and Joseph, M. *Integrating the Inner City: The Promise and Perils of Mixed Income Public Housing Transformation.* (University of Chicago Press 2015).
117 "Black men arrested at Philadelphia Starbucks reach agreements", Yon Pomrenze and Darran Simon, CNN, May 2, 2018 https://www.cnn.com/2018/05/02/us/starbucks-arrest-agreements/index.html.
118 "Central Park birder Christian Cooper on being 'a Black man in the natural world.'" NPR, June 12, 2023 https://www.npr.org/2023/06/12/1181314626/central-park-birder-christian-cooper-on-being-a-black-man-in-the-natural-world.

119 Cooper, C. *Better Living Through Birding: Notes from a Black Man in the Natural World*. (Random House 2023).
120 Low, S. *Why Public Space Matters*. (Oxford University Press 2023) p. 30.
121 Brandon, G. 'A black Yale graduate student took a nap in her dorm's common room. So a white student called police' (May 12, 2018) *CNN*. https://www.cnn.com/2018/05/09/us/yale-student-napping-black-trnd/index.html.
122 Criss D and A Vera. 'Three black people checked out of their Airbnb rental. Then someone called the police on them.' CNN. May 10, 2018. Accessed June 13, 2025. https://www.cnn.com/2018/05/07/us/airbnb-police-called-trnd/index.html.
123 Misra T, "Yes, 311 Nuisance Calls Are Climbing in Gentrifying Neighborhoods", Bloomberg, October 18, 2018.
124 Low, S. *Why Public Space Matters*. (Oxford University Press 2023) p. 254.
125 Goldhagen, S. *Welcome to Your World: How the Built Environment Shapes Our Lives* (Harper 2017) p. 25.
126 Hamann, C. Peek-Asa, C. and Butcher, B. 'Racial disparities in pedestrian-related injury hospitalizations in the United States. BMC Public Health 20. 1459 (2020).
127 National Complete Streets Coalition: Smart Growth America. *Dangerous by Design 2022*. p. 35.
128 National Complete Streets Coalition: Smart Growth America. *Dangerous by Design 2022*. p. 35.
129 Archer, D. Dividing Lines: How Transportation Infrastructure Reinforces Racial Inequality. Norton 2025.
130 National Complete Streets Coalition: Smart Growth America. *Dangerous by Design 2024*. p. 35.
131 National Complete Streets Coalition: Smart Growth America. *Dangerous by Design 2024*.
132 Dumbaugh, E., Stiles, J., Mitsova, D., & Saha, D. (2023). The Most Vulnerable User: Examining the Role of Income, Race, and the Built Environment on Pedestrian Injuries and Deaths. *Transportation Research Record, 2678*(2), 743–752. https://doi.org/10.1177/03611981231175888 (Original work published 2024).
133 Dumbaugh, E., Stiles, J., Mitsova, D., & Saha, D. (2023). The Most Vulnerable User: Examining the Role of Income, Race, and the Built Environment on Pedestrian Injuries and Deaths. *Transportation Research Record, 2678*(2), 743–752. https://doi.org/10.1177/03611981231175888 (Original work published 2024). p. 743.
134 Zipper D. 'The Deadly Myth That Human Error Causes Most Car Crashes'. The Atlantic. November 26, 2021.
135 Steuteville, R. 'Changing street design is critical to reducing bike-ped deaths.' February 14, 2023. Public Square: Congress of New Urbanism. Accessed June 14, 2025 https://www.cnu.org/publicsquare/2023/02/14/changing-street-design-critical-reducing-bike-ped-deaths; Speck J, 'Secretary Pete's safe streets plan won't succeed if engineers continue business as usual'. The Hill. 2 November 2023. Accessed June 14, 2025. https://thehill.com/opinion/finance/3853943-secretary-petes-safe-streets-plan-wont-succeed-if-engineers-continue-business-as-usual/.
136 Davis, V. Inclusive Transportation: A Manifesto For Repairing Divided Communities. (Island Press 2023).
137 Zipper, D. 'Repairing the Roads that Were Built to Divide'. July 12, 2023. Bloomberg City Lab: Transportation.
138 James, W. Loewen. Sundown Towns: A Hidden Dimension of American Racism 254–255 (2006).
139 Center for American Progress. 'Bridging the Divide: A Transportation Plan for the 21st Century,' March 30, 2026. Accessed June 14, 2025. https://www.americanprogress.org/events/bridging-the-divide-a-transportation-plan-for-the-21st-century/.
140 National Academies of Sciences, Engineering, and Medicine 2022. *Racial Equity, Black America, and Public Transportation, Volume 1: A Review of Economic, Health,*

and Social Impacts. Washington, DC: The National Academies Press. https://doi.org/10.17226/26710.

141 Schulman, B. 'Stop Putting Pedestrians to Blame: The Case of Raquel Nelson.' Smart Cities Dive. Accessed June 14, 2025. https://www.smartcitiesdive.com/ex/sustainablecitiescollective/stop-putting-pedestrians-blame-case-raquel-nelson/27369/.

142 National Public Radio (NPR) Staff. 'Child's Death Casts Light On Pedestrian Traffic Woes.' July 30, 2011. Accessed June 14, 2025. https://www.npr.org/2011/07/30/138855279/convicted-suburban-mom-has-city-planners-nervous.

143 'Mall Accused of Racism in a Wrongful Death Trial In Buffalo.' The New York Times. November 15, 1999.

144 Brown, C. Arrested Mobility: Overcoming the Threat to Black Movement. (Island Press 2025).

145 Brown, C., J Rose and S Kling, Arrested Mobility: Barriers to Walking, Biking, and E-Scooter Use in Black Communities in the United States. (Equitable Cities March 2023). p. 22.

146 Campbell, E. 'Ticket to walk: How jaywalking enforcement impacts Washingtonians. (For Transportation Choices Coalition January 23, 2024).

147 Transportation Choices, 'Statewide data on jaywalking stops confirms disproportionate impact on Black and unhoused individuals.' Feb. 9, 2023. https://transportationchoices.org/statewide-data-on-jaywalking-stops-confirms-disproportionate-impact-on-black-and-unhoused-individuals/.

148 PROPEL ATL 'Atlanta "jaywalking" citations show stark racial disparities; no measurable impact on pedestrian safety.' October 30, 2021. Accessed June 14, 2025. https://www.letspropelatl.org/jaywalking_disparities_oct_2021.

149 ProPublica and Florida Times-Union. Sanders T, K. Rabinowitz, and B. Conarck. 'Walking while Black'. November 16, 2017. Accessed June 14, 2025. https://features.propublica.org/walking-while-black/jacksonville-pedestrian-violations-racial-profiling/.

150 Kim, J., Smorodinsky, S., Lipsett, M., Singer, B., Hodgson, A., and Ostro, B. 'Traffic-related air pollution near busy roads: The East Bay children's respiratory health study' (2004) *American Journal of Respiratory and Critical Care Medicine.*

151 American Lung Association. 'Disparities in the Impact of Air Pollution.' Updated November 2, 2023. Accessed June 14, 2025. https://www.lung.org/clean-air/outdoors/who-is-at-risk/disparities.

152 Badger, E. 'Pollution is segregated, too'. (April 15, 2014) *Washington Post.*

153 Uteuova A. 'Like a high-sodium diet: traffic pollution can cause rise in blood pressure – study'. The Guardian. November 29, 2023.

154 Carrington, D. 'Health impact of tyre particles causing 'increasing concern', say scientists.' The Guardian. February 22, 2023.

155 Kilani H, "'Asthma alley': Why minorities bear burden of pollution inequity caused by white people." The Guardian. April 4, 2019.

156 The Economist: Briefing. 'Segregation still blights the lives of African-Americans: There are policies that could improve things a lot.' July 9th, 2020.

157 Doleac, J. Commentary: New evidence that lead exposure increases crime. Brookings Institute. June 1, 2017 https://www.brookings.edu/articles/new-evidence-that-lead-exposure-increases-crime/.

158 Wong, K. V., Paddon, A., Jimenez, A. 'Review of the World Urban Heat Islands: Many linked to increased mortality' (2013) 135 *Energy Resources Technology* 022101.

159 Borden, K. A. and Cutter, S. L. Spatial patterns of natural hazards mortality in the United States. (2008) 7 *International Journal of Health Geography* 64.

160 Meehl, G. A., and Tebali, C. 'More intense, more frequent, and longer lasting heat waves in the 21st century' (2004) 305 *Science* 994; Habeeb, D., Vargo, J., and Stone, B. 'Rising heat wave trends in large US cities' (2015) 76, *Natural Hazards* 1651–1665.

161 Shandas, V., Voelkel, J., Williams, J., and Hoffman, J. 'Integrating satellite and ground measurements for predicting locations of extreme Urban heat' (2019) 7 *Climate* 5.
162 Li, X., Zhou, Y., Asrar, G. R., and Imhoff, M. 'The surface urban heat island response to urban expansion: A panel analysis for the conterminous United States' (2017) *Science of the Total Environment* 605–606, 426–435.
163 Hoffman, J., Shandas, V., and Pendleton, N. 'The effects of historical housing policies on resident exposure to Intra-Urban Heat: A study of 108 US Urban areas' (2020) 8 *Climate* 12.
164 Madrigano, J., Ito, K., Johnson, S., Kinney, P. L., and Matte, T. 'A case-only study of vulnerability to heat wave-related mortality in New York City (2000–2011)' (2015) 123 *Environmental Health Perspective* 672–678.
165 Voelkel, J., Hellman, D., Sakuma, R., and Shanda, V. 'Assessing vulnerability to Urban heat: A study of disproportionate heat exposure and access to refuge by socio-demographic status in Portland, Oregon' (2018) 15 *International Journal of Environmental Research and Public Health* 640.
166 Murage, P., Hajat, S., and Kovats, R. S. Effect of night-time temperatures on cause and age-specific mortality in London' (2017) 1 *Environmental Epidemiology*.
167 Aram, F., Higueras, G., Solgi, E., and Mansournia, S. 'Urban green space cooling effects in cities' (2019) 5 *Heliyon* e01339.
168 Bullard, R, G. Johnson, and A. Torres (eds.) Highway Robbery: Transportation Racism & New Routes to Equity. (South End Press 2004).
169 Kimble, M. City Limits: Infrastructure, Inequality and the Future of America's Highways. (Crown 2024).
170 This is especially problematic, but by no means limited to 'red' states. See for example recent controversy over a Louisiana law that limits the publication of climate data. Wiinter J. 'Chilling and dangerous': Grassroots groups sue over Louisiana law that censors air quality data. The Grist. May 29, 2025.
171 Rehm, D. 'Transportation Secretary Anthony Foxx On The Legacy Of The U.S. Highway System'. March 31, 2016. WAMU 88.5. American University Radio. Accessed June 14, 2025. https://dianerehm.org/shows/2016-03-31/transportation-secretary-anthony-foxx-on-transportation-opportunity-and-the-legacy-of-the-u-s-highway-system.
172 'The American metropolis at century's end: Past and future influences' (20) *Robert Fishman* Housing Policy Debate, Vol. 11, Issue 1. Fannie May Foundation.
173 Raymond, M. The Interstates and the Cities: The U.S. Department of Transportation and the Freeway Revolt, 1966–1973. p. 195.
174 Raymond, M. Race and space in the modern city: Interstate-95 and the black community in Miami. In A. Hirsch and R. Mohl (eds.), *Urban Policy in Twentieth-Century America*. (Rutgers University Press 1993).
175 Gioielli R. 'The Tyranny Of The Map: Rethinking Redining.' The Metropole: The Official Blog of the Urban History Association. Accessed June 14, 2025. https://themetropole.blog/2022/11/03/the-tyranny-of-the-map-rethinking-redlining/.
176 Schindler, S. 'Architectural exclusion: Discrimination and segregation through physical design of the built environment (2015)124 (6) *Yale Law Journal* 1934.
177 Bullard, R., G. Johnson, and A. Torres (eds.) Highway Robbery: Transportation Racism & New Routes to Equity. (South End Press 2004).
178 99% Invisible Podcast. Episode 583. 'The Lost Subways of North America'. 21 May, 2024. Accessed June 14, 2025 https://99percentinvisible.org/episode/the-lost-subways-of-north-america/transcript/.
179 Schmidt, W. 'Racial Roadblock Seen In Atlanta Transit System.' July 22, 1987. The New York Times.
180 Susaneck, A. "Segregation by Design." TU Delft. Centre for the Just City, 2024. Accessed June 14, 2025. https://www.segregationbydesign.com/.

181 Jason, H. 'Secessionist automobility: Racism, Anti-Urbanism, and the politics of automobility in Atlanta, Georgia' (2006) 30 (293) *International Journal of Urban and Regional Research* 300.
182 Schmitt A. 'How Racial Discrimination Shaped Atlanta's Transportation Mess'. Streetsblog USA. February 8, 2017. Accessed June 14, 2025 https://usa.streetsblog.org/2017/02/08/how-racial-discrimination-shaped-atlantas-transportation-mess.
183 Kruse, K. The 1619 Project. "What does a traffic jam in Atlanta have to do with segregation? Quite a lot." The New York Times Magazine. August 14, 2019.
184 Karner A. 'MARTA Rebranding Rooted in a Racialized History of Public Transit.' Atlanta Studies. April 23, 2019. https://doi.org/10.18737/atls20190423.
185 Kruse, K. The 1619 Project. "What does a traffic jam in Atlanta have to do with segregation? Quite a lot." The New York Times Magazine. August 14, 2019.
186 99% Invisible Podcast. Episode 583. 'The Lost Subways of North America'. 21 May, 2024. Accessed June 14, 2025 https://99percentinvisible.org/episode/the-lost-subways-of-north-america/transcript/.
187 Gioielli, R. 'Pruitt-Igoe In The Suburbs: Connecting White Flight, Sprawl, and Climate Change in Metropolitan America. Amerikastudien/American Studies. Volume 65, Issue 2 (2020). pp. 213–233, DOI: https://doi.org/10.33675/AMST/2020/2/9.
188 Rehm, D. 'Transportation Secretary Anthony Foxx On The Legacy Of The U.S. Highway System'. March 31, 2016. WAMU 88.5. American University Radio. Accessed June 14, 2025. https://dianerehm.org/shows/2016-03-31/transportation-secretary-anthony-foxx-on-transportation-opportunity-and-the-legacy-of-the-u-s-highway-system.
189 Ramirez, E. Transportation for America Blog. 'Time to tip the scales in favor of more transportation options.' October 8, 2024. Accessed June 14, 2025. https://t4america.org/2024/10/08/time-to-tip-the-scales-in-favor-of-more-transportation-options/.
190 Boyum, T. "Anthony Foxx: How Transportation Helped Build Systemic Racism." Tying It Together. Podcast. 6 January 2021. Spectrum News 1.
191 McGhee H. *The Sum of Us: What Racism Costs Everyone and How We Can Prosper Together.* (One World 2021).
192 Berman, J. *The Lost Subways of North America: A Cartographic Guide to the Past, Present, and What Might Have Been.* (University of Chicago Press 2023).
193 99% Invisible Podcast. Episode 583. 'The Lost Subways of North America'. 21 May, 2024. Accessed June 14, 2025 https://99percentinvisible.org/episode/the-lost-subways-of-north-america/transcript/.
194 Smart Growth America. *Divided by Design*. 2023. P. 3.
195 Wilson, K. 'One Flat Tire Away From Dropping Out': Why Transit to Community Colleges Matter". Streetsblog, February 21, 2023. Accessed June 14, 2025 https://usa.streetsblog.org/2023/02/21/one-flat-tire-away-from-dropping-out-why-transit-to-community-colleges-matters.
196 Smart Growth America. *Divided by Design*. 2023. P. 4.
197 Smart Growth America. *Divided by Design*. 2023. P. 30.
198 Brown, C. with J'Lin Rose and Samuel Kling. 'Arrested Mobility: Barriers to Walking, Biking, and E-Scooter use in Black communities in the United States.' Equitable Cities. March 2023. P. 9.
199 Hugh, M. Clark, Who Rides Public Transportation, CJI Research Corporation, January 2017.
200 McGhee H. *The Sum of Us: What Racism Costs Everyone and How We Can Prosper Together*. (One World 2021).
201 WorkRise. 'Segregation in the Low-Wage Workforce'. October 24, 2024.
202 Causey, K. "Safer streets, rail crossings, bike paths are key to a better community." May 12, 2023. Palo Alto Online. Accessed June 14, 2025 https://www.paloaltoonline.com/news/2023/05/12/opinion-safer-streets-rail-crossings-bike-paths-are-key-to-a-better-community/.

203 Hazzard, A. 'UMN researcher dives into transportation happiness'. MPR News. October 3, 2022. Accessed June 14, 2025. https://www.mprnews.org/story/2022/10/03/umn-researcher-dives-into-transportation-happiness.

204 Los Angeles Department of City Planning. *Plan for a Healthy Los Angeles: A Health and Wellness Element of the General Plan*. March 2015.

205 Schlossberg, J. 'When air pollution becomes a health equity issue'. UCLA Health. November 1, 2021. Accessed June 15, 2025. https://www.uclahealth.org/news/article/air-pollution-health-equity-los-angeles.

206 Ta-Nehisis, C. *Between The World And Me*. (One World 2015) p. 151.

6 Restorative Architecture and the Politics of Reparations

> Architecture is not passive. Architecture is about listening, seeing, tasting, smelling, and feeling. It impacts us deeply. It changes our brain, and therefore our behavior and our wellbeing. It has the capacity to uplift and to degrade; it can bring people together and it can divide. And it is our collective and individual responsibility to know how our designs affect people working in, living in, passing by, or visiting them. Architecture is my small way of helping to improve people's lives.
>
> *Melissa Farling, Architect.*

6.1 REPARATIVE ARCHITECTURE AND TRANSITIONAL JUSTICE

Transitional justice emerged in the 1980s as an area of practice and scholarship largely focused on addressing the legacy of human rights abuses committed by authoritarian regimes in Latin America and Eastern Europe.[1] Since then, the field has expanded considerably from its original focus on disappearances, torture, and state killings to include socioeconomic rights violations, memorialization, as well as structural and systemic injustices within established democracies, including the U.S.[2] Efforts to formulate a more holistic understanding of the relationship between systemic inequalities and conflict have led to rethinking "transition" as an interim process that links the past and the future, to the concept of "transformation" that implies long-term, sustainable processes embedded in society.[3] Transformative justice has emerged in recent years as a response to perceived shortcomings in the standard toolkit of transitional justice, particularly in relation to addressing structural violence and violations of socioeconomic rights.[4] Like transitional justice, transformative justice has primarily been concerned with addressing historical wrongs in post-conflict and post-authoritarian contexts.[5] Gready and Robins sum up transformative justice as "transformative change that emphasizes local agency and resources, the prioritization of process rather than preconceived outcomes and the challenging of unequal and intersecting power relationships and structures of exclusion at both the local and the global level."[6]

The protests and disturbances, within and beyond the U.S., following the murder of George Floyd led to increased demands that transitional justice engage more explicitly with systemic racism and structural disadvantage.[7] This "travel of ideas" about transitional justice praxis and scholarship[8] can be viewed as something of a counterpoint to 'Foucault's boomerang" (Chapter 1) and is illustrated by the briefing paper issued by the International Center for Transitional Justice (ICTJ) making the

DOI: 10.1201/9781003208129-6

case for applying transitional justice approaches to "dismantle systemic racism" in the U.S.[9] The ICTJ currently partners with both local and national advocates to help advance efforts "to uncover truth, pursue justice, acknowledge and repair harms, and reform institutions within the United States to address systemic racism and injustice." The African American Redress Network (AARN), a U.S.-based transitional justice project, aims to "co-produce transformative efforts alongside communities to advance research and policy making towards a more just society."[10] Pressure is mounting for a national Commission on Truth, Racial Healing, and Transformation, as well as a Commission to Study and Develop Reparations Proposals for African Americans. The National African Americans Reparations Commission (NAARC) was established in 2015 to fight for reparatory justice, compensation, and restoration of African American communities that were "plundered by the historical crimes of slavery, segregation, and colonialism."[11] Joshi has noted that the centuries-long oppression of Black Americans is precisely the kind of "massive human rights violation that necessitates a systemic transitional justice response."[12]

Several important grassroots-based projects have also emerged in various states and cities across the country in recent decades, including the Greensboro Truth and Reconciliation Commission established in 2004 and the Maryland Lynching Truth and Reconciliation Commission established in 2019. Both of these employ a transitional justice framework to address legacy issues including racially motivated lynchings and killings.[13] In 2023 a judge dismissed a lawsuit seeking reparations from the city of Tulsa, Oklahoma, for the 1921 race massacre that left approximately 300 African Americans dead in the Greenwood neighborhood (the "Black Wall Street").[14] Like other local initiatives, advocates had sought to establish the truth about historic killings through the establishment of a formal truth-seeking commission. These initiatives all lie within conventional transitional justice frameworks. What is notable however is that social and economic legacy issues have also been front and center of recent deliberations about restitution of past harms and the need to address the legacy of systemic racism in the U.S. Of particular interest for this work is the way in which efforts to address the legacy of racist policies and practices that shape the built environment in the U.S have, in recent years, also tapped into the transitional justice framework. This includes programs focused on reparations, apologies, truth finding, and memorialization.[15]

The wider debate around reparations for historic harms perpetrated against African Americans received a significant boost following the publication by Ta-Nehisi Coates of "The Case for Reparations" in *The Atlantic* in 2014. Coates highlighted how white wealth was built on two and a half centuries of slavery augmented by discriminatory housing policies and practices. Drawing attention to the role of the Home Owners Loan Corporation (HOLC) and the Federal Housing Association (FHA) (see Chapter 2) in impeding African American physical, residential, social, and economic mobility in the middle decades of the twentieth century, Coates made a compelling case for financial redress. The role of urban renewal and the highway program in perpetuating systemic inequalities through wealth destruction in African American communities has been taken up by several other actors. The California Reparations Task Force for example is examining how best to address the environmental racism caused by highway construction which has led to increased pollution and health issues in Black neighborhoods.[16] The Task Force published a 500-page report for the state legislature as part of the first state-level effort to provide compensation for the legacy of slavery and discrimination in the U.S.[17] These legacy issues have also precipitated debate in mainstream print and broadcast media. In the L.A. Times, Matthew Fleischer suggested that if the U.S. is serious about tearing down monuments to racism a useful starting point would be "the ugly truth about the LA highway system."[18] This reflects growing awareness about how architecture and planning continue to shape patterns of racial disadvantage and demonstrates how transitional justice discourse in the U.S. has moved to embrace the legacy of "infrastructural racism."

These initiatives have been complemented by a number of local initiatives on the part of municipal authorities to address the legacy of discriminatory lending and racial zoning practices by providing reparations in the form of funding for housing grants.[19] One program, launched by the city of Evanston, north of Chicago, was the subject of a Public Broadcasting Service (PBS) documentary entitled "The Big Payback."[20] The city of Asheville, North Carolina, developed a reparations program focused on addressing harms that arose from urban renewal and the displacement of African Americans from land, businesses, and properties without recompense. In that city the East Riverside neighborhood lost over 1,000 homes as well as grocery stores, beauty parlors, barber shops, laundromats, churches, a hotel, a hospital, doctors' offices, and several funeral homes.[21] Following the standard playbook of that period most of those displaced in Asheville moved to segregated public housing projects, disconnected from the wider economic and social life of the city, with the consequent problems of educational underachievement, unemployment, poverty, and social disadvantage. In 2021, the city passed a resolution apologizing for past harms and calling for the establishment of a "community reparations initiative" to report on how best to make amends for past harms. A key goal of the initiative is to increase access to healthcare, transportation, and affordable housing.

In 2024, the state of Virginia established a Commission to study the disruption public college and university expansions had on African American communities.[22] The Commission will work with various institutions of higher education to examine property transactions in majority Black communities to determine what form of compensation would be appropriate. This initiative follows a series of reports by the Virginia Center for Investigative Journalism that revealed how the expansion of Christopher Newport University adversely impacted the predominantly Black neighborhood of Shoe Lane in Newport News.[23] In the early 1960s, as part of a deliberate effort to halt the expansion of a middle-class African American community the Newport News City Council used their powers under eminent domain to seize Black property in order to facilitate the expansion of the college despite protests from civil rights activists and homeowners that there were more suitable and less expensive sites for the development. Records from the period show that property owners were paid 20% less than the value set by an independent appraiser. The city of Newport News and the Congress of New Urbanism (CNU) announced a task force to review decades of property acquisitions and consider possible redress for displaced families.

Some scholars and activists have cautioned against calling these local initiatives "reparations" insisting that they be referred to as "racial equity initiatives."[24] Concerns have arisen that these local and more piecemeal programs instituted at state and municipal levels could undermine efforts to secure a much larger formal reparations program from the federal government that would focus on addressing the racial wealth gap.[25] On the other hand, a national conference on reparations in Atlanta, sponsored by the Decolonizing Wealth Project, argued in favor of local initiatives. The National Coalition of Blacks for Reparations in America (N'COBRA) took a similar view arguing that reparations can be "in as many forms as necessary". N'COBRA aligns its work with initiatives such as the campaign for compensation of Japanese Americans interned in concentration camps in World War II. As outlined in Chapter 2, the federal government, state officials, and local municipalities all played their own role and all benefited from the decimation of African American communities across the twentieth century. Damario-Solomon-Simmons, Director of Justice for Greenwood, which filed the lawsuit against Tulsa, "supports the fight for national reparations but says Tulsa and other local institutions have committed specific harms for which they are culpable."[26] This author would agree.

Chapter 2 provided a detailed analysis of the impact of discriminatory lending practices as well as the substantial amount of wealth that was destroyed as African American neighborhoods across the country were deliberately targeted for destruction as part of the Interstate Highway and urban renewal programs. Individual beneficiaries of these policies, including universities, hospitals, and civic centers ought to share responsibility for redressing harms from these actions but that does not preclude the need for wider programs to address actions committed by federal and state agencies. The difficulty of identifying who or what is responsible for systemic injustices and the challenge of conceiving practical remedies has been mooted as a reason for transitional justice not considering socioeconomic issues – these examples offer some solutions for overcoming these problems.[27]

6.2 MEMORIALIZATION, NARRATIVE JUSTICE, & SYSTEMIC DISADVANTAGE

Another finding of this study is that research on the socioeconomic dimensions of conflict and systemic racism is important for memorialization and narrative justice. A transitional justice approach serves not only to advocate for justice and reparations but also to provide a framework for conversations about conflict and peace.[28] As the documentary evidence uncovered in this text has shown, the acute intersection of socioeconomic and identity-based deprivation within Belfast today can be traced directly to a deliberate conflict management strategy established in the early 1970s. Likewise, the racial wealth gap and current patterns of racial disadvantage in the U.S. can be traced directly to discriminatory housing policies, lending practices, and architectural and planning practices that intentionally obliterated African American communities in the 1950s and 1960s and subsequently excluded those communities from the benefits of spatial mobility. Patterns of concentrated socioeconomic disadvantage, whether in Belfast or Baltimore, are not the result of African American or Catholic indolence, self-segregation, poor choices, lack of work ethic, or any other "ethnic myths." Rather, they were shaped systematically by public and private actors, including architects and planners, assisted by government polices over several decades. Truth commissions and similar initiatives within the transitional justice wheelhouse are one, but by no means the sole mechanism for unpacking false and misleading narratives. Robust research and dissemination of materials that challenge false narratives attributing group inequalities social disadvantage, and violence to the pathologies and failings of particular communities ought to be an essential function of transitional justice praxis.[29] Heather McGhee describes a conversation with a white woman from Tulsa who had only recently learned about the obliteration of the Black Wall Street. The woman described seeing a Black Tulsa neighborhood as a child and wondering why it seemed so poor. "In the absence of history, negative stereotypes filled the void." To paraphrase Miller, scholars and practitioners ought to aim to render visible what has hitherto been invisible in order to help fill these voids.[30]

Important developments in the U.S. in recent years provide examples of how narrative justice might be achieved regarding the racist legacy of discriminatory lending, transportation, and planning practices. Chapter 5 examined how planning decisions in the 1950s helped shape African American exclusion from Deerfield, Illinois, and contribute to the current racial wealth gap. In 2019, Deerfield Public Library launched a series of events that examined the racist history of the town including the decision to refuse planning approval for a development that would have provided housing for African Americans in the late 1950s.[31] The site of the original proposed development, subsequently turned into a park, was renamed in 2020 following a campaign by some members of the local community who wished to atone for the racist actions of the past.[32]

There have also been attempts to reimage or memorialize high-profile racial walls and barriers erected to exclude African Americans from housing developments and comply with federal housing requirements. In some cases these initiatives reflect not only changing views but also the changing demographics of the surrounding communities. For example, the South Miami "segregation wall," as it is known locally, was originally constructed in the 1950s to maintain racial boundaries and prevent African Americans wandering into the nearby white neighborhood. Following the murder of George Floyd, South Miami residents and officials met at the wall to usher in the painting of a South Miami Black History Mural.[33] In the west Texas town of Crane, a wall that was built in the 1940s at the insistence of the FHA to cut off the Black section of the town from a whites-only development had a plaque installed in 2005 identifying the structure as "The Crane Segregation Wall."[34] The Halls Hill community in Virginia celebrated an analogous unveiling of their segregation wall in 2017 after the Halls Hill/High View Park pushed Arlington County to erect a plaque.[35] When parts of the wall collapsed after flooding, some people welcomed it.[36]

In Detroit, neighborhoods on both sides of the notorious Eight Mile wall ordered that the structure be covered with murals.[37] The Michigan State Historic Preservation Office now seeks to place the wall on the National Register of Historic Places. In Lake Worth, Florida, residents have sought to preserve a 1,100-foot-long wall that was built in the 1950s to enclose the African American community of Osborne. Retha Lowe, a City Commissioner in Lake Worth, and one of the first African Americans to move to the white side of the wall in the 1970s argued that the structure should be preserved as a memorial.[38] The barrier was later renamed the "Wall of Unity," providing a canvas for local artists to create 100 10-foot-wide pieces of artwork that include images of Martin Luther King and other prominent figures from the civil rights era. Controversy resurfaced in 2020 when residents and other community stakeholders voted not to include any reference to the Black Lives Matter movement on grounds that such a message would be divisive.[39] This led to public accusations by some that the community leadership was "behind the times." When "All Lives Matter" was painted on the wall in 2021 the incident "almost started a riot."[40] Ongoing disputes around the content of the artwork on the Wall of Unity speak to the difficulties associated with reaching consensus around memorializing structures that were designed to segregate and divide, even when the surrounding communities have become more racially homogenous. Such efforts arguably provide a useful means for local communities to draw attention to the racist legacy of coercive architecture and highlight the important role that physical barriers played in shaping patterns of settlement across the U.S. As outlined in Chapter 2, some scholars have viewed structures such as the Detroit Eight Mile wall as an aberration, or oddity. Local memorialization efforts can help shed light on the systemic nature of this method of promoting exclusion.

In Overtown, Miami, the local community have memorialized their neighborhood through wall art and local museums that seek to celebrate the vibrant "Harlem of the South" deliberately targeted for destruction and obliterated by I-95 (see Section 3.1). The museum and wall art challenge the narrative that the area was torn down because it was a slum and links memorialization of Overtown with other significant events in the civil rights struggle across the country (see Figures 6.1 and 6.2).

Technological advances mean that the legacy of infrastructural racism can now be captured very effectively by digital media. Video conferencing and online film clips have been deployed for example to highlight the history of the Tremé community devastated by the Claiborne Expressway.[41] The emergence of websites like Segregation by Design has also played an important role in allowing contemporary audiences to view images of thriving African American neighborhoods like Sugar Hill, Rondo, Black Bottom, and Sweet Auburn that were devastated by the combined efforts of politicians, architects, planners, and civic leaders in the name of slum

FIGURE 6.1 Four Little Girls wall painting by artist Just Ro that honors the legacy of Addie Mae Collins, Denise McNair, Carole Robertson, and Cynthia Wesley, who were killed when white supremacists bombed the 16th Street Baptist Church in Birmingham, Alabama, in 1963. Photograph by Tim Cunningham, October 2024.

FIGURE 6.2 Judge Lawson Edward Thomas Law Office. Thomas was the first Black attorney in Miami to present a case at trial. He led opposition to the barring of African Americans from area beaches in the 1940s which helped establish an all-Black area on Virginia Key Beach. The building is a Florida Heritage Site but now sits isolated in a sea of urban voids. It is difficult to image this as a busy office with clientele coming and going. Photograph by Tim Cunningham, October 2024.

clearance.[42] The circulation of historic maps that show first class transit networks that were torn up challenges pervading but erroneous narratives that public transportation is inherently alien to the U.S. way of doing things. These resources serve a valuable role in the face of an increasing wave of what Heather McGhee has labelled "miseducation laws" that have emerged in recent years limiting what can be taught in schools and colleges about the history of the country.[43]

6.3 REPARATIVE ARCHITECTURE AND TRANSFORMATIVE JUSTICE

If, as Pablo De Greiff has argued, the future of transitional justice depends upon its ability to occupy a much more explicitly "normative space" that includes non-recurrence and prevention of conflict then one lesson from this research is that it is impossible to achieve this goal without a critical examination of the means by which systemic inequalities intersect with political and civil conflict.[44] This research suggests that a genuinely holistic approach to transitional justice that incorporates the social and economic dimensions of conflict legacy issues cannot ignore the role of planning and architecture. Planners and architects have had prime responsibility for the design and organization of urban space, and "we cannot get away from the fact that our polities are spatial."[45] To date, urban planning and architecture have received fairly scant attention from transitional justice scholarship.[46] The relationships between the planning and design of the urban environment, systemic injustice, and political conflict outlined in this book would suggest that this is an oversight that ought to be addressed.[47]

Some scholars concede the importance of addressing past and present socioeconomic inequalities as a matter of both justice and potential conflict prevention but argue that this should be done "through democratic politics and distributive justice – not through elite bargains and transitional justice … the remedying of socio-economic injustices is a long-term political project."[48] Given that contemporary planning is characterized by reflexive processes that often involve participation and deliberation on the part of numerous stakeholders one might ask whether it is appropriate that transitional justice scholars and practitioners sit on their hands while negotiations about the future shape of contemporary cities and suburbs are carried out by others?

A common theme among those arguing for transformative approaches to transitional justice is the need for participatory frameworks which encourage more engagement with civil society. In this context there is a role for transitional justice proponents to participate in deliberations around future development plans and aim to assist politicians and community leaders as part of the reflexive politics and planning processes that characterize contemporary governance. De Greiff has noted that an important legacy of conflict is the way in which normative expectations are shattered, resulting in "a generalized weakening of agency" that applies to both victims and non-victims alike. He characterizes this weakening of agency by a withdrawal from public spaces, disengagement with social networks, and lack of willingness to make claims to authorities and formal institutions.[49] This diminished agency is also characterized by a lack of civic trust and social capital, all of which combine to make social coordination more difficult, which in turn has consequences for economic development.

This research has highlighted how coercive architecture and planning helped shape the current lack of social capital and diminished agency in areas like the Lower Oldpark in North Belfast, and the Bronx, in New York City. As outlined in Chapter 4, in a context where basic planning principles like connectivity mean "someone can come and attack my house" fear, lack of trust, and a generalized weakening of agency pervades. Rep. Ocasio-Cortez also alludes to this phenomenon when she talks about the "psychic weight" that physical infrastructure imposes upon her community in the Bronx (see Chapter 5). A healthy and functioning city ecosystem requires that diminished agency and lack of trust be repaired and reestablished over time. Partly, this involves establishing trust in institutions like the police, the criminal justice system, and political institutions so that they can be seen to offer an alternative method of resolving disputes through the rule of law rather than through physical force. But this also involves the gradual rebuilding of local communities, physically, socially, economically, and culturally. In other words, helping to support what Ignatieff characterizes as the "subliminal operating system" necessary to preserve some level of peace and stability.[50]

6.4 RESTORATIVE ARCHITECTURE AND THE POLITICS OF REPARATION

Efforts to address the legacy of infrastructural racism in the U.S. by the federal government offer useful insights into the way in which reflexive politics and a thicker understanding of transitional justice can help promote transformation. Anthony Foxx, Secretary for Transportation in the Obama administration, played a significant role in stimulating debate about the need to address the racial legacy of transportation planning in the U.S.[51] Foxx made repairing the damage caused by highways both a federal priority and a personal crusade, aiming to refocus transportation policy on issues like "equity, accessibility, and justice."[52] His attitude toward the divisive and racist consequences of highway planning and design was shaped by personal experience.[53] As outlined in Chapter 2, he described growing up in Charlotte neighborhood in North Carolina, walled in by "two of the busiest interstates on the eastern seaboard … I didn't see the horizon as many people can from their front door. I saw fences and freeways."[54]

Foxx used the term "restorative infrastructure"[55] to describe the kind of architectural and urban design features that reflect a "place making," rather than a "place getting" approach to transportation planning. One such project is a cap over I-71 in Columbus, Ohio, that specifically aims to repair and reconnect communities that were devastated by transportation planning in the past. I-71 had bifurcated the King/Lincoln neighborhood and the city's downtown, leading to social and economic decline as people were cut off from jobs, retail, and business. The vision for the highway "cap" includes a bridge providing homes, businesses, and recreation facilities. A federally funded highway cap in Pittsburgh was developed that includes a three-acre park, with a garden, music venues, food kiosks, and local artwork, aiming to restore connectivity and revitalize the city's Hill District.[56] This area, known as "Little Harlem," was devastated by I-579, which displaced thousands of residents and severed the neighborhood's connection to downtown. The population of the district dropped from 54,000 in 1950 to around 9,500 in 2013. The Pittsburgh cap was funded through the $500 million Transportation Investment Generating Economic Recover (TIGER) program established by his predecessor as Department of Transportation (DOT) Secretary, Ray LaHood. Foxx established another program, Ladders of Opportunity, that also aims to revitalize neighborhoods through access to jobs, education, and health care.

The Reconnecting Communities program, launched by the Biden Administration, builds on this work and aims to remediate the legacy of racial inequalities, segregation, and environmental injustices created by transportation policies of the past. Established under the Infrastructure Investment and Jobs Act (aka Bipartisan Infrastructure Law) the goal of the RCP is to reconnect communities harmed by transportation infrastructure through community-supported planning activities and capital construction projects. Initially funded for $1 billion over five years, the program supports and engages economically disadvantaged communities to "increase affordable, accessible, and multimodal access" to daily destinations like healthcare, employment, grocery stores, schools, places of worship, recreation, and park space." In other words, to begin to knit together some of the urban fabric that was shattered during the 1950s and 1960s.

The RCP program was launched in June 2022 by the Secretary of Transportation Pete Buttigieg, in Birmingham, Alabama.[57] Buttigieg conceded that Birmingham was "as fitting a place as you can find in the country to talk about the history and about the future of transportation equity in America." He acknowledged the role I-65 played in displacing Black neighborhoods in Birmingham and creating "physical barriers, and dead zones that still keep people apart." As with many other cities across the country, urban renewal and the highway program empowered white elites in Birmingham to relocate African Americans and expand downtown on the cleared land.[58] The extension of the University of Alabama Medical Center, for example, was facilitated by the forced relocation of African Americans who were living on adjacent land. Buttigieg delivered a mea culpa for the role that the federal highway program and transportation infrastructure played in segregating communities and concentrating systemic racial inequalities across the country.

> We can't ignore the basic truth that some of the planners and politicians … built them [highways] directly through the heart of vibrant, populated communities, sometimes to reinforce segregation. Sometimes because the people there had less power to resist. And sometimes as part of a direct effort to replace or eliminate Black neighborhoods.[59]

The then Transportation Secretary conceded that neighborhoods and communities across the U.S. remain "constrained by pieces of infrastructure" that were supposed to connect people but which "sometimes have the opposite effect and cut people off from opportunity, sometimes dividing or even destroying whole neighborhoods or communities."[60] Buttigieg also drew attention to the increased risk of illnesses such as asthma that many communities experience by living in proximity to highways. The repentant tone of the launch event was evidenced by the presence of Benard Simelton, President of the Alabama State Conference of the National Association for the Advancement of Colored People (NAACP) who welcomed the initiative and reminded the audience of the social, economic, and political legacy of racist architecture and planning that had denied African Americans access to jobs, healthcare, and polling stations, which had the effect of suppressing Black voting rights as well as social and economic mobility.[61]

A central theme of Reconnecting Communities Pilot (RCP) is the need to repair the legacies of the past through improved planning and design practices that will promote investment, create economic benefits for local communities, and reduce racial disparities.

> The evidence is clear: when people have access to opportunity through physical mobility, the result is more social and economic mobility and prosperity.

The RCP program aims to achieve these objectives through the allocation of Planning Grants and Capital Construction Grants awarded on a competitive basis and dedicated to reconnecting communities that were previously cut off from economic opportunities by transportation infrastructure.

A second thematic area of RCP is that the benefits of the program ought not precipitate "green gentrification," the process whereby infrastructure removal or retrofitting leads to healthier, more walkable, and livable spaces that in turn precipitate an increase in property values and displacement of local communities. At the launch the Transportation Secretary declared,

> When we evaluate applications for this funding, we'll be looking for project sponsors to demonstrate strong community input and buy-in and looking for projects that will improve life for everyone with plans to serve the people who live in these communities already.[62]

Community Land Trusts are one example of anti-displacement measures that ensure long-term residents can remain and enjoy the wider social, economic, and environmental benefits of having their neighborhoods transformed into human-centered places. This aspect of the program has been welcomed by several advocates, including Ben Crowther of the Congress of New Urbanism:

> The repeated emphasis on keeping current residents in place is highly encouraging and will go a long way toward making the Reconnecting Communities program truly reparative, especially in conjunction with program criteria to increase housing supply. It gives Reconnecting Communities the power to turn highways into homes.[63]

The comments from Crowther and others highlight the reparative intention of the RCP, even though it was not explicitly launched as a national reparations program. This reflects the political reality of a measure that was included within the Infrastructure Investment and Jobs Act which required cross-party support to secure congressional approval. Although RCP was widely welcomed by activists and local communities, concerns remain about several aspects of the program that originated as "The Restoring Neighborhoods and Strengthening Communities Program" (also known as the "Highways to Boulevards" initiative). The original proposal was for a $10-billion pilot program to help repair the devastation inflicted by highway builders on predominantly Black, Brown, and poor communities by tearing down highways and rebuilding surrounding neighborhoods to address the needs of marginalized and excluded communities. The intention was that funding would be restricted "to those projects located in regions with a high concentration of low-income residents or residents of color."[64] The political compromises required to ensure Congressional approval broadened the scope of the program beyond tearing down highways to include other measures such as building pedestrian walkways or bridges. This led to concerns that state DOTs could use the program to propose highway expansions with pedestrian bridges. The original initiative was also whittled down from $10 billion to a "disappointingly small" $1 billion federal fund over five years.[65] The $200 million per year of federal funding allocated for RCP should be viewed in the context of the estimated costs of repurposing highways. For example, it is estimated that $500 million is needed to remove just one segment of the Claiborne Expressway that runs through the Tremé neighborhood in New Orleans.

Recognizing local communities face significant challenges in developing their own proposals, the Reconnecting Communities Pilot (RCP) was complemented by a series of other measures including the Thriving Communities program, funded with $25 million through the Consolidated Appropriations Act of 2022.[66] Thriving Communities sought to ensure that disadvantaged communities "adversely or disproportionately affected by environmental, climate, and human health policy outcomes have the technical tools and organizational capacity to compete for federal aid and deliver quality infrastructure projects that enable their communities and neighborhoods to thrive." The USDOT also established a Reconnecting Communities Institute (RCI) to provide technical assistance to build organizational and community capacity in grant recipients and other eligible entities interested in reconnecting communities.[67] RCI was established to serve as DOT's center for learning to restore and reconnect communities that have been harmed, isolated, and cut off from opportunity by transportation infrastructure.[68] Several non-profits, notably Smart Growth America, have also organized briefing sessions and other measures to assist local communities with grant submission.

6.5 RECONNECTING COMMUNITIES: FROM THEORY TO PRAXIS

The RCP program announced $185 million in grant awards for 45 projects for its first phase of fund allocations. Among the successful applicants was a proposal from Buffalo, New York, for $55.6 million to build a new highway cap and tunnel over the Kensington Expressway, a physical barrier that isolates residents on the city's primarily Black east side. The cap will reconnect several east-west roads that were severed by the expressway, create greenspace, safer pedestrian crossings, and make it easier for residents to access community services and amenities like restaurants and cultural facilities. The project has long been a goal of the Restore Our Community Coalition (ROCC), a group of local residents who wish to "remediate the devastation and civic injustice caused by the construction of Route 33 in Buffalo, NY."[69]

Other grantees included the city of Baltimore, which was awarded $2 million to examine options for the removal or redesign of the US 40/Franklin-Mulberry Expressway, known locally as the "Highway to Nowhere" or "the Ditch."[70] The city of Akron, Ohio, received approximately $1 million to create a community-based plan to redevelop its own "road to nowhere." This mile-long section of the city's Innerbelt freeway was closed to traffic in 2017 but still separates "a mostly Black neighborhood from the commercial and cultural life of downtown."[71] The funding aims to reconnect the street grid and possibly add walking trails, a park, public art, and housing. Liz Ogbu, a California-based architect and consultant, who specializes in creating racial and spatial justice, has been working with Akron officials and residents to develop a plan that would address the legacy of displacement and divestment in the neighborhood.[72] An oral history project has also begun to capture the voices of those who had their homes and businesses destroyed by the freeway.

The Baltimore and Akron examples illustrate one legacy feature of the enthusiasm for cutting trenches through U.S. cities that began in the 1950s: a surplus of tarmac in many urban centers across the country. These projects now represent low-hanging fruit with vacant land ripe for economic redevelopment. There are significantly fewer objections from planners and politicians to filling in a highway when it has been decommissioned and doesn't lead anywhere. The work of Ogbu and others highlights the important role of architects and planners in helping local communities develop proposals that are sufficiently robust and technically sound to secure funding from programs like RCP.

The situation is more complex when communities push for highways to be dismantled when they are still operational. RCP grant recipients also include $1.6 million to a group in Tulsa, Oklahoma, which includes the North Peoria Church of Christ which aims to rebuild the historic "Black Wall Street" of Greenwood by removing I-244, which devastated that community. A feasibility study, sponsored by State Representative Regina Goodwin in 2021 had analyzed the impact of I-244 on Greenwood and options for its removal. The study used the city of Syracuse, New York, as a model for the way forward. Plans are currently in place in that city to remove I-81 which divided Syracuse along racial lines in the 1960s and replace the highway with housing and businesses, and reconnect the old street grid. Campaigners in Tulsa wish to see a similar process take place in their city that would repopulate the area now covered by I-244, creating jobs, building affordable housing, boosting businesses, and reconnecting Greenwood with the commercial center.

The Oklahoma Department of Transportation (ODOT) which will ultimately decide the fate of the highway is more circumspect about the prospects of removing this stretch of I-244 and has emphasized the importance of this section of roadway as a part of the wider regional transportation network. Rather than highway removal, the ODOT has focused to date on beautification efforts that make the road more visually appealing, including murals and a walking path. Residents expressed concern that the ODOT may even wish to expand the highway.

The different strategies employed by elements of the local community in Tulsa reflect a sensible division of labor and tactics among local activists. Efforts to secure a formal truth commission to identify details of the 1921 massacre, while thus far unsuccessful (see above) have been complemented by separate efforts to secure funding through RCP in order to repair the damage inflicted by I-244 and urban renewal in later decades. Scholars have sometimes given the impression, even if not explicitly stated, that a broader understanding of transitional justice which incorporates social and economic dimensions involves an "either/or" approach and a zero-sum game. Gready for example argues that given finite resources it is doubtful whether even the most expert truth commissions can assemble sufficient expertise to make credible recommendations on a wider range of civil and political rights and economic and social rights.[73] It has also been argued that reparations programs may generate even more challenges than truth commissions.[74] Local initiatives like these show that many concerns, while legitimate in principle, can be surmounted in practice, as evidenced by local communities, municipalities, and private actors, from Evanston, Illinois, to Newport, Virginia, all of whom have developed reparations for housing and planning policies that concentrated racial exclusion with their own areas.

Another RCP grant recipient, "Reconnect Rondo," based in the Twin Cities in Minnesota developed a plan to revitalize the local area with a land bridge that physically reconnects the Rondo community, including a "lid" constructed on a section of the highway with housing, retail, and other developments.[75] The Rondo neighborhood was devastated by I-94 which split the community in half destroying 700 homes and 300 businesses in the process, with the remaining homes and businesses cut off from social, economic, and cultural opportunity. Reconnect Rondo combines a desire to promote economic, social, and cultural development to address existing housing and other needs, with a desire to secure reparations for the actions of those who deliberately targeted a thriving African American community for destruction:

> … this is more than a land bridge. This is the way to right past wrongs, reweave the community through shared history and create an equitable future we can all be proud of.[76]

The reparative aspects of the project are evident in the formal apology delivered by Chris Coleman, Mayor of St Paul, in 2015.

> We regret the stain of racism that allowed so callous a decision as the one that led to families being dragged from their homes creating a diaspora of the African-American community in the City of Saint Paul. … I apologize, on behalf of the city, to all who call Rondo home … for the acts and decisions that destroyed this once vibrant community.[77]

Coleman declared July 17th "Rondo Remembrance and Reconciliation Day" in St. Paul. The city plans to transform a vacant lot into "Rondo Plaza" which will include monuments to educate groups about the neighborhood that once stood there. In an interview with this author Keith Baker (Executive Director of Reconnect Rondo) emphasized the importance of technical assistance to address the "weakened agency" identified by De Grieff, as well as the difficulties inherent in relying solely on "democratic politics and distributive justice."[78] Baker, a former professional transportation planner, pointed out that state DOTs in the U.S. typically plan transportation networks decades in advance, developing detailed proposals that sit on the shelf until sufficient funding arrives to enable them to be acted upon. This framework means that by the time funding arrives and proposals go out to formal consultation the key decisions have already been taken and it is very difficult for local communities (or anyone else) to reshape development plans. Aware of this dynamic, Reconnect Rondo developed their own proposals some years ago and were better placed than many other communities across the country to secure funding from the RCP program. Taking a similar approach architects and planners worked with African American community members in Syracuse for many years to develop detailed proposals to address the legacy of I-81.[79] After an extended period of community planning and several legal challenges the city of Syracuse will finally move ahead with their plan.

Syracuse and Rondo illustrate the inherent complexities associated with efforts to address spatial inequalities and the need for local communities to secure the services of technocrats who can assist with putting inclusive and restorative community visions into practice. From a theoretical perspective this highlights the importance of ensuring that "diminished agency" of local communities is addressed and alternative visions for a city are presented that

> cannot be disregarded as totally foreign, permanently outside and of no concern to the system itself.[80]

The involvement and participation of professionals who understand how municipal and regional transportation planning systems operate and have the technical expertise to assist in the development of realistic proposals is essential to help repair systemic inequalities woven into the urban fabric. The Reconnect Rondo initiative and indeed the wider RCP program also aligns political efforts to redress the legacy of racist policymaking with initiatives for creating more walkable and people-friendly cities. Efforts to repair communities that have been divided and isolated by design in the U.S. have been boosted by the significant rise in community activism across the country as a new wave of "Freeway Fighters" join in efforts to address the legacy of infrastructural racism. Many cities have discovered the financial benefits of using vast tracts of tarmac that are presently devoted to the movement and storage of vehicles for other purposes, notably housing, retail, and businesses. Much U.S. infrastructure is coming near the end of its shelf-life and must be replaced or repaired which has concentrated minds about the economic costs of hulking highways. In addition, emerging scientific evidence highlighting the impact of vehicle pollution and the role of urban highways in promoting climate change has also led to increased focus on the environmental and health costs of auto centric planning and urban design. These assorted movements allow those working to promote racial equality to cut with the grain of other activists who wish to create more healthy and inclusive spaces.

These positive developments ought to be considered within the context of what David Zipper, Senior Fellow at the MIT Mobility Initiative, refers to as the "Unstoppable Appeal of Highway Expansion" in the U.S.[81] The budget for RCP and similar programs is dwarfed by the overall transportation budget. Technical design manuals used by transportation engineers still favor wider roads which are good for keeping traffic moving but not so good for everyone else. In her recent book, Veronica O. Davis offers "a manifesto for repairing divided communities" that recommends an update of infrastructure design manuals.[82]

Urban and state political dynamics also mitigate against efforts to reconnect and repair communities that were divided by transportation infrastructure. A defining characteristic of every recent U.S. presidential election is the way in which maps across the country show blue cities with voting patterns turning a deeper shade of red the further one goes from the urban core. This is true whether one is looking at the electoral map of Vermont, Texas, California, or Florida. These maps highlight the very real divide that exists between more conservative rural and suburban voters and the more progressive voting patterns of city dwellers. This difference plays out in several ways in state versus city policymaking. In recent years, Republicans have intensified their efforts to limit the power of city governments as part of a wider attack on cities by state authorities.[83] Texas, for example, has sought to limit the power of cities to promote labor rights and other measures.[84] Efforts to remediate systemic racism and "arrested mobility" for African Americans take place within a political system where entire cities like Los Angeles are viewed as "pathology" with the deployment of heavily armored troops "dramatizing on a local scale the well-worn American custom of invading a foreign city inhabited by non-white people.'[85]

6.6 TRANSFORMATIVE JUSTICE AND REPARATIVE ARCHITECTURE

Notwithstanding the range of challenges facing those who wish to address the legacy of systemic and structural racism in the U.S., there are some valuable lessons that can be drawn for those elsewhere while avoiding the trap of "isomorphic mimicry."[86] In this context it is worth noting that many of the positive initiatives outlined above in the U.S. reflect a global move towards 'architectures of spatial justice.'[87] In Canada, the non-profit Architects Against Housing Alienation based in Toronto have called for "Reparative Architecture" to address the long history of racist urban planning that forcibly displaced and disenfranchised Black communities in that country.[88] The group demands that the state provide reparations by funding Black-led community land trusts for the creation of affordable housing and commercial space for the Black residents who were displaced or are at risk of being displaced by state policies. In Syria, Marwa al-Sabouni has also highlighted the relationship between the shape of the built environment and conflict transformation, championing the creation of 'architectures of belonging.'[89]

As outlined in Chapter 4, efforts to adopt a more holistic approach to transformative justice that takes account of the social and economic legacies of conflict needs to recognize the volatile relationship that exists between territory, identity, and inequality in places like North Belfast or poor majority Black, Brown, or white communities in the U.S. Although the nature of conflict differs between the two contexts the need for remedies is similar. In practice, this means accepting political compromises like those that took place around the Girdwood redevelopment and recognizing "… the dangers that lurk in our own certainties and convictions and avoid them as best we can…."[90] Accepting solutions that may cut against the grain of established human rights and equality norms related to housing need does not mean that transitional justice ought to be completely absent from deliberations around the post-conflict reconstruction of cities. As with the U.S., the diminished agency of local communities and among elected politicians is a legacy of conflict and historic exclusion that requires the active involvement of scholars and practitioners to help address the spatial legacy of communities still confined within concrete.

As noted above, the narrative role of transitional justice is a crucial component. The Belfast "Westlink" did not, as many have claimed, "fortuitously" serve as a barrier around the north and west of the city but rather was designed specifically to impede lateral movement across the city. There were several consequences of this including concentrating social and economic inequalities and patterns of spatial disadvantage and undermining social cohesion. There were also significant consequences for the mental and physical health of entire communities forced to live in fixed molds that diminished their capacity to access essential services and jobs and engage in the kind of informal social interactions that are essential for human flourishing.

Politicians who are making decisions about planning, health, housing, and transportation policy need to be aware of the devastating consequences of several decades of coercive architecture that was specifically built to segregate and divide communities across the city. In Belfast, politicians, media commentators, and scholars have yet to get off first base regarding the impact of divisive architecture. It is notable that in the U.S., most mainstream planning and architecture scholars, as well as politicians, including senior officials in the Biden administration, have now recognized and acknowledged the role of infrastructural racism in shaping patterns of inequality and disadvantage – as evidenced by the launch of Reconnecting Communities Program (RCP). This is important because a first step in any process of repair involves recognition of the problem in the first place. Hostility to the RCP by the Trump administration, as evidenced by the suspension of federal funding for many parts of the program, demonstrates the fragility of top-down political interventions. The reality however is that regardless of how the Trump administration proceeds with RCP, Reconnect Rondo, the Claiborne Avenue Alliance, and the plethora of architects and planners who work with them and similar initiatives across the U.S. will continue to campaign to address the legacy of infrastructural racism. Local initiatives like those pioneered by Evanston and Deerfield will continue regardless of who is in the White House.

In Belfast, new economic developments, notably, the Ulster University Belfast Campus and the widening of the M2 at the York Street Interchange, have led to concerns from local communities similar to those around the Crumlin Road Gaol/ Girdwood Barracks. The political, social, economic, cultural, and environmental consequences of several decades of coercive architecture and planning expand beyond communities in the north and west of the city. In December 2024, the number one issue for politicians, policy makers, local communities, and social media influencers across the north of Ireland was the chronic traffic congestion that clogged movement around the city. Government officials went so far as to recommend that people work more from home. Irate phone calls on local radio and interviews with furious commuters filled the airwaves as visitors to Belfast wishing to do their Christmas shopping sat in vehicles alongside those wishing to get home from work.

As outlined in Chapter 5, commentators in the U.S. have quite rightly made the connection between infrastructural racism, traffic congestion, and climate change. The same legacy is apparent in Belfast, but the debate around the issue is quite different. Everyone in the north of Ireland is affected by the legacy of conflict-driven transportation policies that ensured neither airport in Belfast is connected directly to a rail link, and the huge swathes of tarmac that constitute the grey donut around the city fill up with vehicles at peak periods, bringing the city to a standstill. The legacy of decades of transportation planning focused on impeding rather than facilitating free movement, and a democratic deficit that weakened agency among elected officials and local communities mean that the region has yet to begin to recognize the connection between these problems.

Local communities, local politicians, and, indeed, the new planning department of Belfast City Council are all currently engaged in deliberative planning processes that will shape the future city, for better or worse. There is clearly a role here for technical expertise and outside knowledge about the social consequences of new development in the city. A key finding of this research is that a genuinely thicker understanding of transitional justice that incorporates socioeconomic dimensions ought to constitute a more deliberative and *reflexive* form of theory and practice in the field. Such an approach would recognize both the autopoietic nature of modern governance and provide a framework for analyzing the complex relationship that exists between social and economic inequalities and political conflict. It would also avoid the criticism that transitional justice is insufficiently attentive to context, formulaic, and technocratic,[91] providing space for the range of stakeholders who have an important role in helping to understand, and address, systemic and structural conflict legacy issues. The Belfast-based Participation and the Practice of Rights Project is currently engaged with MA Architecture students from Queen's University Belfast as part of the global Change Stories project to explore sustainable urban development in the city. This is the kind of structural coupling between architecture and rights activism that ought to be encouraged.

A more reflexive form of transitional justice is one located on the interface between transitional and transformative justice; it is deliberative, with means and ends determined at the level of the "local." Such an approach fits with other developments characterized as transitional justice "from below."[92] Reflexive transitional justice also fits with the demand from conflict transformation scholarship that prescriptive and mimetic approaches to transition ought to be jettisoned in favor of those that are context-specific, participatory, and bottom-up – what Paul Ledarach has called "elicitive."[93] Gready and Robins argue that the processes likely to emerge from such an emancipatory approach to transition will necessarily be hybrid in nature, dependent upon informal as well as formal governance, grounded in the cultures and the context from which they emerge, and more or less hybridized with global discourses and mechanisms.[94] It is in this "hybridized space" that a genuinely more holistic understanding of transformative justice might hope to succeed on the streets of Belfast and Baltimore.

NOTES

1 Arthur, P. 'How "transitions" reshaped human rights: A conceptual history of transitional justice'. *Human Rights Quarterly* 31, no. 2 (2009): 321–367.
2 Newman, M. *Transitional Justice: Contending with the Past*. Polity Press, 2019; Leyh, B. M., 2023. Unable to see the forest for the trees: Transitional justice and the United States of America. In *Transitional Justice in Aparadigmatic Contexts* (pp. 86–105). Routledge; King, D.S. and Page, J.M., 2020. Towards transitional justice? Black reparations and the end of mass incarceration. In *Rethinking Reconciliation and Transitional Justice After Conflict* (pp. 123–142). Routledge.
3 Lambourne, W. Transformative justice, reconciliation and peacebuilding. In S. Buckley-Zistel (ed.), *Transitional Justice Theories*, p. 20. Taylor & Francis Group, 2013.
4 Evans, M. (ed), *Transitional and Transformative Justice: Critical and International Perspectives*. Routledge, 2019 (*passim*).
5 Evans, M. (ed), *Transitional and Transformative Justice: Critical and International Perspectives*, p. 37. Routledge, 2019 (*passim*).
6 Gready, P. and Robins, S. 'From transitional to transformative justice: A new Agenda for practice'. *International Journal of Transitional Justice* no. 8 (2014) (8): 340.
7 "Racism, Ethnicity, and Transitional Justice", Global Initiative for Justice, Truth and Reconciliation. https://gijtr.org/wp-content/uploads/2023/04/Racism-Ethnicity-and-TJ-final_2023.pdf.
8 McEvoy, K. 'Travel, dilemmas and nonrecurrence: Observations on the 'Respectabilisation' of transitional justice'. *International Journal of Transitional Justice* 12, no. 2 (2018): 185–193.
9 Ladisch, V. and Roccatello, A. 'The color of justice: Transitional justice and the legacy of slavery and Racism in the United States' April (2021) *International Center for Transitional Justice.*
10 The African-American Redress Network. Accessed June 15, 2025 https://redressnetwork.org.
11 National African-American Reparations Commission. Accessed June 15, 2025 https://reparationscomm.org/about-naarc/.
12 Yuvraj, J. Racial Transitional Justice in the United States (June 1, 2022) Race and National Security (Matiangai Sirleaf ed., 2023). Available at https://papers.ssrn.com/sol3/papers.cfm?abstract_id=4088738.
13 Maryland Lynching Truth and Reconciliation Commission. Accessed June 15, 2025. https://msa.maryland.gov/lynching-truth-reconciliation/ See also Fox, N. and Cunningham, D. 'Transitional Justice in public and private: Truth commission narratives in Greensboro'. *International Journal of Transitional Justice* 16, no. 2 (2022): 235–253.

14 Christensen, K. 'The tulsa race massacre, transitional justice, and the ongoing fight against impunity' (August 25, 2021) *Stanford Lawyer,* https://law.stanford.edu/stanford-lawyer/articles/the-tulsa-race-massacre-transitional-justice-and-the-ongoing-fight-against-impunity/.
15 Quarcoo A. and M. Husakovic, 'Racial Reckoning in the United States: Expanding and Innovating on the Global Transitional Justice Experience.' October 26, 2021. Accessed June 15, 2025 https://carnegieendowment.org/research/2021/10/racial-reckoning-in-the-united-states-expanding-and-innovating-on-the-global-transitional-justice-experience?lang=en.
16 Taylor Jr., Otis R. (2023, March 13). America's Highway System Is a Monument to Environmental Racism and a History of Inequity. KQED.
17 United States Government Accountability Office. Report to Congressional Committees. Highway Infrastructure. 'Better Alignment with Leading Practices Would Improve DOT's Reconnecting Communities Pilot Program. May 2023.
18 Fleischer, M. 'Want to tear down insidious monuments to racism and segregation? Bulldoze L.A. freeways.' L.A. Times. June 24, 2020.
19 Mock, B. 'What it actually means to pass local "Reparations"' (April 15 2021), *Bloomberg City Lab.*
20 Alexander, E. and W. Dow. 'The Big Payback.' (PBS Documentaries 2022.).
21 Mock B 'What it Actually Means to Pass Local "Reparations"' April 15 2021, Bloomberg City Lab.
22 Kellam, B., 'Virginia lawmakers approve commission to examine universities' Displacement of Black Communities, (14 March 2023) *Propublica.*
23 Kellam, B. and Hansen, L., 'Erasing The "Black Spot": How a virginia college expanded by uprooting a Black neighborhood, virginia center for investigative journalism, September 5, 2023.
24 Mock, B., "The case against local reparations" (July 2023) *Bloomberg City Lab.*
25 Darity, W. and Mullen, A. *The Black Reparations Project: A Handbook for Racial Justice*. (University of California Press 2023).
26 Mock, B. 'The case against local reparations' (July 2023) *Bloomberg City Lab.*
27 Dancy, G. and Wiebelhaus-Brahm, E. 'Bridge to human development or vehicle of inequality? Transitional justice and economic structures'. *International Journal of Transitional Justice* 9, no. 1 (2014): 51–69.
28 Dancy, G. and Wiebelhaus-Brahm, E. 'Bridge to human development or vehicle of inequality? Transitional justice and economic structures'. *International Journal of Transitional Justice* 9, no. 1 (2014): 51–69.
29 Steinberg, S. *The Ethnic Myth: Race, Ethnicity and Class in America* (Beacon Press 2001).
30 Miller, Z. 'Effects of invisibility: In search of the 'Economic' in Transitional justice' (2008) 2 *The International Journal of Transitional Justice* 266–291.
31 Deerfield Public Library. Accessed June 15, 2025. The Fight to Integrate Deerfield. https://www.deerfieldlibrary.org/fid/.
32 Eric DeGrechie, "Park Board Votes to Rename Mitchell Park", June 18, 2020. Patch.
33 Hines, B. 'How South Miami transformed a dividing wall from Jim Crow era into a unifying force.' February 24, 2024. Miami Herald.
34 Evelyn Rossler Stroder, (2017) *The Wall That Failed: How One Community Navigated the Shifting Sands of Integration in West Texas*. 329–330, 455–463.
35 Henry, J. 'Arlington community remembers segregation wall.' WUSA9. March 2, 2017.
36 Blitz, M. 'Once There Was a Segregation Wall in Arlington'. Arlington Magazine. June 1, 2020.
37 Schwartz, M. and L. Baker. 'In Detroit, A Colorful Mural Stands As A Reminder Of The City's 'Segregation Wall.' NPR. July 22, 2017.
38 Jane, V. S. 'Lake Worth woman makes dividing wall symbol of unity' (February 6, 1994) *Palm Beach Post.*

39 Lake Worth Beach's Wall of Unity sets aside discord as refurbishing project begins, The Palm Beach Post, Jorge Milian, January 16, 2021.
40 Jorge, M. 'Lake Worth Beach's Unity Wall finds the controversy it was trying to avoid' (January 21, 2021) *The Palm Beach Post*.
41 Amy, S. Guerilla in the room. In R. Reft, A. De Lucas, and R. Retzlaff (eds.), *Justice and the Interstates: The Racist Truth About Urban Highways* (Island Press 2023).
42 Susaneck, A. "Segregation by Design." TU Delft. Centre for the Just City, 2024. Accessed June 14, 2025. https://www.segregationbydesign.com/.
43 Susaneck, A. "Segregation by Design." TU Delft. Centre for the Just City, 2024. Accessed June 14, 2025. https://www.segregationbydesign.com/.
44 De Greiff, P. 'The future of the past: Reflections on the present state and prospects of transitional justice'. *International Journal of Transitional Justice* 14 (2020): 254.
45 Collier, P. *The Future of Capitalism: Facing The New Anxieties* (Harper Collins 2018) 61.
46 Mihai, M. 'Architectural transitional justice? political renewal within the scars of a violent past' (2018) 12 *International Journal of Transitional Justice*. The Historical Urbanism Project, which is an interdisciplinary project looking at the role of the design of the urban environment in promoting conflict transformation, is a welcome addition to the limited attention transitional justice has given to planning architecture to date, see https://brandonhamber.blogspot.com/2019/01/historical-urbanism-project.html.
47 Further evidence of the important role that urban planning and architecture can play in precipitating and preventing conflict is provided by the large body of scholarship from urban theorists, political scientists, and others who have focused attention on this issue and who have been cited in this book.
48 Waldorf, L. 'Anticipating the past: Transitional justice and socio-economic wrongs'. *Social and Legal Studies* 21, no. 2 (2012): 171.
49 De Greiff, P. Articulating the links between transitional justice and development: Justice and social integration. In P. De Greiff and R. Duthie (eds.) *Transitional Justice and Development: Making Connections* (International Center for Transitional Justice 2009).
50 Ignatieff, M. *The Ordinary Virtues: Moral Order in Divided World* (Harvard University Press 2017) p. 46.
51 Center for American Progress. "Secretary Foxx Discusses A Transportation Plan for the 21st Century." Accessed June 13, 2025 https://www.youtube.com/watch?v=QqtJgWaU0zM&t=494s.
52 Ross-Brown, S. 'Transportation Secretary Foxx Moves to Heal Scars of Urban Renewal.' The American Prospect. September 30, 2016.
53 Boyum, T. "Anthony Foxx: How Transportation Helped Build Systemic Racism." Tying It Together. Podcast. 6 January 2021. Spectrum News 1.
54 Rehm, D. 'Transportation Secretary Anthony Foxx On The Legacy Of The U.S. Highway System'. March 31, 2016. WAMU 88.5. American University Radio. Accessed June 14, 2025. https://dianerehm.org/shows/2016-03-31/transportation-secretary-anthony-foxx-on-transportation-opportunity-and-the-legacy-of-the-u-s-highway-system.
55 Rehm, D. 'Transportation Secretary Anthony Foxx On The Legacy Of The U.S. Highway System'. March 31, 2016. WAMU 88.5. American University Radio. Accessed June 14, 2025. https://dianerehm.org/shows/2016-03-31/transportation-secretary-anthony-foxx-on-transportation-opportunity-and-the-legacy-of-the-u-s-highway-system.
56 Ross-Brown, S. 'Transportation Secretary Foxx Moves to Heal Scars of Urban Renewal.' The American Prospect. September 30, 2016.
57 U.S. Department of Transportation. Transcript: Secretary Pete Buttigieg Remarks on Launch of Reconnecting Communities Program – Birmingham, Alabama. Thursday, June 30, 2022. Accessed June 15, 2025 https://www.transportation.gov/briefing-room/transcript-secretary-pete-buttigieg-remarks-launch-reconnecting-communities-program.

58 Connerly, C. *The Most Segregated City in America: City Planning and Civil Rights in Birmingham, 1920–1980*. (University of Virginia Press 2005).
59 U.S. Department of Transportation. Transcript: Secretary Pete Buttigieg Remarks on Launch of Reconnecting Communities Program – Birmingham, Alabama. Thursday, June 30, 2022. Accessed June 15, 2025 https://www.transportation.gov/briefing-room/transcript-secretary-pete-buttigieg-remarks-launch-reconnecting-communities-program.
60 Transcript: Secretary Pete Buttigieg Remarks on Launch of Reconnecting Communities Program – Birmingham, Alabama. June 30, 2022.
61 Transcript: Secretary Pete Buttigieg Remarks on Launch of Reconnecting Communities Program – Birmingham, Alabama. June 30, 2022.
62 Transcript: Secretary Pete Buttigieg Remarks on Launch of Reconnecting Communities Program – Birmingham, Alabama. June 30, 2022.
63 Wilson, K. 'Four Things Advocates Need to Know About the 'Reconnecting Communities' Program' (June 30, 2022) *Streetsblogusa.*
64 Wilson, K. 'Senate considering $10B for highway removal' (January 11, 2021) *Streetsblog. Accessed June 14, 2025* https://usa.streetsblog.org/2021/01/11/senate-considering-10b-highway-removal-bill.
65 Wilson, K. 'Four things advocates need to know about the 'Reconnecting Communities' Program' (June 30, 2022) *Streetsblogusa*. Accessed June 14, 2025 https://usa.streetsblog.org/2022/06/30/four-things-advocates-need-to-know-about-the-reconnecting-communities-program.
66 U.S. Department of Transportation. Thriving Communities Program. Accessed June 15, 2025. https://www.transportation.gov/grants/thriving-communities.
67 U.S. Department of Transportation. Reconnecting Communities Institute. Accessed June 15, 2025. https://rciconnect.org.
68 U.S. Department of Transportation. Past, Present and Future of Reconnecting Communities. February 16, 2023. Accessed June 15, 2025. https://storymaps.arcgis.com/stories/dc61e3bcecd44464b4bfef8299d9d6ce.
69 Restore Our Community Coalition (ROCC). Accessed June 15, 2025. https://roccbuffalo.org/?p=2873.
70 Cassie, R. 'Road to Ruin', Baltimore. February 2023. https://www.baltimoremagazine.com/section/businessdevelopment/highway-to-nowhere-baltimore-expressway-demolished-black-neighborhoods/.
71 Maher, K. 'Black community cut off by Akron's 'Road to Nowhere' Seeks to Undo Damage' (September 20, 2023) *Wall Street Journal.* https://www.wsj.com/us-news/black-community-cut-off-by-akrons-road-to-nowhere-seeks-to-reconnect-a58cb054?mod=Searchresults_pos1&page=1.
72 City of Akron. Reconnecting Our Community: Reimage the Innerbelt, Together. Accessed June 15, 2025 https://www.akroninnerbelt.com.
73 Gready, P. *The Era of Transitional Justice: The Aftermath of the Truth and Reconciliation Commission in South Africa and Beyond* (Routledge 2010).
74 Dancy, G. and Wiebelhaus-Brahm, E. 'Bridge to human development or vehicle of inequality? Transitional justice and economic structures' (2014) 9 (1) *International Journal of Transitional Justice* 54.
75 Reconnect Rondo. Accessed June 15, 2025 https://reconnectrondo.com/landbridge/.
76 Reconnect Rondo. Accessed June 15, 2025 https://reconnectrondo.com/landbridge/.
77 Constantini, A. 'Rondo neighborhood gets apologies for I-94.' KARE11. July 17, 2015. Kare11. Accessed June 15, 2025 https://www.kare11.com/article/news/local/rondo-neighborhood-gets-apologies-for-i-94/89-105454642.
78 Interview with Keith Baker, Executive Director, Reconnect Rondo. (November 2024).
79 Interview with Marc Wouters, Architect. (March 2024).
80 Carson, W. *The Other Price of Britain's Oil: Safety and Control in the North Sea*, (Martin Robertson 1981) 301 in J Paterson, *Behind the Mask: Regulating Health and Safety in Britain's Offshore Oil and Gas Industry* (Ashgate 2000) 226 (emphasis in original).

81 Zipper, D. 'The unstoppable appeal of highway expansion' (September 28, 2021) *Bloomberg CityLab.*

82 Zipper, D. Repairing the Roads That Were Built to Divide. (July 12, 2023) *Bloomberg CityLab.*

83 'Republicans intensify their assault on city governments.' June 3rd, 2023. The Economist.

84 Holder, S. 'Texas Wrests Power From Local Governments With Sweeping New Law.' (June 28, 2023) *Bloomberg CityLab.*

85 Nguyen, V. 'The Struggle for Los Angeles'. (June 14, 2025) Financial Times.

86 De Greiff, P. 'The future of the past: Reflections on the present state and prospects of transitional justice' (2020) 14 *International Journal of Transitional Justice* 251–259.

87 Cuff, D. Architectures of Spatial Justice. (MIT Press 2023.).

88 Architects Against Housing Alienation. Accessed June 15, 2025 https://aaha.ca/en/demand/reparative-architecture.

89 al-Sabouni, M. Building for Hope: Towards an Architecture of Belonging. (Thame and Hudson 2021).

90 Ignatieff, M. *The Ordinary Virtues: Moral Order in Divided World*, (Harvard University Press 2017) 137.

91 De Greiff, P. 'The future of the past: reflections on the present state and prospects of transitional justice' (2020) 14 *International Journal of Transitional Justice* 254.

92 McEvoy, K. 'Beyond legalism: Towards a thicker understanding of transitional justice' (2007) 34 (4) *Journal of Law and Society.* See also De Greiff, P. 'The future of the past: Reflections on the present state and prospects of transitional justice' (2020) 14 *International Journal of Transitional Justice* 257.

93 Ledarach, P. 'Building mediative capacity in deep-rooted conflict', (2002) 26 (1) *Fletcher Forum on World Affairs* 91–101.

94 Gready, P. and Robins, S. 'From transitional to transformative justice: A new agenda for practice' (2014) (8) *International Journal of Transitional Justice* 351.

Index

Pages in *italics* refer to figures, pages followed by "n" refer to notes.

O

P

R

S

T

U

Made in the USA
Monee, IL
13 September 2026

62979535R10166